AIP CONFERENCE PROCEEDINGS 206

ATOMIC PROCESSES IN PLASMAS
GAITHERSBURG, MD 1989

EDITORS:
YONG-KI KIM
NATIONAL INSTITUTE OF
STANDARDS AND TECHNOLOGY

RAYMOND C. ELTON
NAVAL RESEARCH LABORATORY

American Institute of Physics   New York

Authorization to photocopy items for internal or personal use, beyond the free copying permitted under the 1978 US Copyright Law (see statement below), is granted by the American Insitute of Physics for users registered with the Copyright Clearance Center (CCC) Transactional Reporting Service, provided that the base fee of $2.00 per copy is paid directly to CCC, 27 Congress St., Salem, MA 01970. For those organizations that have been granted a photocopy license by CCC, a separate system of payment has been arranged. The fee code for users of the Transactional Reporting Service is: 0094-243X/87 $2.00.

© 1990 American Institute of Physics.

Individual readers of this volume and non-profit libraries, acting for them, are permitted to make fair use of the material in it, such as copying an article for use in teaching or research. Permission is granted to quote from this volume in scientific work with the customary acknowledgment of the source. To reprint a figure, table or other excerpt requires the consent of one of the original authors and notification to AIP. Republication or systematic or multiple reproduction of any material in this volume is permitted only under license from AIP. Address inquiries to Series Editor, AIP Conference Proceedings, AIP, 335 E. 45th St., New York, NY 10017.

L.C. Catalog Card No. 90-55265
ISBN 0-88318-769-8
DOE CONF 891052

Printed in the United States of America.

ATOMIC PROCESSES IN PLASMAS

Contents

Preface vii

MEMORIAL

In Memory of Clarence F. Barnett "Barney" October 7, 1923–June 11, 1989 ix

ATOMIC PROCESSES

Electron-Impact Ionization and Recombination in Positive Ions 3
    D. C. Griffin and M. S. Pindzola
Relativistic and Quantum Electrodynamic Effects in Highly-Charged Ions 19
    Yong-Ki Kim
State Selective Charge Transfer Relevant for Plasma Physics 40
    R. Hoekstra, F. J. de Heer, and R. Morgenstern
Effects of Electric Fields on Autoionization and Implications
for Dielectronic Recombination 48
    R. R. Jones and T. F. Gallagher

DATABASES

The ALADDIN Atomic Physics Database System 63
    Russell A. Hulse
The Opacity Project: Progress and Methods 73
    Yan Yu
Update on the OPAL Opacity Code 82
    F. J. Rogers, C. A. Iglesias, and B. G. Wilson

LOW-DENSITY PLASMAS

A Discussion of Some Less Well Accounted for Atomic Processes
Responsible for XUV Emission from Magnetically Confined Fusion
Plasmas 95
    Michael Finkenthal
Laser Optogalvanic and Fluorescence Spectroscopy in Glow Discharge
Plasmas 107
    J. E. Lawler and E. A. Den Hartog
A Plasma Spectroscopic Study of Molecular Hydrogen and Hydrocarbons
in a Tokamak: Techniques and Problems 116
    Takashi Fujimoto and WT-III Group
Atomic Processes and Spectroscopic Techniques Applied to Fusion
Plasma Diagnostics 122
    R. J. Fonck

Modeling of Impurity Emissions from Tokamak Plasmas 135
    B. C. Stratton

Liquid and Solid Atomic Ion Plasmas 152
    J. J. Bollinger, S. L. Gilbert, D. J. Heinzen, W. M. Itano,
    and *D. J. Wineland*

Critical Quantities for Solar System Science in Atomic and Molecular Reactions 163
    D. E. Shemansky

DENSE PLASMAS

Problems in Line Broadening and Ionization Lowering 177
    J. Davis and M. Blaha

Large Scale Calculations of the Structure and Dynamics of Dense Matter 193
    Stephen M. Younger

Spectroscopic Analysis of Hot Dense Laser Produced Plasmas 204
    C. F. Hooper, Jr., R. C. Mancini, D. P. Kilcrease,
    and L. A. Woltz

Atomic Processes in Plasmas under Ultra-Intense Laser Irradiation 217
    G. T. Schappert, D. E. Casperson, J. A. Cobble,
    J. C. Comly, L. A. Jones, G. A. Kyrala, K. J. LaGattuta,
    P. H. Y. Lee, G. L. Olson, and A. J. Taylor

Spectroscopy of Dynamic, Nonequilibrium and Turbulent Plasmas in Pulsed-Power Systems 232
    Y. Maron

XUV LASERS

A Quest for More Efficient X-Ray Lasers: Atomic Parameters Involved and Recent Experiments 257
    R. C. Elton, E. A. McLean, J. A. Stamper, C. K. Manka,
    and B. H. Ripin

Improved Performance of Recombination Balmer-α Laser by Short Pulse Irradiation 267
    Y. Kato, H. Azuma, K. Yamakawa, M. Nishio, T. Tachi, H. Shiraga,
    E. Miura, H. Takabe, *K. Nishihara*, S. Nakai, S. A. Ramsden,
    G. J. Pert, M. H. Key, S. J. Rose, and C. P. J. Barty

Harmonic Generation in Rare Gases: Single Atom Response and Propagation Effects 277
    L.-A. Lompré, A. L'Huillier, M. Ferray, and G. Mainfray

Author Index 289

Note: The author who delivered the talk is designated by italics.

PREFACE

This volume contains 23 of the 31 invited talks presented at the Seventh American Physical Society Topical Conference on Atomic Processes in Plasmas held in Gaithersburg, Maryland on October 2–5, 1989. The conference was attended by about 200 participants from all over the world. This conference was a sequel to the one held in Santa Fe, New Mexico in September 28–October 2, 1987 (AIP Conference Proceedings No. 168). As before, the main topics were atomic processes in low-density plasmas, such as in magnetic fusion devices and in astrophysical applications, and in dense plasmas, such as in laser produced ones.

The editors would like to dedicate this volume to Clarence F. Barnett, a pioneer who envisioned the need for close interaction among atomic and plasma physicists and initiated one of the original conferences of this series. We dearly miss his foresight and inquisitiveness that stimulated us to think deeper and sharper. David Crandall kindly prepared the memorial article.

In this volume, there are two articles reviewing atomic structure and collision theories for highly-charged ions, and two articles reporting experiments on charge transfer and field ionization. Atomic databases play important roles in plasma modeling, and the current status of three such databases are reported.

We are fortunate to have six articles on atomic processes in low-density plasmas. The field of unconventional plasmas confined in an ion trap is reviewed by Wineland in an article coauthored with his colleagues (see Bollinger et al.). A review of atomic and molecular processes for solar physics is also presented.

Articles on dense plasmas include a review of the Los Alamos "Very Bright Light Source" and a computer simulation of dense matter. Quests for more efficient x-ray lasers are discussed in two articles; and an impressive experiment that produced very high-order harmonics is reported in another article in the section on XUV lasers.

We hope that this volume will contribute to the interdisciplinary interaction between atomic and plasma physicists, which has been very fruitful in providing ideas and means for advanced technologies.

We are grateful to the National Institute of Standards and Technology, the Lawrence Livermore National Laboratory, and the Office of Fusion Energy, U.S. Department of Energy for providing financial support for the conference and the publication of these proceedings.

Yong-Ki Kim
Gaithersburg, Maryland

Raymond C. Elton
Washington, D.C.

March, 1990

In Memory of
Clarence F. Barnett
"Barney"
October 7, 1923–June 11, 1989

Clarence Barnett originated the conference on "Atomic Processes in High Temperature Plasmas" in 1977 and was well known for promoting communication between atomic and plasma physics. He was associated with the Oak Ridge National Labortories for his entire professional career. The photo shows him at work near the beginning of that career in 1943 when he worked on ion sources for isotope separation for the secret Manhattan Project.

Barney's contributions to physics included basic cross-section measurements, development of plasma devices and diagnostics, ion source improvements, and selection and evaluation of atomic data for fusion. He is widely known for the "Redbooks" of atomic data and continued to work on them up until he passed away.

Barney received various recognitions for his work including the DOE's Distinguished Associate Award. He is best known to many participants in this conference as a friendly and energetic colleague who loved to challenge us to reach beyond the comfortable limits of our accomplishments for more meaningful results from our research.

This photo is from his "official" retirement in 1985. His spirit will continue to encourage us in the future.

Atomic Processes

ELECTRON-IMPACT IONIZATION AND RECOMBINATION IN POSITIVE IONS

D. C. Griffin
Department of Physics, Rollins College,
Winter Park, Fl. 32789

M. S. Pindzola
Department of Physics, Auburn University,
Auburn, Al. 36849

ABSTRACT

Theoretical calculations of electron-impact ionization and recombination in positively charged ions are reviewed and compared to cross section measurements. In the case of ionization, the contributions of excitation autoionization and resonant recombination followed by double autoionization are considered. The excitation-autoionization contributions can dominate the total ionization cross section even for highly ionized species. In addition, we present the results of fully relativistic calculations of ionization cross sections obtained from lowest-order QED theory for H-like and He-like uranium. Finally, we consider the contribution of dielectronic recombination (DR) to electron-impact recombination. Comparison of theoretical calculations of DR for He-like and Li-like carbon and oxygen ions with recent high-resolution, merged-beam measurements, using the electron cooler at Aarhus, seem to indicate that total DR cross sections can be accurately determined from the isolated-resonance approximation. The effects of electric fields on DR are briefly discussed.

INTRODUCTION

Electron-impact ionization and recombination are important processes in both laboratory and astrophysical plasmas. In this paper, we give a brief review of recent work on the theoretical calculations of these processes in multiply charged atomic ions. We begin by considering electron-impact ionization with an emphasis on the contribution from indirect processes involving doubly excited states. As an example, we discuss the ionization of highly ionized Na-like ions, and compare results from theoretical calculations with crossed-beam measurements of the ionization cross sections for these ions. Then we consider the ionization of H-like and He-like uranium, and present results of ionization cross section calculations which employ fully relativistic wavefunctions and include contributions from lowest-order QED. Finally we discuss electron-impact recombination with an emphasis on dielectronic recombination (DR), and compare theoretical calculations of DR for He-like and Li-like ions with results from high-resolution merged-beam experiments.

THE CONTRIBUTION OF INDIRECT PROCESSES TO ELECTRON-IMPACT IONIZATION

A number of processes can contribute to the total electron-impact ionization cross section of multiply charged ions. In addition to direct ionization of an ion with atomic number Z and N bound electrons

$$e^- + X(Z,N) \longrightarrow X(Z,N-1) + e^- + e^- , \qquad (1)$$

we have inner-shell excitation followed by autoionization

$$e^- + X(Z,N) \longrightarrow X^{**}(Z,N) + e^- \qquad (2)$$
$$\hookrightarrow X(Z,N-1) + e^- ,$$

and resonant recombination to a doubly excited state of the (N+1)-electron ion followed by sequential autoionization of two electrons

$$e^- + X(Z,N) \longrightarrow X^{**}(Z,N+1) \qquad (3)$$
$$\hookrightarrow X^{**}(Z,N) + e^-$$
$$\hookrightarrow X(Z,N-1) + e^- .$$

In addition, we can have resonant recombination to a doubly excited state of the (N+1)-electron ion followed by the simultaneous emission of two electrons (auto-double ionization). This process has been calculated in Li-like ions,[1] and has been measured in the case of Li-like C^{3+};[2] however, its contribution to the total ionization cross section is small and it will not be considered here.

For now we only consider the first two processes of direct ionization and excitation autoionization. In an independent-processes approximation, the total-ionization cross section is then given by

$$\sigma_{tot}(i) = \sum_f \sigma_{ion}(i \to f) + \sum_j \sigma_{exc}(i \to j) B_j^a , \qquad (4)$$

where $\sigma_{ion}(i \to f)$ is the direct-ionization cross section from the initial configuration i to the final configuration f of the (N-1)-electron ion and $\sigma_{exc}(i \to j)$ is the inner-shell excitation cross section from configuration i to a particular doubly excited autoionizing level j. B_j^a is the autoionization branching ratio from level j and is given by

$$B_j^a = \frac{\sum_m A_a(j \to m)}{\sum_m A_a(j \to m) + \sum_k A_r(j \to k)}, \quad (5)$$

where $A_a(j \to m)$ is the autoionizing rate from level j to the continuum channel m, and $A_r(j \to k)$ is the radiative rate from level j to a final bound level k.

Using this approximation, numerous calculations of ionization cross sections have been made. In our work, the direct-ionization cross sections are calculated using a configuration-average distorted-wave ionization program, the energy levels and autoionizing and radiative rates are determined using a modified version of Cowan's atomic structure programs,[3] and the inner-shell excitation cross sections are calculated using our multi-configuration, intermediate-coupling distorted-wave excitation program. Bound-state and continuum radial wavefunctions are calculated in a potential which includes the Darwin and mass-velocity relativistic corrections.[4] Details of the calculational methods employed in these programs are given in Ref. 5.

In addition to ionization from the ground-state configuration of a given ion, one must also consider ionization from possible long-lived metastable states. Indeed, comparisons of theoretical calculations with crossed-beam experiments using the ECR ion source at Oak Ridge National Laboratory for ions in the magnesium isoelectronic sequence,[6] as well as ions in the iron[7] and nickel[8] isonuclear sequences indicate that this source preferentially populates low-lying, long-lived states. Whether this is true in other laboratory and astrophysical plasmas is still an unanswered question.

However, as an example of the contribution of excitation autoionization to the total ionization cross section, we shall consider the case of Na-like ions, where no such metastable states exist. Then for ionization from the $2s^2 2p^6 3s$ ground-state configuration, we include direct ionization out of the 3s, 2p, and 2s subshells and inner-shell excitations to the $2s^2 2p^5 3sn\ell$ and $2s2p^6 3sn\ell$ configurations which can then autoionize to the $2s^2 2p^6$ configuration of the neon-like ion. In Fig. 1, we show an energy-level diagram for the case of Ni^{17+}.

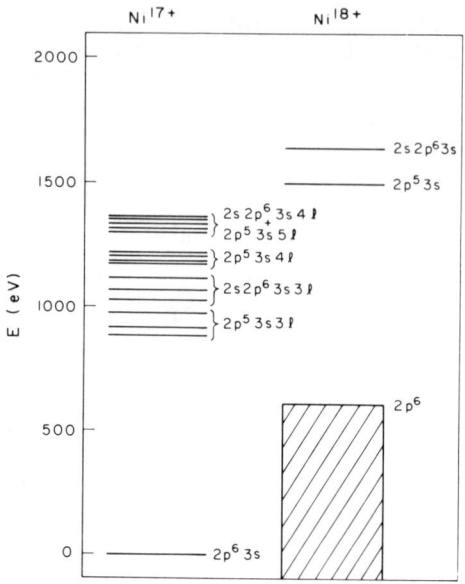

Fig. 1. Energy-level diagram showing the $2s^2 2p^5 3snl$ and $2s2p^6 3snl$ configurations of Ni^{17+} relative to the ionization thresholds for ionization from the 3s, 2p, and 2s subshells.

Configuration interaction within the intermediate autoionizing levels can be very important. Therefore, in addition to the doubly excited configurations shown, which can be populated through single-electron excitations, one must also include other configurations which mix with these particular doubly excited configurations. For example, the $2p^5 3s3d$ configuration strongly interacts with the $2p^5 3p^2$ configuration. This type of interaction has very little effect on the excitation cross section, but can change the autoionizing rates by as much as three orders of magnitude.[5]

Shown in Fig. 2 is the calculated ionization cross section for this ion. As can be seen, the contributions from excitation autoionization dominate the total cross section near threshold, even though the average branching ratio for autoionization is only about 0.58. For many highly ionized species, excitation autoionization contributions are very large, depending on the details of the atomic structure involved. For example, Reed et al. have recently shown that inner-shell excitations especially of the form $2p^6 3s \rightarrow 2p^5 3s3p$ contribute significantly to the total ionization cross section even in Na-like Au^{68+}.[9] This makes extrapolations along isoelec-

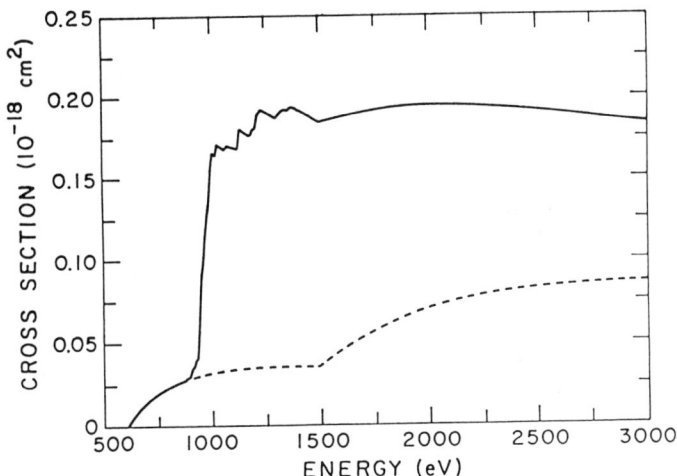

Fig. 2. Calculated ionization cross section for Ni^{17+}. Solid curve, total ionization cross section; dashed curve, direct-ionization contribution only.

tronic, and even more so, isonuclear sequences[7,8] very difficult.

We now consider resonant recombination double autoionization (RRDA), which is the process described in Eq. 3. In order for this to contribute to the total-ionization cross section, the doubly excited state of the (N+1)-electron ion must autoionize to a doubly excited autoionizing state of the N-electron ion which, in turn, must autoionize to a bound state of the (N-1)-electron ion. The cross section for RRDA is given by

$$\sigma_{rrda} = \sum_j \sigma_{rr}(i \to j) B_j^{da}, \qquad (6)$$

where $\sigma_{rr}(i \to j)$ is the energy-averaged resonant-recombination cross section from level i of the N-electron ion to level j of the (N+1)-electron ion. By the principle of detailed balance, $\sigma_{rr}(i \to j)$ can be written in terms of the rate $A_a(j \to i)$ for the reverse process of autoionization from level j to level i (in atomic units) as

$$\sigma_{rr}(i \to j) = \frac{2\pi^2}{\Delta \varepsilon k_i^2} \frac{2J_j+1}{2(2J_i+1)} A_a(j \to i), \qquad (7)$$

where J_i and J_j are the total angular momenta of the

levels i and j, respectively, and $\Delta\varepsilon$ is an energy bin width larger than the largest resonance width. Finally, B_j^{da} is the branching ratio for double autoionization from level j and is given by

$$B_j^{da} = \sum_d \left[\frac{A_a(j\rightarrow d)}{\sum_k A_a(j\rightarrow k) + \sum_n A_r(j\rightarrow n)} \times \frac{\sum_f A_a(d\rightarrow f)}{\sum_f A_a(d\rightarrow f) + \sum_m A_r(d\rightarrow m)} \right], \quad (8)$$

where the sum over d includes autoionization only to levels of the N-electron ion which themselves are autoionizing, while the sum over k includes autoionization to all lower levels of the N-electron ion, whether they be bound or autoionizing, and $A_r(j\rightarrow n)$ is the radiative rate from level j to a bound level n.

Contributions from RRDA are difficult to include in calculations of the total-ionization cross section because of the very large number of levels involved in the determination of the branching ratio for autoionization. In Fig. 3, is shown our calculation of

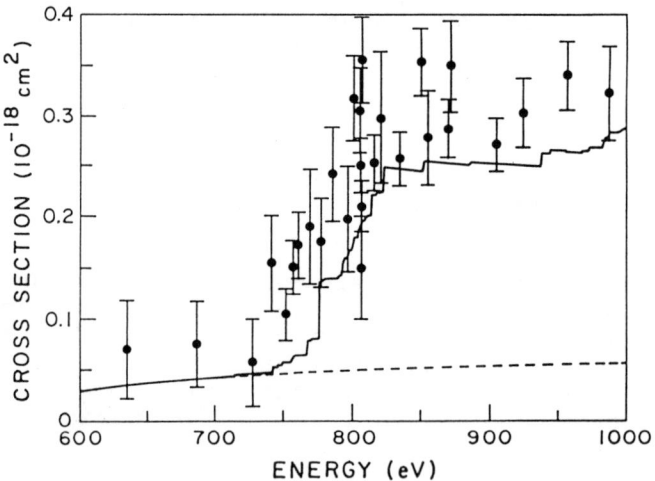

Fig. 3. Calculated ionization cross section for Fe^{15+}. Solid curve, total-ionization cross section; dashed curve, direct-ionization cross section only. Experimental points from Ref. 10.

the ionization cross section for Fe^{15+} which does not include contributions from RRDA,[5] in comparison to the crossed-beam measurements of Gregory et al.[10] A similar level of agreement between theory[5] and experiment[11] is also found for Ti^{11+} and Cr^{13+}. The differences are primarily due to the contributions from the RRDA process. This has been verified recently by an extremely large calculation of the resonant contributions in Fe^{15+} by Chen et al.[12] They find that their calculation of RRDA through doubly excited states of the form $2p^53s3\ell n\ell$ and $2p^53s4\ell n\ell$, when added to the contributions from direct ionization and excitation autoionization, agree well with the experimental measurements.[10]

ELECTRON-IMPACT IONIZATION OF RELATIVISTIC SYSTEMS IN LOWEST-ORDER QED

We also consider the electron-impact ionization of very highly ionized species for which the contributions of excitation autoionization are negligible, but relativistic effects should be large. In lowest-order QED, the direct scattering amplitude in atomic units is given by

$$S_d = i \int d^4x \int d^4y \int \frac{d^4q}{(2\pi)^4} \left\{ \bar{\Psi}_f(x) \gamma_\mu \Psi_i(x) \right\} \frac{e^{-iq(x-y)}}{q^2 + i\varepsilon}$$
$$\times \left\{ \bar{\Psi}_e(y) \gamma^\mu \Psi_b(y) \right\}, \quad (9)$$

where $\Psi(x)$ are the Dirac bispinors, q is the 4-component momentum transfer, the Lorentz gauge has been chosen for the photon propagator, and γ^μ represents the 4 x 4 matrices:

$$\gamma^0 = \begin{bmatrix} 1 & 0 \\ 0 & -1 \end{bmatrix} \quad \gamma^i = \begin{bmatrix} 0 & \sigma^i \\ -\sigma^i & 0 \end{bmatrix} \quad i = 1, 2, 3 \quad (10)$$

and σ^i represents the ith component of the Pauli-spin matrices. A similar expression holds for the exchange scattering amplitude, except that the wavefunction for the scattered electron Ψ_f and the wavefunction for the ejected electron Ψ_e are interchanged. The 12-dimensional integrals may be reduced using partial-wave expansions for the incident, scattered, and ejected wavefunctions. The bound-state wavefunction, Ψ_b is obtained from the

Dirac-Fock approximation, and the radial partial waves are computed in relativistic distorted-wave potentials. Using standard methods of angular algebra, one obtains expressions for the direct ionization cross section which include both magnetic and retardation effects.[13]

A program to calculate the ionization cross section based on these methods has been written, and applied to K-shell ionization in H-like and He-like uranium.[13] The results of these calculations were compared to the relativistic channeling measurements of Claytor et al.[14] for K-shell ionization in these ions, as well as with non-relativistic calculations and relativistic calculations without retardation and magnetic effects. The measured cross sections at about 1.7 times threshold are significantly higher than the calculated cross sections, while the difference between the various calculated cross sections are quite small at this electron energy.[13]

For H-like uranium, a comparison of the QED results, with the cross section obtained using our non-relativistic ionization program, and also with the cross section obtained using a Dirac-Fock ionization code, which does not include magnetic and retardation effects,[15] is shown in Fig. 4. We see that the calculated cross sections are all in close agreement up to about

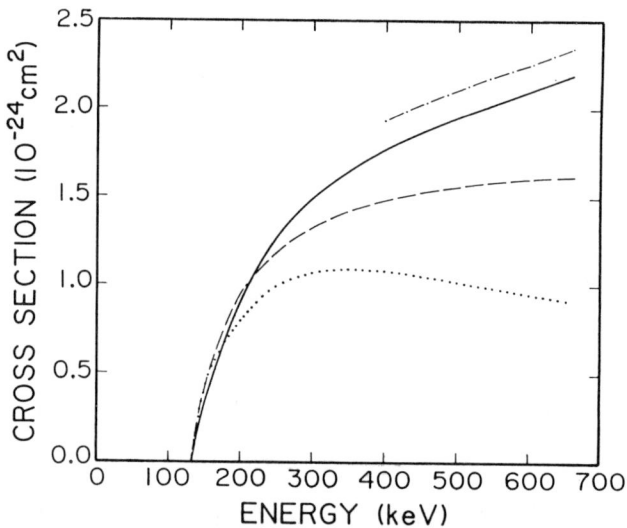

Fig. 4. Electron-impact ionization cross section for H-like uranium. Solid curve, QED theory; dashed curve, Dirac-Fock calculation without retardation and magnetic effects; dotted curve non-relativistic theory; chain curve, plane-wave theory of Scofield (Ref. 16).

200 keV, but show significant departures at higher energies. We also compare our QED cross section with the plane-wave QED results of Scofield.[16] The Scofield calculation was made for K-shell ionization of neutral uranium, so that his cross section was divided by two to obtain the curve shown in Fig. 4. The agreement with the Scofield calculation would be somewhat better if it had been adjusted for the difference between the ionization potential for K-shell electrons in neutral uranium (116.3 keV) and that for hydrogen-like uranium (132.2 keV).

ELECTRON-IMPACT RECOMBINATION

Finally we consider electron-impact recombination of an N-electron ion. The first process is three-body recombination

$$e^- + e^- + X(Z,N) \longrightarrow X(Z,N+1) + e^- . \quad (11)$$

It is just the reverse of electron-impact ionization, and due to its relatively low probability, it is only important at high electron densities. Next we have radiative recombination

$$e^- + X(Z,N) \longrightarrow X(Z,N+1) + h\nu . \quad (12)$$

This is just the reverse of photoionization, and it dominates the recombination cross section at relatively low electron energies; however, its importance decreases rapidly with energy. Finally, we have the indirect process of dielectronic recombination (DR)

$$e^- + X(Z,N) \longrightarrow X(Z,N+1)^{**} \longrightarrow X(Z,N+1) + h\nu . \quad (13)$$

This process involves resonant recombination to a doubly excited state of the (N+1)-electron ion followed by radiative decay to a bound state. Of course, the doubly excited state can also autoionize and provide a resonance contribution to the elastic or excitation cross section of the N-electron ion. DR is the dominant recombination process at relatively high energies, and we shall now consider it in more detail.

In the isolated-resonance approximation, the DR cross section from an initial level i of the N-electron ion is given by

$$\sigma_{dr}(i) = \sum_j \sigma_{rr}(i \to j) B_j^r , \qquad (14)$$

where again, $\sigma_{rr}(i \to j)$ is the resonant recombination cross section from level i of the N-electron ion to level j of the (N+1)-electron ion and is given by Eq. (7). B_j^r is the branching ratio for radiative decay to a bound state of the (N+1)-electron ion and is given by

$$B_j^r = \frac{\sum_n A_r(j \to n)}{\sum_k A_a(j \to k) + \sum_m A_r(j \to m)} , \qquad (15)$$

where the sum over n is over all lower bound levels of the (N+1)-electron ion, while the sum over m is over all lower levels of the (N+1)-electron ion, bound and autoionizing.

As an example, let us consider dielectronic recombination associated with the 2s→2p excitation in Li-like C^{3+}

$$e^- + C^{3+}(2s) \longrightarrow C^{2+}(2pn\ell) \longrightarrow C^{2+}(2sn\ell) + h\nu$$
$$\longrightarrow C^{2+}(2pn'\ell') + h\nu' . \qquad (16)$$

The radiative transitions involving the Rydberg electron ($n\ell \to n'\ell'$) can dominate for low Rydberg states, but their importance falls off rapidly with the principal quantum number n. On the other hand, the radiatve rates for the transitions involving the core electron (2p→2s) are nearly independent of the Rydberg electron. An energy-level diagram for this system is shown in Fig. 5.

The low lying resonances such as 2p4d have relatively large cross sections because of the size of the radiative rates for transitions in which the Rydberg electron is the active electron; however, these resonances are rather widely spaced. On the other hand, the cross sections for the closely spaced high-Rydberg states are smaller, but they continue to contribute to the cross section until we reach rather high values of the principal quantum number, n, where the resonant-recombination cross section (which falls of approximately as $1/n^3$) becomes so small that the states become essentially closed to recombination. Thus, the DR cross section associated with such $\Delta n = 0$ excitations are dominated by high Rydberg states.

External electric fields can have a pronounced effect on dielectronic recombination. First of all such

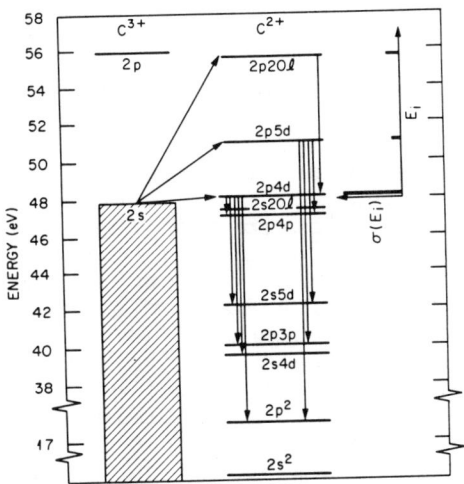

Fig. 5. An energy level diagram illustrating several DR transitions associated with the 2s→2p excitation in C^{3+}. A schematic diagram of the DR cross section resulting from transitions through the 2p4d, 2p5d, and 2p20ℓ configurations of C^{2+} is inset on the far right.

fields can ionize electrons in high Rydberg states, and thereby decrease the DR cross section. Secondly, fields mix high Rydberg states with different values of the orbital angular momentum ℓ. For DR associated $\Delta n = 0$ transitions such as the 2s→2p excitation in C^{3+}, the rates for resonant recombination (compared to the radiative rates) are high for small values of ℓ and small for high values of ℓ; therefore, in the absence of a field, the high angular momentum states are essentially closed to recombination. However, external fields mix these states in such a way as to open up more channels for recombination and thereby enhance the cross section.[17]

In DR experiments, the fields in the interaction region are relatively small, while the fields in the analyzing region are very large, typically of the order of 10 kV/cm. Thus the smaller fields in the interaction region enhance the DR cross section, and then after recombination has occurred, the fields in the interaction region strip off the high Rydberg states. The effects of field ionization can be estimated with reasonable accuracy by using a semiclassical formula to determine a maximum value of the principal quantum number n_m, above which field ionization will occur.[18]

We determine the effects of field mixing in the interaction region by employing eigenvectors for the doubly excited states which are obtained by diagonalizing a Hamiltonian matrix that includes the Stark matrix elements as well as the internal electrostatic and spin-orbit terms.[19] Finally, the DR experiments do not measure the cross section, but rather the product of the electron velocity and the cross section convoluted with an electron velocity distribution function $<v\sigma>$ - a sort of rate coefficient which is a function of electron energy. Thus we combine our cross section with the experimental velocity distribution to make a theoretical prediction of the measurement.

The first DR measurements on a series of Li-like and Na-like ions were done by Dittner et al. at ORNL.[20,21] In this set of experiments, the field in the interaction region was expected to be about 30 V/cm. Although the agreement between experiment and our theoretical calculations[19] was quite good for the Na-like ions,[20] the measured rates were larger than the calculated rates for the Li-like ions (including the C^{3+} case considered above) even for the maximum field enhancement of the theoretical cross sections.[21] Furthermore, the velocity distribution in these experiments was sufficiently wide that a comparison between experiment and theory was impossible for low Rydberg states, which are not affected by electric fields.

Andersen and Bolko have now completed measurements of DR for C^{3+} and O^{5+} using the Aarhus EN-tandem accelerator and a beamline equipped with an electron cooler.[22] The electric field in the interaction region is expected to be less than 5V/cm and the electron distribution is very narrow, so that individual low Rydberg states are resolved.

In Fig. 6 are shown their measurements for C^{3+} in comparison with our most recent calculations for various small fields in the interaction region[23]. The agreement between experiment and theory in the region of the high Rydberg states for energies between 7.0 and 8.0 eV is excellent for a field of about 3 V/cm. However, the calculated rate for the peak between 0 and 0.6 eV, which is due to the 2p4d and 2p4f resonances, is nearly twice the measured rate. Since the cross section for resonant recombination varies as the inverse of the electron energy (see Eq. 7), and these resonances lie so close to zero, this discrepancy is most likely due to errors in the calculated positions of the doubly excited levels for these configurations. The agreement with experiment for O^{5+} is not quite as good, but still very reasonable.[22,23]

Andersen et al.[22,24] have also measured DR from the $1s2s(^3S)$ and $1s2s(^1S)$ metastable states of He-like C^{4+} and O^{6+}. However, comparison of these measurements with

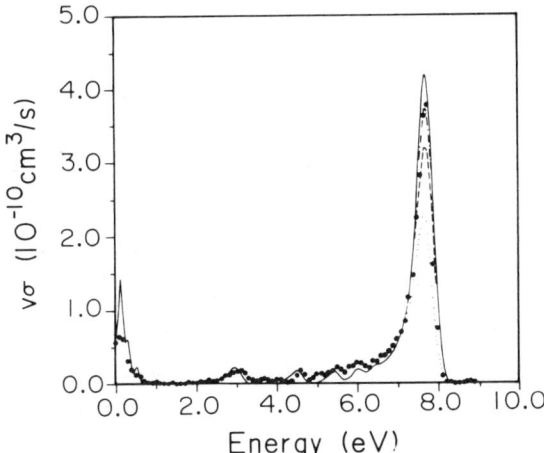

Fig. 6. Theoretical DR rate coefficients for C^{3+} (as measured in the Aarhus electron-ion merged-beam experiment) as a function of electron energy. Dotted curve, E = 0 V/cm; dashed curve, E = 1 V/cm; dot-dash curve, E = 3 V/cm; and solid curve, E = 5 V/cm. The calculations include all resonances up to and including n_m = 35. The experimental points are from Ref. 22.

theoretical calculations are complicated by the fact that the relative populations of the incident ion beam in these states is not known. The dominant DR transitions are

$$e^- + 1s2s(^3S) \longrightarrow 1s2p(^1P)n\ell \longrightarrow 1s^2n\ell + h\upsilon \quad (17)$$

and

$$e^- + 1s2s(^1S) \longrightarrow 1s2p(^1P)n\ell \longrightarrow 1s^2n\ell + h\upsilon . \quad (18)$$

The $1s2s(^1S)n\ell$ and $1s2p(^3P)n\ell$ doubly excited states are also populated in transitions from $1s2s(^3S)$, but they only weakly radiate to bound states, and then only for $\ell=1$. The competing autoionizing transitions from $1s2p(^1P)n\ell$ are

$$1s2p(^1P)n\ell \longrightarrow 1s^2 + e^- , \quad (19)$$

and for high enough values of n

$$1s2p(^1P)n\ell \longrightarrow 1s2s(^3S) + e^-$$
$$ \longrightarrow 1s2s(^1S) + e^- \qquad (20)$$
$$ \longrightarrow 1s2p(^3P) + e^- .$$

Theoretical calculations of these transitions in C^{4+} and O^{6+} have recently been completed.[25] Effects from electric fields in the interaction region are expected to be very small for these cases since the dominant radiative transitions $1s2p(^1P)n\ell \rightarrow 1s^2 n\ell$ have rates that are competive with the autoionizing rates.[25] In order to compare the calculations with the measurements, the ratio of the $1s2s(^3S)$ to $1s2s(^1S)$ populations were varied until the best agreement with experiment was obtained. In the case of C^{4+} this ratio was found to be 18, while in the case of O^{6+} it was 70.

The results for C^{4+}, plotted on an arbitrary scale, are shown in Fig. 7. As can be seen, there is excellent agreement between theory and experiment with regard to the relative heights of the various peaks. In the case of O^{6+}, the agreement is not as good, but still quite

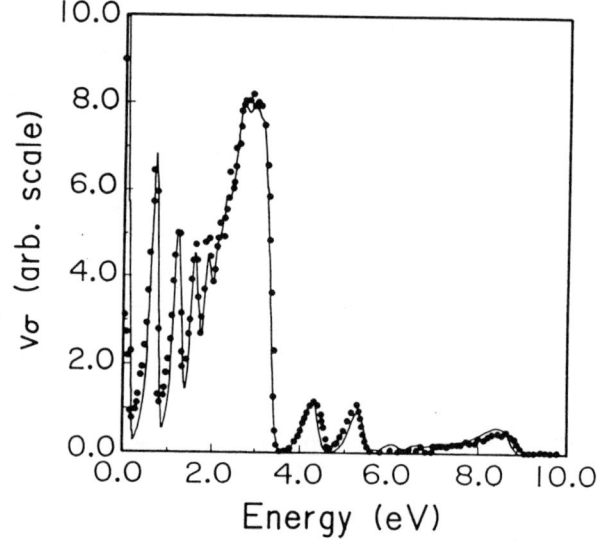

Fig. 7. Comparison of theoretical and experimental rate coefficients (as measured in the Aarhus experiment) for DR from the metastable states of C^{4+}, on an arbitrary scale. Solid curve, theory with ratio of the population fractions of $1s2s(^3S)$ to $1s2s(^1S)$ equal to 18; solid circles, experimental points from Ref. 22.

acceptable, especially in light of the sensitivity of the calculation to the positions of the various resonant states.[25] By comparing the theoretical calculations with the absoulute values of the measured rates, it is also possible to determine the population of the $1s2s(^3S)$ states which give the best agreement with experiment. In the case of O^{6+}, this turns out to be 20%, which is quite reasonable. However, in the case of C^{4+}, a value of 70% is obtained; one would expect about the same population of $1s2s(^3S)$ in both ions,[25] and the reason for this apparent discrepancy is not understood at this time.

CONCLUSIONS

Indirect process play an important role in electron ion-collisions. Perturbation methods in which multi-step processes are determined from branching ratio calculations, seem to give reasonably good agreement with total cross-section measurements. However, high-resolution ionization experiments,[2,26] as well as high-resolution DR experiments using electron coolers[22,24] and ion storage rings,[27] are beginning to provide more complete tests of the theoretical methods for calculating electron-impact ionization and recombination. As these experiments move to more highly ionized species, the contributions from relativistic effects, perhaps including QED corrections, may be revealed.

ACKNOWLEDGEMENTS

The authors acknowledge the contributions of their collaborators Chris Bottcher, Nigel Badnell, and David Moores. We thank Mau Chen, Kennedy Reed, and David Moores for providing us with the results of their ionization calculations for Au^{68+} and Fe^{15+} prior to publication, and Don Gregory for providing ionization data on Ti^{11+} and Cr^{13+} prior to publication. This work was supported by the Office of Fusion Energy, U. S. Department of Energy, under Contract No. DE-AC05-84OR21400 with Martin Marietta Energy Systems, Inc. and Contract No. DE-FG05-86ER53217 with Auburn University.

REFERENCES

1. M. S. Pindzola, and D. C. Griffin, Phys. Rev. <u>A36</u>, 2628 (1987).
2. A. Müller, G. Hofmann, K. Tinschert, and E. Salzborn, Phys. Rev. Letts., <u>61</u>, 1352 (1988).
3. R. D. Cowan, The Theory of Atomic Structure and Spectra (California Press, Berkeley, 1981).
4. R. D. Cowan and D. C. Griffin, J. Opt. Soc. Am. <u>66</u>, 1010 (1976).

5. D. C. Griffin, M. S. Pindzola, and C. Bottcher, Phys. Rev. A36, 3642 (1987).
6. M. S. Pindzola, D. C. Griffin, and C. Bottcher, Phys. Rev. A33, 3787 (1986).
7. M. S. Pindzola, D. C. Griffin, and C. Bottcher, Phys. Rev. A34, 3668 (1986).
8. D. C. Griffin and M. S. Pindzola, J. Phys. B21, 3253 (1988).
9. K. J. Reed, M. H. Chen, and D. L. Moores (private communication).
10. D. C. Gregory, L. J. Wang, K. Rinn, and F. W. Meyer, Phys. Rev. A35, 3256 (1987).
11. D. C. Gregory (private communication).
12. M. H. Chen, K. J. Reed, and D. L. Moores (private communication).
13. M. S. Pindzola, D. L. Moores, and D. C. Griffin, Phys. Rev. A40 (1989).
14. N. Claytor, B. Feinberg, H. Gould, C. E. Bemis, Jr., J. Gomez del Campo, C. A. Ludemann, C. R. Vane, Phys. Rev. Lett. 61, 2081 (1988).
15. M. S. Pindzola and M. J. Buie, Phys. Rev. A37, 3232 (1988).
16. J. H. Scofield, Phys. Rev. A18, 963 (1978).
17. D. C. Griffin, M. S. Pindzola, and C. Bottcher, Atomic Excitation and Recombination in External Fields, edited by M. H. Nayfeh and C. W. Clark (Harwood Academic, New York, 1985)
18. C. Bottcher, D. C. Griffin, and M. S. Pindzola, Phys. Rev. A34, 860 (1986).
19. D. C. Griffin, M. S. Pindzola, and C. Bottcher, Phys. Rev. A33, 3124 (1986).
20. P. F. Dittner, S. Datz, P. D. Miller, P. L. Pepmiller, and C. M. Fou, Phys. Rev. A33, 124 (1986).
21. P. F. Dittner, S. Datz, P. D. Miller, P. L. Pepmiller, and C. M. Fou, Phys. Rev. A35, 3668 (1987).
22. L. H. Andersen and J. Bolko Phys. Rev. A40, ####, (1989).
23. D. C. Griffin, M. S. Pindzola, and P. Krylstedt, Phys. Rev. A40 (1989).
24. L. H. Andersen, P. Hvelplund, H. Knudsen, and P. Kvistgaard, Phys. Rev. Lett. 62, 2656 (1989).
25. N. R. Badnell, M. S. Pindzola, and D. C. Griffin, accepted, Phys. Rev. A.
26. A. Müller, G. Hofmann, B. Weissbecker, M. Stenke, K. Tinschert, M. Wagner, and E. Salzborn, Phys. Rev. Lett. 63, 758, (1989).
27. G. Kilgus, J. Berger, P. Blatt, M. Grieser, D. Habs, B. Hochadel, E. Jaeschke, D. Krämer, R. Neumann, G. Neureither, W. Ott, D. Schwalm, M. Steck, R. Stokstad, E. Szmola, and A. Wolf, submitted to Phys. Rev. Lett.

RELATIVISTIC AND QUANTUM ELECTRODYNAMIC EFFECTS IN HIGHLY-CHARGED IONS

Yong-Ki Kim
National Institute of Standards and Technology
Gaithersburg, Maryland 20899

ABSTRACT

Recent spectroscopic data clearly indicate that both relativistic and quantum electrodynamic (QED) effects are indispensable in understanding the spectra of highly-charged ions. QED effects are discernible even in spectra involving M- and N-shell electrons. Current understanding and theoretical capability in predicting relativistic and QED corrections in atomic structure are reviewed.

1. INTRODUCTION

Recent advances in technologies that depend on hot and/or dense plasmas, such as high-temperature fusion devices and powerful lasers, have amplified the need for better understanding of and predictive capability for the properties of highly-charged ions. For a precise prediction or verification of atomic properties, such as energy levels and transition probabilities, nonrelativistic theories based on independent particle models are inadequate. We must supplement such theories with improvements to account for many-body effects known as electron correlation, relativistic corrections, which arise mainly from replacing the Schrödinger Hamiltonian by the Dirac Hamiltonian, and QED corrections, which have no nonrelativistic counterpart.

Relativistic and QED corrections affect every bound electron in hydrogenic ions as well as heavy neutral atoms. Changes caused by these corrections to the hydrogenic energy levels predicted by the nonrelativistic Schrödinger theory are illustrated schematically in Fig. 1. In neutral and lightly-charged atoms, electron correlation effects—effects due to mutual screening of bound electrons—are the most serious source of uncertainty in theoretical descriptions of such atoms.

Relativistic effects become significant only for atoms with moderate to high nuclear charge Z. The effective nuclear charge, Z_{eff}, seen by valence electrons in neutral atoms of $Z \approx 20$–50 is rather low, $Z_{eff} < 5$, and valence-shell properties are not seriously affected by relativity. The inner shells of such atoms, however, are significantly affected by relativity, while the influence of correlation effects is markedly reduced.

On the other hand, valence shells of heavy atoms are affected both by relativity and correlation, while QED corrections—often called the Lamb shift or radiative corrections—gain importance in inner shells of heavy atoms.

In this article, theoretical methods for relativistic and QED corrections are reviewed and typical predictions based on such theories are compared with experimental data to indicate current theoretical capabilities.

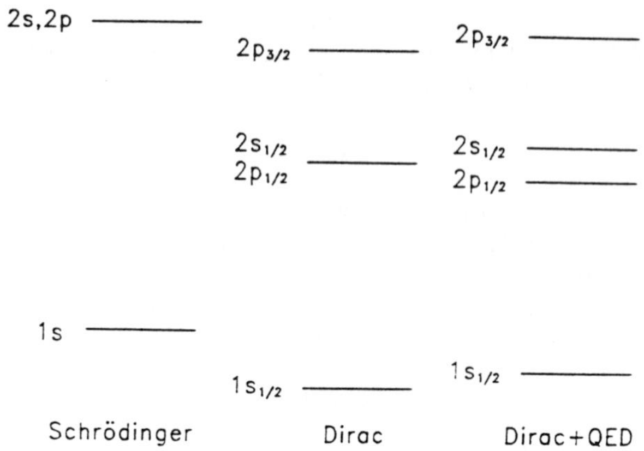

Figure 1. Schematic energy levels of hydrogen-like ions.

2. MULTICONFIGURATION DIRAC-FOCK (MCDF) METHOD

2.1. Hartree-Fock Wave Function

One of the most successful and useful approaches to describing an atom is the Hartree-Fock (HF) method,[1] which is based on the Schrödinger Hamiltonian of an atom:

$$H_{nr} = \sum_j [\mathbf{p}_j^2/2m_0 - Ze^2/r_j] + \sum_{j>k} e^2/r_{jk}, \quad (1)$$

where the summations are over all bound electrons, $\mathbf{p}_j$ is the momentum of the jth electron, e is the electronic charge, m_0 is the electron (rest) mass, r_j is the distance between the nucleus and the jth electron, and r_{jk} is the distance between the jth and kth electrons. A HF wave function, Ψ_{HF}, is a linear combination of Slater determinants, which are, in turn, antisymmetrized products of one-electron orbitals, $u_{n\ell m}$:

$$u_{n\ell m} = r^{-1} P_{n\ell}(r) Y_{\ell m}(\theta, \phi), \quad (2)$$

where n is the principal quantum number, $P_{n\ell}$ is the radial function, $Y_{\ell m}$ is the spherical harmonics with the orbital angular momentum ℓ and its projection m. A HF wave function can be made to represent a specific orbital angular momentum $\mathbf{L}$ and spin angular momentum $\mathbf{S}$ by choosing appropriate combinations of ℓ, m and two-component spin functions on the RHS of Eq. (2). The radial functions are determined by coupled equations that are obtained by applying the variational principle to the expectation value of H_{nr}, i.e.,

$$\delta \langle \Psi_{HF} | H_{nr} | \Psi_{HF} \rangle = 0, \quad (3)$$

with a subsidiary condition that the radial functions with the same ℓ be orthogonal to each other. A single Slater determinant is adequate to describe the HF wave function for a closed-shell atom, such as the rare gases. However, more determinants with different m_ℓ and m_s are needed to represent an open-shell atom in most cases.[2]

The radial functions can be numerically tabulated as functions of r, or approximated by linear combinations of simple analytic functions of r, known as the basis functions. The latter, commonly called analytic HF wave functions,[3] are compact and convenient for applications but their quality depends on the choice of the basis functions.

Much of the numerical effort in calculating HF wave functions is spent in satisfying the Pauli exclusion principle. The reason for using a Slater determinant instead of a straightforward product of u's is to satisfy this requirement. The use of the Slater determinant introduces terms in the Hartree-Fock equation for the radial functions known as exchange terms. To save numerical effort, one often omits the exchange terms (Hartree approximation) or replaces them with simpler expressions (e.g., Hartree-Slater approximation, average-configuration approximation). Using the unaltered exchange terms in a Hartree-Fock calculation requires a new set of radial functions to be calculated for each level. This is sometimes referred to as the optimized-level calculation. In principle, the optimized-level calculation yields the most reliable wave functions, but the resulting wave functions for different levels are not necessarily orthogonal to each other unless the total angular and spin quantum numbers are different.

Although the total energy predicted by a Hartree-Fock wave function is accurate within a percent or two, predictions for transition energies, i.e., the difference between two total energies, are too crude (a few to 10%) to be of any practical value. Traditionally, the difference between the total energy predicted by a single-configuration HF wave function (with minimum number of one-electron orbitals) and the corresponding experimental value, after adjusting for relativistic corrections, is defined as the *correlation energy*.[4]

For a theoretical prediction of energy levels to be reliable, this correlation energy, along with relativistic and QED corrections, must be determined to very high accuracy. There are many competing methods to improve on the correlation energy. A popular method is to use more than the minimum number of one-electron orbitals to form Ψ with more flexibility. These extra orbitals are called correlation orbitals.

2.2. Configuration Interaction and Multiconfiguration Wave Functions

One can choose several combinations of orbitals that lead to the same total **L**, **S** and parity as correlation orbitals.. For instance, the minimum set of orbitals needed to describe the ground state of Be is $1s^2\ 2s^2$. We can also include a 2p correlation orbital by adding the configuration $2p^2$, i.e.,

$$\Psi(\text{Be}) = a\psi(2s^2) + b\psi(2p^2), \qquad (4)$$

where a and b are configuration mixing coefficients and the K-shell core is omitted for brevity. The wave function $\psi(2p^2)$ is chosen to be an eigenfunction of

$L^2 = S^2 = 0$. Equation (4) is an example of a multiconfiguration wave function. The mixing coefficients a and b are determined by sandwiching Eq. (1) with Eq. (4) to build a Hamiltonian matrix and diagonalizing the matrix. A multiconfiguration wave function is further classified according to the choice of ψ's.

If ψ's are preselected by another method, e.g., hydrogenic functions with a Z_{eff}, and kept constant thereafter, then it is called a *configuration-interaction* (CI) wave function. If the coupled Hartree-Fock equations are iteratively solved with different values of a and b such that an optimum set of radial functions and mixing coefficients are simultaneously obtained, then the solution is called a *multiconfiguration Hartree-Fock* (MCHF) wave function.[1]

In general, MCHF wave functions are more efficient in representing electron correlation than CI wave functions if both contain the same number of configurations, but the former is numerically more difficult to calculate. One can combine the MCHF method with the CI method by using an MCHF wave function to represent the leading correlation effect and then adding a large number of correlation orbitals using the CI method to evaluate the rest of the correlation correction.

As is shown in the next section, most of the formalism for the nonrelativistic HF, CI, and MCHF wave functions is applicable to their relativistic counterparts.

2.3. Dirac-Fock and Multiconfiguration Dirac-Fock Wave Functions

The choice of the terms in the nonrelativistic Hamiltonian, H_{nr}, is well justified from the classical mechanics, but its extension to the relativistic case is by no means trivial. Formal solutions of the Dirac Hamiltonian,

$$H_D = \vec{\alpha} \cdot \mathbf{p}c + \beta m_0 c^2 - Ze^2/r, \qquad (5)$$

where c is the speed of light, $\vec{\alpha}$ and β are 4×4 Dirac matrices, have both positive and negative energy eigenvalues.[5] Hence, when a relativistic, many-body Hamiltonian, H_r, is formed by replacing the one-electron operators in H_{nr} by Eq. (5), viz.,

$$H_r = \sum_j [\vec{\alpha}_j \cdot \mathbf{p}_j c + \beta_j m_0 c^2 - Ze^2/r_j] + \sum_{j>k} e^2/r_{jk}, \qquad (6)$$

the formal solutions of H_r include infinitely many combinations of degenerate states where, for instance, one electron has its energy in the positive continuum while another in the negative continuum such that the total energy is still the same as that for a solution in which both electrons are bound.

This dilemma is called the Brown-Ravenhall "disease" or the continuum dissolution.[6] This does not imply that atoms do not have stable bound states; this is simply a *mathematical* difficulty associated with the solutions of a particular Hamiltonian operator. One can introduce formal projection operators into H_r to allow only the positive-energy solutions.[7] In practice, numerical procedures to obtain relativistic eigenfunctions of H_r are limited to functional forms appropriate only to the positive-energy solutions.

In analogy to the nonrelativistic case, Eq. (2), the Dirac Hamiltonian has eigensolutions of the form:

$$\psi_{n\kappa m} = r^{-1} \begin{pmatrix} P_{n\kappa}(r)\chi_{\kappa m}(\theta,\phi) \\ iQ_{n\kappa}(r)\chi_{-\kappa m}(\theta,\phi) \end{pmatrix}, \tag{7}$$

where κ is the Dirac quantum number (e.g., $\kappa = -1$ for an s electron, $+1$ for a $p_{1/2}$ electron, -2 for a $p_{3/2}$ electron, $+2$ for a $d_{3/2}$ electron, and so on), $P_{n\kappa}$ is the large-component radial function, $Q_{n\kappa}$ is the small-component radial function, and $\chi_{\kappa m}$ is a 2-component angular function,

$$\chi_{\kappa m} = \sum_{\sigma=\pm 1/2} C(\ell s j, m-\sigma, \sigma) Y_{\ell, m-\sigma}(\theta,\phi)\varphi_\sigma, \tag{8}$$

with $s = 1/2$,

$$\varphi_{\frac{1}{2}} = \begin{pmatrix} 1 \\ 0 \end{pmatrix}, \quad \varphi_{-\frac{1}{2}} = \begin{pmatrix} 0 \\ 1 \end{pmatrix}, \tag{9}$$

and $C(\ell s j; m - \sigma, \sigma)$ is the Clebsch-Gordan coefficients. The Dirac quantum number κ is related to the angular quantum number j by $j = |\kappa| - 1/2$.

The bound-state eigenvalues of the Dirac Hamiltonian for a hydrogenic system are given by

$$E_{n\kappa} = m_0 c^2 \left[\frac{1}{\sqrt{1+(\alpha Z/n^*)^2}} - 1\right], \tag{10}$$

where α is the fine-structure constant, and

$$n^* = n - |\kappa| + \sqrt{\kappa^2 - \alpha^2 Z^2}. \tag{11}$$

Note that $E_{n\kappa}$ depends on n and j unlike the nonrelativistic case, which depends only on the principal quantum number. The rest mass is subtracted out in Eq. (10) to be consistent with the nonrelativistic definition of energy levels. In the limit $\alpha \to 0$, the nonrelativistic formula for the hydrogenic energy level is recovered:

$$E_{n\kappa} \cong -\frac{Z^2}{n^2}\left(1 + \frac{\alpha^2 Z^2}{4n\kappa^2}\right)\left(\frac{m_0 c^2 \alpha^2}{2}\right). \tag{12}$$

The last factor on the RHS of Eq. (12) is the Rydberg constant (= 13.606 eV). Equation (12) clearly indicates that the relativistic corrections to the hydrogenic energy levels is to lower the levels for all κ, that is, all ℓ and j. Also, the relativistic expectation values of r are smaller than the nonrelativistic counterparts. These properties of the Dirac orbitals are modified by mutual screening of bound electrons in a many-electron heavy atom, such that the one-electron energy parameters and $\langle r \rangle$ values may exhibit opposite trends from that predicted by the Dirac theory for a one-electron ion. For instance, the DF value of the binding energy of the 4f orbital in Hg is about 10% less than the value predicted by the HF wave function, while the relativistic value of $\langle r \rangle_{4f}$ is about 2% larger than the corresponding nonrelativistic value.[8]

In analogy with the nonrelativistic case, one can construct Slater determinants of Dirac orbitals, make linear combinations of the Slater determinants to form eigenfunctions of $\mathbf{J}^2$, take the expectation value of H_r, apply the variational principle, and derive Dirac-Fock equations for the radial functions. The radial functions can be either numerical or analytic, though care must be taken in selecting the appropriate type of basis functions in the latter case.[9,10]

In reality, the amount of numerical work to calculate Dirac-Fock wave functions far exceeds that for Hartree-Fock wave functions because each relativistic orbital has two radial functions, the large and small components, and for each ℓ, there are two distinct j's except for an s orbital.

The use of Dirac orbitals, Eq. (7), requires jj coupling, which is not convenient to represent neutral or lightly-charged atoms. Combinations of jj configurations are needed not only to cope with electron correlation but also to represent intermediate coupling. For instance, the 1s2p state of He actually consists of four levels, ^{3}P, $J = 0, 1,$ and 2 and ^{1}P, $J = 1$. In terms of relativistic orbitals, the $J = 0$ level consists of 1s2p$_{1/2}$; the $J = 2$ level consists of 1s2p$_{3/2}$; and the two $J = 1$ levels are combinations of the 1s2p$_{1/2}$ and 1s2p$_{3/2}$ configurations. To distinguish the 1s2p ^{3}P$_1$ and ^{1}P$_1$ levels in a relativistic calculation, a multiconfiguration calculation that mixes the 1s2p$_{1/2}$ and 1s2p$_{3/2}$ configurations must be performed. Such a wave function is called a multiconfiguration Dirac-Fock (MCDF) wave function.[11] In this example, the combination that results in the lower total energy will correspond to the ^{3}P$_1$ level, and the upper total energy to the ^{1}P$_1$ level. The analogous nonrelativistic calculation must mix ^{3}P and ^{1}P wave functions through the spin-orbit interaction. In general, the total number of configurations needed to account for electron correlation in jj coupling is far greater than the corresponding nonrelativistic case, imposing more numerical burden on relativistic calculations.

2.4. Breit Operator

Relativity modifies the interaction between two bound electrons. One can derive the leading correction to the Coulomb repulsion between two bound electrons known as the Breit operator,[12] but the resulting expression is gauge dependent. Although the Lorentz gauge (also called the Feynman gauge) leads to a simpler expression, Sucher[13] has shown that the Coulomb gauge is more consistent with the perturbation theory with which the Breit operator is derived. The Breit operator in the Coulomb gauge is given by:

$$H_B(\omega) = -e^2 \sum_{j>k} \{ \frac{\vec{\alpha}_j \cdot \vec{\alpha}_k}{r_{jk}} \exp(i\omega_{jk} r_{jk}) \\ - \frac{(\vec{\alpha}_j \cdot \nabla_j)(\vec{\alpha}_k \cdot \nabla_k)}{\omega_{jk}^2 r_{jk}} [\exp(i\omega_{jk} r_{jk}) - 1] \}, \quad (13)$$

where

$$\omega_{jk} = |\epsilon_j - \epsilon_k|/c, \quad (14)$$

with ϵ_j and ϵ_k being the one-electron eigenvalues of the jth and kth electron, respectively. The ω-dependent exponential factors in Eq. (13) are called the "frequency"-dependent part of the Breit operator. There is some ambiguity in identifying ϵ, which is a one-electron energy in a many-electron atom. The first term on the RHS of (13) represents the magnetic interaction between two bound electrons, and the second term arises from the retardation of the Coulomb repulsion. The differential operators in the second term apply only to the exponential factor after it, but not to wave functions when an expectation value of $H_B(\omega)$ is taken.

The frequency-dependence becomes significant only for very high values of Z. One can safely use, in most applications, a simpler form of $H_B(\omega)$ that is independent of ω, thus avoiding the ambiguity associated with it. This frequency-*independent* form of the Breit operator, H_B, is obtained by taking the limit $\omega \to 0$ in Eq. (13):

$$H_B = -e^2 \sum_{j>k} \{\frac{\vec{\alpha}_j \cdot \vec{\alpha}_k}{r_{jk}} + \frac{[(\vec{\alpha}_j \cdot \nabla_j)(\vec{\alpha}_k \cdot \nabla_k)r_{jk}]}{2}\}. \tag{15}$$

Here again, the differential operator applies only to the r_{jk} in the bracket but not to the wave function. The contribution to the total energy from the magnetic interaction term is positive—i.e., reduces the binding energy—while the contribution from the retardation term is negative and an order of magnitude less than the contribution from the magnetic interaction.

In using $H_B(\omega)$ or H_B, one can either include it in the main Hamiltonian in a variational calculation or use it as a perturbation after wave functions have been determined using H_r only. Including the Breit operator in a variational calculation requires much longer computer time with slight, improvements in energy expectation values. Only deep inner-shell orbitals of heavy atoms are affected by including the Breit operator in a variational calculation.

In terms of the Feynman diagrams, the Breit interaction represents the exchange of one transverse virtual photon between two bound electrons.[5] (The exchange of one longitudinal virtual photon results in the usual Coulomb repulsion.) The study of contributions from Feynman diagrams that correspond to the exchange of two or more virtual photons is the subject of current interest.

2.5. Z-Expansion Theory

Trends in various atomic properties, such as energy levels, transition probabilities, excitation cross sections and ionization cross sections, can easily be identified if the properties are studied along an isoelectronic sequence. Although *ab initio* results may not be numerically accurate, their qualitative behavior as a function of Z is often correct and hence useful in verifying experimental results. The leading terms of many atomic properties, energy levels being one of them, have a definite Z dependence.

It can be shown that the nonrelativistic total energy of an atom, E_{nr}, can be expanded in a descending power series of Z[14]:

$$E_{nr} = (m_0 c^2 \alpha^2/2)[E_2 Z^2 + E_1 Z + E_0 + E_{-1}/Z + E_{-2}/Z^2 + \ldots], \tag{16}$$

where E_2 is the sum of the hydrogenic energies, $E_{hyd} = -Z^2/n^2$ rydberg, and the remaining terms arise from the Coulomb repulsion between bound electrons and mutual screening of the electrons, i.e., electron correlation. The values of E_i, $i \leq 1$, in Eq. (16) depend not only on the quantum numbers of individual electrons but also on the total quantum numbers such as **L**, **S** and seniority.

A nonrelativistic Hartree-Fock wave function that consists of one configuration, such as the first term on the RHS of (4), leads to the correct E_2 but only part of the rest of E_i. To obtain as accurate values of E_i as possible, one must introduce correlation orbitals, i.e., go beyond a single-configuration wave function. For two-electron atoms, one can use the Hylleraas-type wave functions, which contain terms that depend explicitly on the interelectronic separation, r_{ij}. For more complex atoms, it is common to use multiconfiguration wave functions.

The values of E_2, E_1 and E_0 for a large number of ions, including configurations for excited states, are available in the literature.[15] As is clear from Eq. (16), however, the knowledge of only the first three coefficients in the Z expansion is insufficient to determine the energy levels of neutral atoms and lightly-charged ions. For highly-charged ions, the first three coefficients in Eq. (16) are adequate, provided that relativistic and QED corrections are also included since these corrections are substantial for such ions.

In contrast to Eq. (16), relativistic corrections for an atom arising from the Dirac Hamiltonian, Eq. (5), are given by:

$$E_{Dirac} = E_{hyd}[1 + E^{(2)}(\alpha Z)^2 + E^{(4)}(\alpha Z)^4 + \ldots], \quad (17)$$

where the coefficients $E^{(\mu)}$ are determined from the power series expansion of the Dirac energy formula, Eq. (10). For highly-charged ions, such an expansion converges very slowly, often requiring expansions up to $E^{(8)}$ or more to maintain high accuracy.

For a many-electron atom, Eqs. (16) and (17) can be combined to yield a double series in descending powers of Z and ascending powers of $(\alpha Z)^2$ by replacing E_{hyd} in Eq. (17) by E_{nr} in Eq. (16). The resulting Z dependence in such a combination is far more complicated than those seen either in Eq. (16) or (17) alone.

The expectation value of the Breit operator, H_B, has the following Z dependence:

$$E_B \equiv \langle \Psi | H_B | \Psi \rangle$$
$$= (m_0 c^2 \alpha^2)(\alpha Z)^2 [aZ + b + c/Z + \ldots], \quad (18)$$

where a, b, c, etc. are constants that depend on the quantum numbers of the total wave function. In general, the magnitude of E_B is very small compared to the expectation values of the Coulomb repulsion. Higher-order relativistic corrections to the Breit operator, e.g., two-photon exchange Feynman diagrams, are expected to have a leading term of the order of α compared to the leading term of E_B.

3. QUANTUM ELECTRODYNAMIC CORRECTIONS

Corrections to the energy levels resulting from the Dirac Hamiltonian are referred to as the quantum electrodynamic (QED) corrections. In this sense, the Breit operator is also a QED correction, but it is common to treat the Breit operator separately because Breit developed his theory decades before the modern QED theory reproduced his result.[12] Most QED effects are concerned with the interaction of each bound electron with the field surrounding it through the exchange of virtual photons. These effects are classified according to the corresponding Feynman diagrams, and the magnitude of each effect diminishes as the number of virtual photons increases. The Feynman theory provides systematic ways to formulate matrix elements representing various energy corrections, but the resulting formulas do not easily avail themselves to numerical evaluation.

In the early days of QED theory, perturbation expansions in powers of αZ were used to obtain numerical results, but applications to heavy atoms had to wait until a pioneering work by Mohr,[16] who developed nonperturbative method to evaluate QED corrections.

Leading corrections to the total energies that result from the expectation values of H_r and H_B are discussed below. The existing QED formalism provides recipes to evaluate these corrections for a single electron bound to a point nucleus, but the extension of such recipes to a many-electron atom with a realistic model of the nucleus is nontrivial and the subject of current research.

3.1. Nuclear-Size Correction

The field that a bound electron is subject to consists of the Coulomb field provided by the nucleus and that by other bound electrons, i.e., mutual screening. In simpler theories, one can treat the nucleus as a point charge, as is normally done in nonrelativistic atomic structure theory. Relativistic and QED effects, however, are sensitive to the charge distribution near the origin such that the point-nucleus model leads to unrealistic numerical results. The departure from the point-nucleus result is known as the "finite-nucleus" or the "nuclear-size" correction. Although the Fermi distribution is a sensible model for a nucleus for our purpose, some nuclear parameters are not known well enough. Fortunately, numerical results are not very sensitive to the details of nuclear models—except for the heaviest elements in the periodic table—as long as an extended nucleus is used instead of a point nucleus.

The Dirac energy levels given in Eq. (10) correspond to those for a point nucleus. The use of an extended nucleus reduces the binding energy of each electron reflecting the diluted concentration of the nuclear charge. Most of the popular computer codes for relativistic wave functions have options for different nuclear models and provide appropriate solutions of the Dirac equation with matching eigenvalues. When an extended nucleus is used, the well-known logarithmic divergence of the relativistic wave functions with $j = 1/2$ (or $|\kappa| = 1$) disappears and bound solutions are obtained for $Z > 137$, contrary to the Dirac eigenfunctions for a point nucleus. The use of an extended nucleus leads to bound-state solutions for superheavy atoms with $Z > 150$ created by colliding heavy ions with heavy atoms.

The nuclear-size correction is more important for $s_{1/2}$ and $p_{1/2}$ electrons than others and it rapidly increases for very heavy ions, becoming the most important correction to the Dirac eigenvalue of the 1s electron for $Z > 100$.

3.2. Self-Energy Corrections

The largest QED correction is the self-energy correction, which arises from a bound electron emitting a virtual photon and absorbing the same photon while interacting with the nucleus [see Fig. 2 (a)]. The self energy (i) may reduce or increase the binding energy of an electron; (ii) is the largest in magnitude for an s electron and diminishes as ℓ increases; (iii) is slightly smaller in magnitude than energy shifts due to the Breit operator for inner-shell electrons; and (iv) has the following Z dependence for a hydrogenic ion with a point nucleus:

Figure 2. Feynman diagrams for (a) the self energy and (b) the vacuum polarization.

$$E_{SE} = m_0 c^2 \alpha (\alpha Z)^4 F_{n\kappa}(\alpha Z)/\pi n^3, \qquad (19)$$

where $F_{n\kappa}(\alpha Z)$ contains $\log(\alpha Z)$ dependence for an s electron, while its αZ dependence becomes weaker as ℓ increases. The self energy for an s electron will be smaller by a factor of $\alpha \log(\alpha Z)$ than the leading relativistic corrections to eigenvalues arising from the Dirac eigenvalues, Eq. (10). For electrons with $\ell \geq 1$, self energy corrections will be smaller by a factor of α.

The $F_{n\kappa}(\alpha Z)$ functions for the 1s, 2s, $2p_{1/2}$, and $2p_{3/2}$ electrons for all values of Z are known for a hydrogenic ion with a point nucleus.[16] One can estimate the self-energy corrections for the electrons with $n > 2$, $\ell \leq 1$ by scaling the results for $n = 2$ based on the n^{-3} dependence in Eq. (19).

As is shown in Table I, the self-energy correction for low-lying states amounts to 80–90% of all QED corrections, the net sum of which are referred to as the Lamb shift. Because of this dominance of the self energy, its modification due to the screening by other bound electrons becomes a critical issue overshadowing other smaller QED corrections. This is another subject of current interest. For instance, one can establish the limits of the screening effect by comparing the self energy of the 2s level of the Zn-like uranium ion, U^{62+}, using the bare nuclear charge ($Z = 92$) and the fully screened charge ($Z_{eff} = 62$):

Table I. QED and other relativistic corrections (in eV) to the Dirac eigenvalues for a point nucleus, Eq. (10) [from W. R. Johnson and G. Soff, Atom. Data Nucl. Data Tables **33**, 405 (1985)]. (FORTRAN notation is used, e.g., 6.2099E-03=6.2099×10^{-3})

Corrections*	1s	2s	$2p_{1/2}$	$2p_{3/2}$
Argon	Z = 18			
Self Energy	1.2169E+00	1.6339E-01	-4.2930E-03	6.2099E-03
SE, nucl size	-3.5334E-05	-4.4167E-06	0	0
SE, sum	1.2168E+00	1.6338E-01	-4.2930E-03	6.2099E-03
Uehling Poten	-8.5542E-02	-1.0825E-02	-4.8584E-05	-8.8334E-06
UP, nucl size	0	0	0	0
Wichmann-Kroll	2.8267E-04	3.5334E-05	0	0
VP, sum	-8.5260E-02	-1.0790E-02	-4.8584E-05	-8.8334E-06
Higher QED	4.5934E-04	5.7417E-05	1.3250E-05	-4.4167E-06
Other relativ	2.8267E-04	3.9750E-05	-4.4167E-06	-4.4167E-06
QED, sum	1.1323E+00	1.5269E-01	-4.3328E-03	6.1922E-03
Dirac, nucl size	9.0454E-03	1.1483E-03	4.4167E-06	0
Xenon	Z = 54			
Self Energy	5.1035E+01	7.7353E+00	9.2300E-02	7.4556E-01
SE, nucl size	-5.7240E-02	-8.5860E-03	0	0
SE, sum	5.0978E+01	7.7267E+00	9.2300E-02	7.4556E-01
Uehling Poten	-7.3525E+00	-1.0271E+00	-4.9370E-02	-5.3663E-03
UP, nucl size	2.5758E-02	3.9353E-03	0	0
Wichmann-Kroll	1.6600E-01	2.2538E-02	1.4310E-03	3.5775E-04
VP, sum	-7.1608E+00	-1.0006E+00	-4.7939E-02	-5.0085E-03
Higher QED	3.7206E-02	4.6508E-03	1.0733E-03	-3.5775E-04
Other relativ	1.7172E-02	2.8620E-03	-3.5775E-04	-3.5775E-04
QED, sum	4.3872E+01	6.7336E+00	4.5077E-02	7.3983E-01
Dirac, nucl size	3.2341E+00	4.6436E-01	1.5026E-02	0
Uranium	Z = 92			
Self Energy	3.5962E+02	6.6368E+01	9.6361E+00	8.9007E+00
SE, nucl size	-6.4140E+00	-1.1845E+00	-1.7180E-01	0
SE, sum	3.5320E+02	6.5183E+01	9.4643E+00	8.9007E+00
Uehling Poten	-9.7995E+01	-1.7310E+01	-2.9960E+00	-1.2659E-01
UP, nucl size	4.3162E+00	8.3189E-01	8.7409E-02	0
Wichmann-Kroll	4.6779E+00	7.7462E-01	1.8989E-01	2.1099E-02
VP, sum	-8.9000E+01	-1.5703E+01	-2.7187E+00	-1.0549E-01
Higher QED	3.1347E-01	3.9183E-02	9.0423E-03	-3.0141E-03
Other relativ	1.2056E-01	2.1099E-02	-3.0141E-03	-3.0141E-03
QED, sum	2.6464E+02	4.9540E+01	6.7516E+00	8.7891E+00
Dirac, nucl size	1.9379E+02	3.6817E+01	4.3041E+00	0

* SE = self energy; nucl size = nuclear size correction; Uehling Poten = UP = Uehling Potential (vacuum polarization); Wichmann-Kroll = higher order correction to UP; VP = vacuum polarization; Other relativ = Other relativistic corrections.

$$\frac{[Z^4 F_{2s}(\alpha Z)]_{Z=62}}{[Z^4 F_{2s}(\alpha Z)]_{Z=92}} = 19.5\%, \tag{20}$$

which suggests that the screening effect on the 2s level of a uranium ion with 30 bound electrons should be somewhere between 0 and 80%, though common sense dictates that the actual screening will be closer to the lower limit than the upper limit.

The value of $F_{n\kappa}$ for a $2p_{1/2}$ electron is negative for low Z but switches sign between $Z = 40$ and 50, while the value for a $2p_{3/2}$ electron remains positive for all Z. A preliminary study of the $F_{n\kappa}$ for the $3p_{1/2}$ electron shows a similar change of sign.[17] In the vicinity of the Z values where these sign changes take place, the general statement made in the preceding paragraph about the magnitude of the self energy correction will not be valid. Moreover, such "crossing points" are likely to be shifted by corrections, such as the screening and the choice of the nuclear charge distribution, to the self-energy values based on a point nucleus. The self-energy values for the hydrogen atom are known for $\ell > 1$ states;[18] for the 3d and 4f states, the self energy is again negative when $\kappa > 0$, and positive when $\kappa < 0$. Further study is needed to find if the self energies for the positive κ (i.e., $j = \ell - 1/2$) states switch signs or not as Z increases.

A few approximate methods are used in relativistic wave function codes to estimate the screening of the self energy,[19,20] but none of them are based on a rigorous QED formalism. First of these determines a Z_{eff} by matching the expectation value of r obtained using a Dirac-Fock (or any other approximate) orbital with a hydrogenic Dirac orbital (for a point nucleus) whose Z value produces the same expectation value. Then, Mohr's self-energy correction for a point nucleus is interpolated for this Z_{eff} to obtain a "screened" self-energy correction. This method produces screened results consistent with those calculated by Johnson and coworkers[21] for the 1s electron of heavy atoms, but sometimes produces a "negative" screening, i.e., $Z_{eff} > Z$ for other electrons.

The second approximation is to use the charge density of an orbital integrated to a short distance from the origin, typically 0.3 Compton wavelength $(= \alpha \times \text{Bohr radius})$. The ratio of this integral computed using a Dirac-Fock orbital and that obtained using the corresponding hydrogenic orbital (again, for a point nucleus) is used to scale down the self-energy correction calculated by Mohr for a bare nuclear charge.[16]

The third approximation is based on Welton's interpretation of the self energy as fluctuations in the classical trajectory of a bound electron due to the nuclear field.[22] The second and third approximations produce similar results for *transition energies*, i.e., differences in total energies, for ions with low to moderate Z, though absolute magnitudes of the screening corrections for individual electrons are different.

All of these approximations, however, should be treated only as educated guesses until verified by a more rigorous procedure. Screening by a few bound electrons in a high-Z ion can be treated as a perturbation to the unscreened self energy, but the example of U^{62+} mentioned above suggests that a perturbative approach should be used with caution as the number of bound electrons increases.

3.3. Vacuum Polarization

In the QED theory, a vacuum consists of any number of virtual electron-positron pairs, and they become virtually-bound pairs in the field of a nucleus. These virtually-bound pairs can interact with the real bound electrons by exchanging virtual photons [see Fig. 2 (b)]. Energy shift due to this type of interaction is called vacuum polarization. The energy shift due to vacuum polarization can also be expressed in the same form as the self energy, Eq. (19) with a different $F_{n\kappa}$. There is no leading $\log(\alpha Z)$ dependence in the $F_{n\kappa}$ for vacuum polarization, and hence the leading term of vacuum polarization is of the order of $\alpha^3 Z^2$ compared to the nonrelativistic energy level, which makes the vacuum polarization correction smaller by a factor of α than the leading relativistic corrections to nonrelativistic energy eigenvalues.

Fortunately, Uehling has derived an effective potential,[23] known as the Uehling potential, to represent the leading term of the vacuum polarization. The main part of the energy correction due to the vacuum polarization then can be calculated as an expectation value of the Uehling potential using a relativistic one-electron orbital. The screening by other bound electrons are indirectly included by using a self-consistent wave function, such as the Dirac-Fock orbitals.

The expectation values of the Uehling potential for all electrons are negative, i.e., increase electron binding energies, but their magnitudes are an order of magnitude smaller than the self energy of low-lying states (except for the "crossing points" of $p_{1/2}$ electrons mentioned in the preceding section). For superheavy ions, vacuum polarization competes with self energy in magnitude. Corrections of the order of α and $(\alpha Z)^2$ compared to the Uehling potential[24] are also included in the computer codes for relativistic wave functions developed by Desclaux and coworkers,[20] but these refinements are still dwarfed by the uncertainties in the screening corrections for the self energy.

3.4. Higher-Order QED Corrections

Higher-order QED corrections that involve the exchange of more virtual photons and/or more virtual electron-positron pairs than the corrections discussed so far can be identified and classified according to the corresponding Feynman diagrams, but their numerical evaluation is very difficult because applications of the standard recipes of QED will require multiple summations over a complete set of unperturbed states, e.g., hydrogenic wave functions for all bound, positive continuum and negative continuum states. At present, uncertainties in the screening of the leading part of the self energy and uncertainties in representing electron correlation in a many-electron atom far exceed expected magnitudes of these higher-order QED corrections.

5. OTHER THEORETICAL METHODS

Although the Dirac-Fock and MCDF methods described earlier are the most popular theoretical methods to obtain fully relativistic wave functions and corresponding atomic properties, there are other theoretical methods which are less accurate but conceptually simpler, or more accurate but less flexible. The

Pauli approximation,[25] which retains only the leading parts of various relativistic effects when appropriate expressions are expanded in powers of αZ, is an example of the former, while the relativistic many-body perturbation theory (MBPT) is typical of the latter.[26,27]

4.1. Pauli Approximation

Both the Dirac Hamiltonian, Eq. (5), and the Breit operator, Eq. (15), can be expanded in powers of αZ. When only those terms up to the order of $(\alpha Z)^2$ compared to the nonrelativistic terms are kept, the resulting expressions contain only those operators familiar in nonrelativistic quantum mechanics. The Pauli approximation refers to this process of retaining only the leading relativistic corrections as well as the resulting operators that go beyond the terms in the nonrelativistic Hamiltonian, Eq. (1). The standard procedure in the Pauli approximation is to calculate energy corrections due to these extra operators as perturbations using the nonrelativistic wave functions obtained by solving the Schrödinger equation. Although the total number of operators involved in the Pauli approximation is larger than those used in the Dirac-Fock method, nonrelativistic wave functions can be used in the Pauli approximation along with well-developed computational tools available in nonrelativistic atomic structure theory.

A variant of the Pauli approximation uses some of the extra one-electron operators, such as the mass-correction term and the Darwin term, in the main Hamiltonian along with the usual terms in H_{nr}, to introduce changes in wave functions due to such operators. The remaining operators in the Pauli approximation are still treated in the first-order perturbation. Operators that are dependent on j are treated as perturbations while operators independent of j are moved into the unperturbed Hamiltonian. This division allows one-electron orbitals to be classified by ℓ and s but not by j, thus retaining the simplicity of nonrelativistic quantum numbers. Small-component radial functions are unnecessary in the Pauli approximation, resulting in a substantial savings of computational resources. The saved resources can then be directed, for instance, toward a better representation of electron correlation.

For instance, the computer codes developed by Cowan,[2] which are widely used by spectroscopists and plasma modelers, are in this category. His one-electron radial functions carry nonrelativistic quantum numbers but are solutions of Hartree-Fock type equations that include some relativistic operators from the Pauli approximation. In his codes, j-dependent operators are introduced as perturbations (although these operators are of the same order of magnitude as those included in the unperturbed Hamiltonian), and the total wave function is properly identified according to J, L, and S.

Relativistic energy corrections obtained using the Pauli approximation are in excellent agreement with those obtained from the corresponding Dirac-Fock or MCDF calculations for ions with low and moderate Z (about 50 or lower). For heavier ions, one must be careful to include terms of $(\alpha Z)^n$, $n \geq 4$, that result from H_D. Other properties that depend on the size of radial functions, such as transition probabilities, may not be predicted well by the Pauli approximation unless changes in radial functions are also accounted for in the Pauli approximation.

4.2. Dirac-Slater Method

Both in the Hartree-Fock and Dirac-Fock methods, most of the computational effort is spent to satisfy the "Fock" part, or the exchange terms in the relevant self-consistent field (SCF) equations. Moreover, the exchange terms form a nonlocal potential which imposes inconveniences on the resulting solutions, such as gauge-dependent transition probabilities and not forming an orthonormal set unless the exchange terms are frozen, i.e., core orbitals are excluded from the SCF process. To overcome these shortcomings, Slater[28] has proposed to replace the exchange terms with a local potential built from the charge density of the bound electrons. The Slater approximation can be used both in relativistic and nonrelativistic SCF equations, and the resulting solutions are known as the Hartree-Slater or Dirac-Slater wave functions. Such wave functions are approximations to the Hartree-Fock or Dirac-Fock wave functions.

Dirac-Slater wave functions can be calculated quickly, the solutions form an orthonormal set, and they are in general more reliable than the wave functions obtained by completely ignoring the exchange terms. With these advantages, Dirac-Slater wave functions represent a sensible compromise between accuracy and convenience for mass production of atomic data, such as oscillator strengths and excitation cross sections, and for applications where a complete set of eigenfunctions are needed.

For highly-charged ions, differences between the Dirac-Fock and Dirac-Slater wave functions diminish since the nuclear potential, which is treated in the same way in both methods, dominates and the exchange terms are less important.

4.3. Many-Body Perturbation Theory

The many-body perturbation theory (MBPT) is a powerful method to calculate atomic energy levels and other properties. In this method, the Hamiltonian is divided into H_0 and H', where H_0 is chosen so that it is possible to generate a complete set of eigenfunctions, covering both discrete and continuous eigenvalues. Often a complete set that consists of only discrete eigenvalues are used.[27] The perturbation energy, i.e., the expectation value of H', is then calculated using the perturbation theory to as many orders as practical. The correction to the unperturbed wave function is also calculated using the standard perturbation procedure. If MBPT calculations include enough orders of perturbation, electron correlation will be well represented resulting in very accurate theoretical results. Usually, two-electron operators in the Hamiltonian, be it relativistic or nonrelativistic, are replaced by approximate one-electron operators to form H_0, and the difference between the exact and approximate potentials is treated as the perturbation, H'. Relativistic effects are included by using relativistic Hamiltonians, such as H_r and H_B, and relativistic eigenfunctions of the unperturbed Hamiltonian.

In practice, two approaches are used in the MBPT. The first approach is to calculate corrections to unperturbed value of the desired property using the standard procedure for the first order perturbation, second order perturbation, and so on.[27] This is conceptually straightforward, but the actual numerical implementation is not. Diagrams, called Bethe-Goldstone diagrams, that resemble Feynman diagrams symbolically describe individual processes, such as the excitation of a core electron to a virtual excited state, and contributions from a very large number of diagrams must be evaluated. Formulas become increasingly complex as the order of perturbation is raised, making it numerically difficult to go beyond the third-order perturbation for an atom with three or more electrons. A relativistic version of the MBPT theory must also include contributions from the negative-energy continuum, or use projection operators or an equivalent method to exclude the negative-energy continuum.

The second approach in MBPT is to convert expressions for the perturbation corrections that include summations over complete sets into equivalent differential equations.[26] These equations are classified according to the number of bound electrons excited into virtual states, e.g., one-body equation, two-body equation, etc. Solving these equations iteratively is equivalent to summing a certain class of diagrams to all orders of the perturbation. The existence of negative-energy solutions in the Dirac equation again poses difficulties in the relativistic version of this approach.

In principle, the MCHF or MCDF method discussed earlier can achieve similar numerical accuracy in accounting for the electron correlation as that achieved by the MBPT method. However, the MBPT may offer an advantage in interfacing with higher order QED corrections since both MBPT and QED are perturbation theories. On the other hand, the MCHF and MCDF methods accommodate closed-shell and open-shell atoms equally well, while currently available versions of relativistic MBPT require further refinements before they can be applied to atoms with two or more open-shell electrons.

5. NUMERICAL RESULTS

5.1. Energy Levels

The energies of the $3s^2\ ^1S_0 \rightarrow 3s3p\ ^3P_1$ transition for a few Mg-like ions are listed in Table II. The theoretical values are given as percentage of experimental values. In the nonrelativistic Hartree-Fock (HF) calculation, the J value of the ^3P state is not identified, while the Dirac-Fock (DF) wave function was calculated for $J = 1$.

The MCDF wave function for the ground state included, in addition to the Ne-like core, $3s^2$, $3p^2_{1/2}$, $3p^2_{3/2}$, $3d^2_{3/2}$ and $3d^2_{5/2}$ configurations, and the triplet state included $3s3p_{1/2}$, $3s3p_{3/2}$, $3p_{1/2}3d_{3/2}$, $3p_{3/2}3d_{3/2}$ and $3p_{3/2}3d_{5/2}$ configurations, using the combinations of Slater determinants that correctly project $J = 0$ for the ground state and $J = 1$ for the triplet state.

The comparison of the HF and the DF results clearly shows that the relativistic effect begins to be significant for the Cu^{17+} ion and beyond, while electron correlation is dominant at low Z. Although the MCDF result agrees very well with experiment for $Z \geq 29$ in this example, the usual accuracy achieved by correlating only the valence shell is of the order of 0.2–1.0%. The remainder comes from the correlation corrections involving core orbitals and

Table II. Energies of the $3s^2\ ^1S_0 \rightarrow 3s3p\ ^3P_1$ transition in Mg-like ions.

Method	Mg	Ca^{8+}	Cu^{17+}	Mo^{30+}
Experiment(cm^{-1})	21870	144640	289400	525020
Hartree-Fock(%)	68.2	88.1	87.2	82.3
Dirac-Fock(%)	68.6	90.3	93.3	95.7
MCDF(%)	96.2	99.3	99.7	99.99

highly excited states, including continuum.

Fine structure separations within the same configuration, e.g., between $2p_{1/2}$ and $2p_{3/2}$ of a lithium-like or boron-like ion, can be calculated to a higher accuracy because some of the correlation effects in the lower and upper states cancel each other.[29] Since this separation is expected to vary smoothly as a function of the nuclear charge Z, one can identify and correct any irregularities in experimental data by plotting the difference between theory and experiment as a function of Z.

The $3p_{1/2}$–$3p_{3/2}$ separation in Al-like ions[30] is shown in Fig. 3, where the theory for Curve A did not include QED corrections for the $n = 3$ electrons ($3s^23p$), while that for Curve B did include the corrections. The wave functions used in this calculation included relativistic equivalent of the $3s^23p$, $3p^3$, $3s3p3d$ and $3p3d^2$ configurations to represent most of the electron correlation in the ground state of aluminum-like ions. Although the absolute values of the theoretical transition energies are still subject to uncertainties from additional electron correlation and the approximate nature of the QED corrections included, the Z dependence of the difference between theory and experiment is so smooth that one can easily correct minor deviations in the experimental data and interpolate missing experimental data.

For instance, the trend seen in the low-Z part of Curve B in Fig. 3 suggests that the experimental value for the Ag^{34+} at $Z = 47$ is too large, though the expected correction to the experimental value is still within experimental error limits.

This complementary use of experimental and theoretical transition energies as is demonstrated in Fig. 3 has become a very powerful tool for spectroscopists to reduce uncertainties in their experimental data and to obtain reliable predictions by interpolation of the differences between theory and experiment and then adding the differences to appropriate theoretical values. We now have several examples where comparisons of theory and experiment along an isoelectronic sequence clearly favor theoretical results that include "educated guesses" of QED corrections even for the N-shell ($n = 4$) electrons, particularly for transitions that involve s or p electrons.[31,32]

As was mentioned earlier, uncertainties in the current theoretical capability to evaluate energy levels by introducing electron correlation among valence electrons and by including leading relativistic and QED corrections are of the order of 0.2–1.0% for ions. Attempts to reduce these uncertainties further require a substantial increase in computational efforts, quickly reaching the point of rapidly diminishing return, in particular, for ions with two or more valence

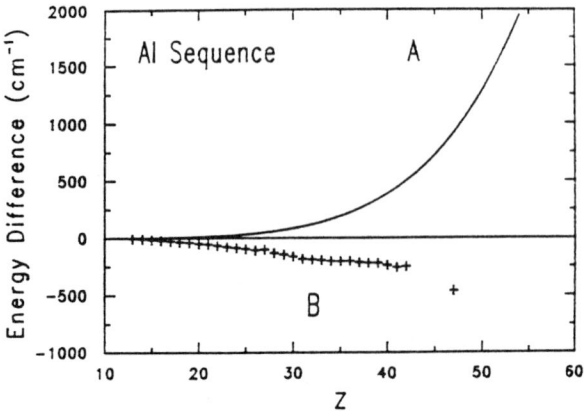

Figure 3. The difference between theoretical and experimental fine-structure splitting, $3p_{1/2}$–$3p_{3/2}$, in the ground state of Al-like ions.[30]

electrons outside a closed shell.

5.2. Transition Probabilities

Many nonrelativistically forbidden transitions become weakly allowed, or sometimes strongly allowed transitions as a result of spin-orbit interaction, commonly referred to as the intermediate coupling. For instance, the np^6 $^1S_0 \to np^5(n+1)s$ $^3P_1$ transition in rare gases is strongly forbidden in low-Z atoms because of the different spins whereas the triplet state mixes heavily with the dipole-allowed $np^5(n+1)s$ $^1P_1$ level as Z increases and becomes an allowed transition at high Z. For xenon,[33] the "triplet" transition is stronger than the "singlet" transition! Selection rules also follow those in jj coupling and J (ignoring the hyperfine interaction with the nuclear spin) becomes the only good quantum number at high Z.

Another example of a transition probability being changed by relativity is the decay of the 2s level to the 1s level in a one-electron ion. For low-Z ions, decay by emitting two photons, known as the two-E1 decay, is the dominant mode but one-photon M1 decay takes over as Z increases[34] since the transition rate for the latter depends on $(\alpha Z)^{10}$ while the rate for the former depends on $(\alpha Z)^6$.

Transition probabilities strongly depend on transition energies, which, in turn, depend on correlation, relativistic and QED corrections. Departures from nonrelativistic predictions of transition probabilities are most noticeable in weak transitions (because of relativistic coupling) and in cases where transition energies are affected strongly by relativity (e.g., splitting of $np_{1/2}$ and $np_{3/2}$ levels). For highly-charged ions, branching ratios based on nonrelativistic considerations will not hold also because the radial functions with the same ℓ but different j have different Z dependence and hence the transition matrix elements will not follow the nonrelativistic ratio as Z increases.

Table III presents an example of such a departure from the nonrelativistic ratio.[35] In Cu-like ions, the nonrelativistic ratio of the electric dipole f values for the 4s → $4p_{3/2}$ and 4s → $4p_{1/2}$ transitions is 2:1. The ratio of the relativistic f values, however, increases as Z increases because (i) the excitation energy in the definition of the f value increases faster for the $4p_{3/2}$ transition than for the $4p_{1/2}$ transition due to the higher j value of the former, and (ii) the $4p_{1/2}$ radial orbital shrinks faster than the $4p_{3/2}$ orbital does. The change in the excitation energies affects the ratio far more than the change in the size of the orbitals.

High accuracy achieved in the theoretical predictions of energy levels discussed in the preceding section is not found in the predictions of transition probabilities except in isolated cases. MCDF wave functions that introduce correlation only in the valence shell will be able to predict transition probabilities to an accuracy of 5 ~ 10% for moderately- or highly-charged ions. Highly correlated wave functions are necessary to achieve a higher accuracy for the transition probabilities of neutral or lightly-charged ions.

QED corrections will have little direct effects on transition probabilities since such corrections are applied after the wave functions are calculated. However, the change in transition energies due to QED corrections will indirectly affect transition probabilities through the latter's dependence on the transition energy.

Table III. Electric dipole oscillator strength f for the 4s → 4p transition of Cu-like ions.[35]

Ion	Z	$f(4p_{1/2})$	$f(4p_{3/2})$	Ratio†
Ge^{3+}	32	0.316	0.654	2.07
Xe^{25+}	54	0.173	0.474	2.74
Pb^{53+}	82	0.109	0.565	5.18

† Ratio = $f(4p_{3/2})/f(4p_{1/2})$.

Relativistic and correlation effects that change transition probabilities will also affect collision cross sections of highly-charged ions for the same reasons described above. In general, relativity makes the orbital sizes of a highly-charged ion smaller and hence most collision cross sections are reduced by relativity. For more details, the reader is referred to the review on collision cross sections by D. Griffin in this proceedings.

6. CONCLUSION

Accurate spectroscopic data on highly-charged ions provide stringent tests of the validity of atomic structure theory that accounts for electron correlation, relativistic and QED corrections. Comparisons with experimental data clearly indicate that QED corrections are necessary along with the other corrections to achieve agreements of the order of ~ 0.5%. To achieve an accuracy of ~ 0.1% or better, extensive improvements are needed in theory to account for a large number of small corrections–such as higher-order corrections to the Breit interaction, introduction of correlation that involve the core and continuum states, and a rigorous calculation of screening effect in the self energy. Meanwhile, the Z dependence of the difference between theory and experiment along an isoelectronic sequence is smooth, and hence the difference can be used to remove irregularities in experimental data or to interpolate them for ions without reliable data.

Relativity also affects transition probabilities mostly through changes in excitation energies but also to a minor extent due to changes in the size of radial functions. Intermediate coupling can transform forbidden transitions into allowed transitions as Z increases.

Further progress in relativistic atomic structure theory and the availability of ever increasing computing power will eventually lead us to an unprecedented predictive capability not only in energy levels but also in transition probabilities of highly-charged ions produced in hot and/or dense plasmas.

ACKNOWLEDGEMENTS

I am greatly indebted to Drs. J. P. Desclaux and P. Indelicato for sharing MCDF and related codes, which were indispensable in learning much of the relativistic effects discussed here; to Drs. P. J. Mohr and A. W. Weiss for elucidating many subtle aspects of the QED and correlation corrections; and to Drs. J. Reader, J. Sugar, and V. Kaufman for helping me to understand complicated experimental data. This research was supported in part by the Office of Fusion, the Department of Energy and also by the Lawrence Livermore National Laboratory.

REFERENCES

1. C. F. Fischer, *The Hartree-Fock Method for Atoms* (John Wiley and Sons, New York, 1977).
2. R. D. Cowan, *The Theory of Atomic Structure and Spectra* (Univ. California Press, Berkeley, 1981).
3. C. C. J. Roothaan and P. S. Bagus, *Methods in Computational Physics* (Academic Press, New York, 1963), Vol. II, p. 47.
4. P.-O. Löwdin, Adv. Chem. Phys. **2**, 207 (1959).
5. M. E. Rose, *Relativistic Electron Theory* (John Wiley and Sons, New York, 1961).
6. G. E. Brown and D. G. Ravenhall, Proc. Roy. Soc. London, A**208**, 552 (1951).
7. J. Sucher, in AIP Conf. Proc. No. 136, p. 1 (1985).
8. J. P. Desclaux and Y.-K. Kim, J. Phys. B **8**, 1177 (1975).
9. Y.-K. Kim, Phys. Rev. **154**, 17 (1967).
10. S. P. Goldman, Phys. Rev. A **37**, 16 (1988).
11. J. P. Desclaux, in AIP Conf. Proc. No. 136, p. 162 (1985).
12. G. Breit, Phys. Rev. **34**, 553 (1929).
13. J. Sucher, in AIP Conf. Proc. No. 189, p. 337 (1989).

14. H. T. Doyle, Adv. Atom. Molec. Phys. **5**, 337 (1969).
15. E. P. Ivanova and U. I. Safronova, J. Phys. B **8**, 1591 (1975); U. I. Safronova and V. S. Senashenko, *Theory of Spectra of Multicharged Ions* (Energoatomizdat, Moscow, 1984) (in Russian).
16. P. J. Mohr, Phys. Rev. A **26**, 2388 (1982) and references therein.
17. P. J. Mohr and Y.-K. Kim, to be published.
18. S. Klarsfeld and A. Marquet, Phys. Lett. **43B**, 201 (1973).
19. I. P. Grant, B. J. McKenzie, P. H. Norrington, D. F. Mayers, and N. C. Pyper, Comput. Phys. Commun. **21**, 207 (1980); I. P. Grant, in AIP Conf. Proc. No. 168, p. 78 (1988).
20. J. P. Desclaux, Comput. Phys. Commun. **9**, 31 (1975); in AIP Conf. Proc. No. 136, p. 162 (1985).
21. A. M. Desiderio and W. R. Johnson, Phys. Rev. A **3**, 1267 (1971); K. T. Cheng and W. R. Johnson, Phys. Rev. A **14**, 1943 (1976).
22. T. A. Welton, Phys. Rev. **74**, 1157 (1948); see also P. Indelicato, O. Gorceix and J. P. Desclaux, J. Phys. B **20**, 651 (1987).
23. E. A. Uehling, Phys. Rev. **48**, 55 (1935).
24. G. Källen and A. Sabry, Danske Vidensk. Selsk. Mat.-Fis. Medd. **29**, 17 (1955); E. H. Wichmann and N. M. Kroll, Phys. Rev. **101**, 843 (1956); G. Soff and P. J. Mohr, Phys. Rev. A **38**, 5066 (1988).
25. H. A. Bethe and E. E. Salpeter, *Quantum Mechanics of One- and Two-Electron Atoms* (Springer-Verlag, Berlin, 1957).
26. I. Lindgren, in AIP Conf. Proc. No. 189, p. 3 (1989) and references therein.
27. W. R. Johnson, in AIP Conf. Proc. No. 189, p. 209 (1989); S. A. Blundell, W. R. Johnson, and J. Sapirstein, Phys. Rev. A **41**, 1698 (1990) and references therein.
28. J. C. Slater, *Quantum Theory of Atomic Structure* (McGraw-Hill, New York, 1960), Vol. II, Appendix 22.
29. Y.-K. Kim and K.-N. Huang, Phys. Rev. A **26**, 1984 (1982).
30. Y.-K. Kim, J. Sugar, V, Kaufman, and M. A. Ali, J. Opt. Soc. Am. B5, 2225 (1988).
31. J. F. Seely, Phys. Rev. A **39**, 3682 (1989).
32. J. F. Seely, J. O. Ekberg, C. M. Brown, U. Feldman, W. E. Behring, J. Reader, and M. C. Richardson, Phys. Rev. Lett. **57**, 2924 (1986).
33. J. Geiger, Z. Phys. A**282**, 129 (1977).
34. R. Marrus and P. J. Mohr, in Adv. Atom. Molec. Phys. **14**, 181 (1978).
35. K. T. Cheng and Y.-K. Kim, Atom. Data Nucl. Data Tables **22**, 547 (1978).

STATE SELECTIVE CHARGE TRANSFER RELEVANT FOR PLASMA PHYSICS

R. Hoekstra[*,‡], F.J. de Heer[‡] and R. Morgenstern[*]

[*] Kernfysisch Versneller Instituut, Zernikelaan 25,
9747 AA Groningen, The Netherlands
[‡] FOM-Institute for Atomic and Molecular Physics,
P.O. Box 41883, 1009 DB Amsterdam, The Netherlands

ABSTRACT

We present the general trends in comparing experimental and theoretical electron capture cross sections for both dominantly and non-dominantly populated states in collisions of fully stripped carbon or oxygen ions on atomic hydrogen. The experimental results have been obtained by means of Photon Emission Spectroscopy. With this technique we have also measured the HeII($n=4 \rightarrow n=3$) emission profiles along the ion beam axis in collisions of He^{2+} on atomic hydrogen. We have succeeded for the first time to identify from the emission profiles the separate contributions from the various quasi-degenerate $4l$-states of He^+ produced in the electron capture process.

INTRODUCTION

Charge transfer processes in slow collisions ($v \leq 1$ a.u.) of fully stripped ions with atomic hydrogen are of fundamental interest[1] and play an important role in the diagnostics of fusion plasmas, e.g. in connection with the injection of atomic hydrogen beams[2,3]. Whereas a lot of theoretical work[4-9] has been done, experimental results are still scarce and mainly restricted to total one-electron capture measurements, see e.g. ref.10. This scarceness of experimental work is mainly due to difficulties in producing both the slow fully stripped ions and the atomic hydrogen with respectively sufficient intensity and density.
Recently we have measured a number of absolute line emission cross sections[11-13] for C^{6+} and O^{8+} colliding on atomic and molecular hydrogen. Not only transitions resulting directly or via cascades from the dominantly populated states ($n=4$ for C^{6+} on H and $n=5$ for O^{8+} on H) were measured but also transitions resulting from non-dominantly populated states, $n=5$, 7 and 8 for C^{6+} on H and $n=6$ for O^{8+} on H. The latter ones provide a stringent test for theory Especially the $\Delta n=1$ transitions between high-n states, yielding light in the visible spectral region are important for future plasma diagnostics[2,3]. Because they allow fibre optic connections between spectrometers and fusion reactor. Unfortunately, our measured emission cross sections for such $\Delta n=1$ transitions, like CVI($8 \rightarrow 7$) and CVI ($7 \rightarrow 6$) (with the numbers referring to the principal quantum numbers n involved in the transitions) cannot be used directly for plasma diagnostics since in a tokamak the l-states within one n-shell are mixed[2,3]. Therefore it is important to deduce experimentally all the l-state electron capture cross

sections separately. Due to the quasi-degeneracy of the l-levels in the resulting hydrogenic ions, light emission from different l-levels cannot be distinguished spectroscopically and hence the measurements have to be compared with the sum of the various theoretical contributions. In this way detailed sum comparisons have been made[11-13], which showed that for the high-n states there were considerable differences. This motivated us to concentrate our experimental work on determining the $4l$ cross sections in collisions of He^{2+} on H (the dominant channel is $n=2$). The He^{2+}-H system has been chosen, because the HeII ($4 \rightarrow 3$) is probably the most important for plasma physics, the number of l-states is not too large, so the separate contributions can be determined and there are calculations[8,9] available for comparison.

In the next sections we will first briefly describe the experimental setup and technique. Thereafter the general tendencies in comparing theory and experiment for collisions of C^{6+} and O^{8+} with atomic hydrogen will be shown. Finally we will present the results of identifying contributions from the degenerate He^+ ($4l$) states.

EXPERIMENT

The fully stripped ions have been produced by the ECR ion source installed at the KVI. In the experimental setup, schematically shown in Fig. 1, the ions crossed a pure molecular or

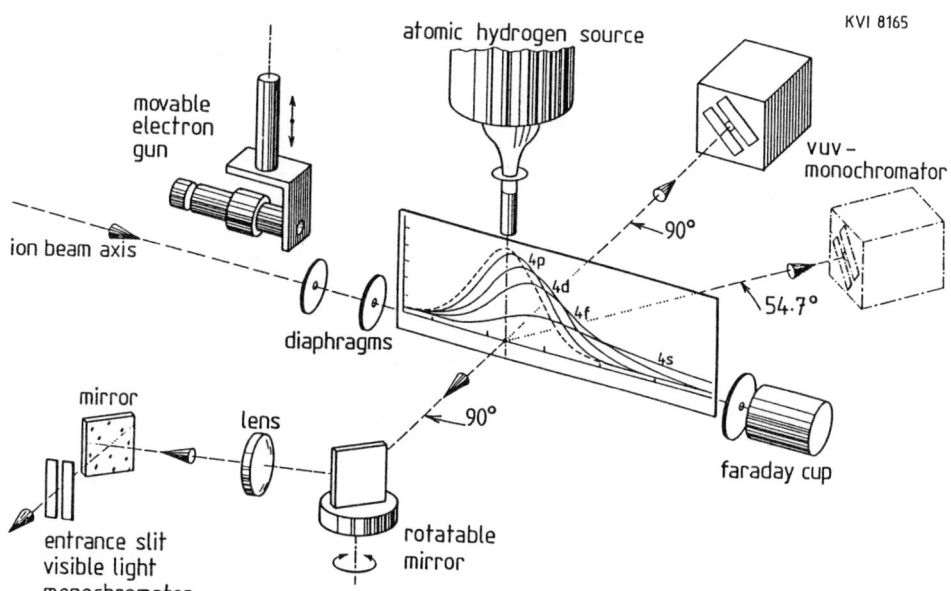

Fig. 1. Schematical view of the experimental set-up. Typical emission profiles of the HeII($4l$) states are included. The target density profile is indicated by the dashed curve.

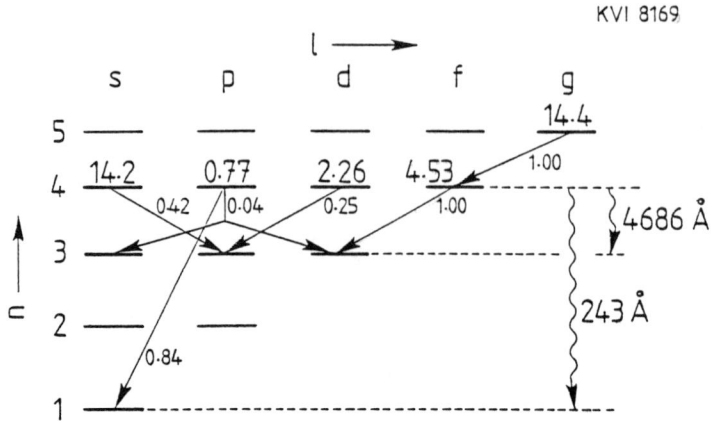

Fig. 2. Energy level scheme of HeII. The relevant transitions are shown together with the branching ratios and the lifetimes (in nsec) of upper states.

a partly dissociated hydrogen beam produced by a Slevin-type radio frequency discharge source. The absolute density profiles of molecular and atomic hydrogen target beams were determined by observation of electron impact induced atomic (Balmer-β) and molecular radiation. The radiation was observed with a monochromator equipped with an imaging lens system which enabled the measurement of the target density along the beam axis. In that way[14] we found a dissociation degree of some 70% in the center of the target. The photon emission, resulting from the decay of C^{5+*}- or O^{7+*} states was observed with the above mentioned spectrometer for visible (300 - 650 nm) light and a grazing incidence vacuum spectrometer for the VUV (10 - 80 nm) radiation. Both spectrometers are calibrated absolutely on wavelength and sensitivity[15]. All CVI and OVIII lifetimes involved are sufficiently short for complete decay to occur within the viewing ranges of the spectrometers.

To deduce the separate contributions of the $He^{+}(4l)$-states to the HeII(4 → 3) transitions we have exploited the fact that the lifetimes of the states are different and long enough that no complete decay occurs within the viewing range of the spectrometer. Fig. 2 shows the energy level scheme of the HeII together with the lifetimes and branching ratios of the states relevant for our experiment. Radiation from short lived levels, e.g. 4p is mainly concentrated on the collision center, whereas radiation from long lived states, e.g. 4s is still emitted further downstream the ion beam axis, see Fig. 1. Hence the determination of emission profiles along the beam axis gives information on the l-states. This method, often used in collision experiments on static targets has only once been used in combination with a beam target, namely to measure capture into 3l-states in H^{+}-Li collisions[16]. He^{2+} colliding on atomic hydrogen is more intricate since (I) more l-states have to be included, (II) the lifetimes of the states are closer to each other, (III) the target has two components, namely H and H_2.

Neglecting cascades the emission profiles P_{4l} along the ion beam axis are described by

$$P_{4l}(z) = \frac{1}{v\tau_{4l}} \int_0^z T(z') e^{-(z'-z)/v\tau_{4l}} dz'$$

where v is the velocity of the ions, τ_{4l} the lifetime of state $4l$, z the position along the ion beam axis and T(z') the target density profile. The measured signals, S(z) are equal to

$$S(z) = K \sum_l \beta(4l \rightarrow 3) P_{4l}(z) \sigma(4l)$$

with K an absolute calibration constant, $\beta(4l \rightarrow 3)$ the branching ratio for decay to $n=3$ and $\sigma(4l)$ the electron capture cross sections. By measuring the signals at some 25-30 positions along the ion beam axis, it is possible to deduce the cross sections from a least squares fit. To increase the accuracy of this deconvolution procedure the contribution of the $4p$, which is relatively small due to the small branching ratio (Fig. 2) has first been subtracted from the signals ($4p$ determined from the $4p \rightarrow 1s$ transition).

RESULTS FOR C^{6+} AND O^{8+}

In collisions of C^{6+} and O^{8+} on H the electron capture process populates dominantly $n=4$ for C^{6+} and $n=5$ for O^{8+}, and to a lesser extent $n=5$ for C^{6+} and $n=6$ for O^{8+}. For emission lines in the vuv spectral region, originating from $n \leq 4$ and $n \leq 5$ for carbon and oxygen ions, respectively, there is good agreement with the most elaborate calculations using a Molecular Orbital[4] (MO), an Atomic Orbital[5] (AO) and an AO-MO[6] approach. As an example of this good agreement between theory and experiment Fig. 3a shows the results for the $4 \rightarrow 2$ transition in C^{5+*} produced in C^{6+}- H collisions. Furthermore, from this figure it can be seen that previous measurements by Dijkkamp et al.[17] agree clearly within error bars with our new results[12,13].

However, for transitions from $n=5$ and $n=6$ for respectively fully stripped carbon and oxygen ions colliding on atomic hydrogen, the experimental results strongly support the molecular orbital calculations[4]. The general feature, that the MO calculations predict cross sections for the first n-state above the dominant one, which are larger than the AO and AO-MO calculations, and the fact that our results show a preference for the MO calculations is shown in Fig. 3b on the basis of the CVI($5 \rightarrow 2$) line. It should be noted that recent extended AO calculations for C^{6+} on H by Fritsch[9] are also in fair agreement with the photon emission cross sections.

Finally Fig. 3c shows the results for two transitions resulting from non-dominantly populated high-n states, namely $n=7$

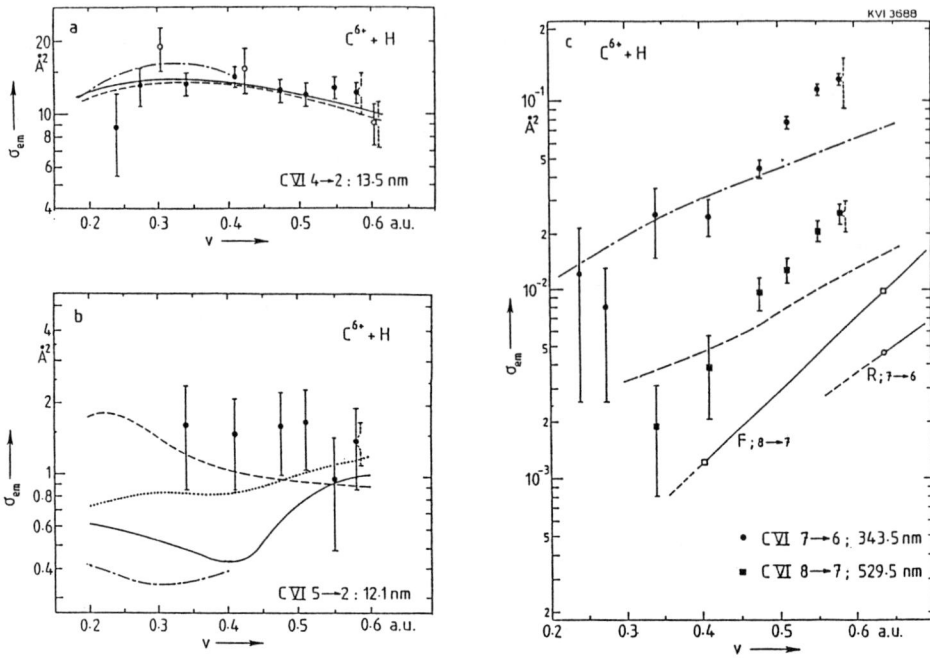

Figs. 3a-c. Comparison of theoretical and experimental emission cross sections. The shown solid and dashed (at the highest velocity) error bars are the relative and absolute experimental errors. Experiment: open symbols ref.[17], closed symbols this work[12,13]; Theory: Fig. 3a and b, dashed curve MO[4], solid curve AO[5], dashed-dotted curve AO-MO[6] and dotted curve extended AO[9]; Fig. 3c, F; 8→7 AO[7], R; 7→6 UDWA[8], dashed curves extended AO[4].

and $n=8$ in CVI. As mentioned before these transitions, yielding light in the visible spectral range are particular interesting for future plasma diagnostics since they allow the use of fibre optic connections between the spectrometer and the fusion reactor[2,3]. However, it can be seen that for these lines there are large differences with the corresponding calculations, an AO calculation[7] for charge transfer into $n=8$ and the unitarized distorted wave approximation[8] (UDWA) for electron capture into $n=7$. Especially the generally known inappropriateness of UDWA theory at $v \leq 1$ a.u. for calculating electron capture into non-dominantly populated high-n states is clearly demonstrated. The recent extended AO calculations by Fritsch[9] are in better agreement with the experimental data than the earlier ones regarding the absolute values, but in somewhat poorer agreement regarding the velocity dependence. The difference between the two AO calculations is that the older calculations[7] only included the dominant $n=4$ capture channels and the $n=8$ states, whereas the recent ones also include all the n-states in between. Hence in the older calculations the capture into $n=8$ is treated as a one step perturbation from the $n=4$ level. Therefore the larger

$n=8$ cross sections predicted by the extended AO calculations[9] indicate the importance of *ladder climbing* processes (with $n=5$, 6 and 7 the rungs of the ladder) in the population of high-n states.

RESULTS FOR He$^+$(4l) CAPTURE

The HeII (4 → 3) emission profile for 5 keV/amu He^{2+} colliding on atomic hydrogen is shown in Fig. 4a together with the results of the least squares deconvolution. It can be seen that the measurements are well described by the fit. Since the deconvolution results[18] differ from theory, Fig. 4b shows a convolution of the theory in order to demonstrate that the theoretical prediction is not just another solution for the fit. The experimental results are some 25% larger than theory. Scaling the theory up with 25% it is seen that down streams along the ion beam axis the theory decreases somewhat faster than experiment. Comparing the fit and the theory it can be seen that the 4s, 4p and 4d are relatively stronger populated than predicted. As a typical example of the comparison between experimental and theoretical 4l cross sections Fig. 5 shows the capture cross sections for the He$^+$ (4f) state. This figure shows that the experimental results increase smoothly with energy, whereas the theory shows a structure around 4 kev/amu. Such a structure is present in all theoretical 4l cross sections and absent in the experimental results. However considering that the 4l cross sections are

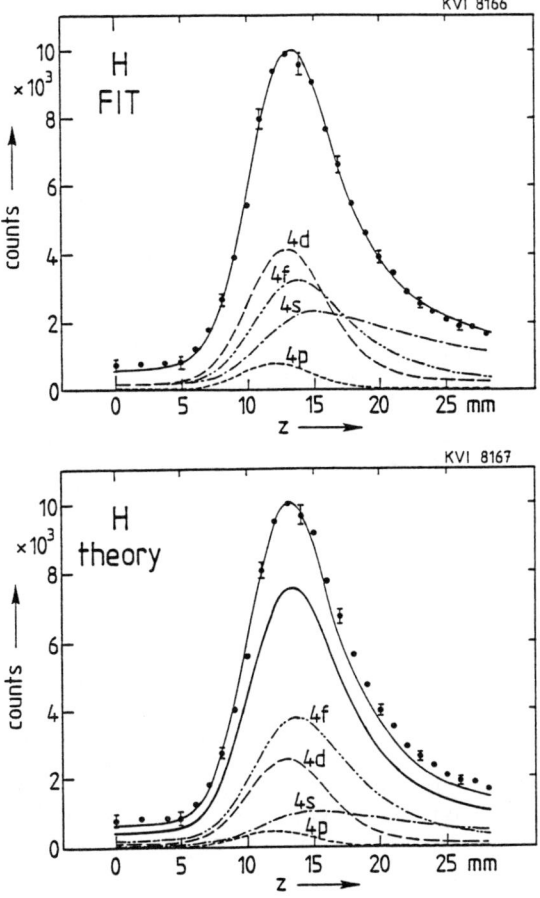

Figs. 4a-b. The HeII(4 → 3) emission profile for 5 keV/amu He^{2+} colliding on H. Fig. 4a shows the least squares fit results (deconvolution) and Fig. 4b the convolution of the AO calculations by Fritsch[9]. The thin solid line gives the scaled AO results.

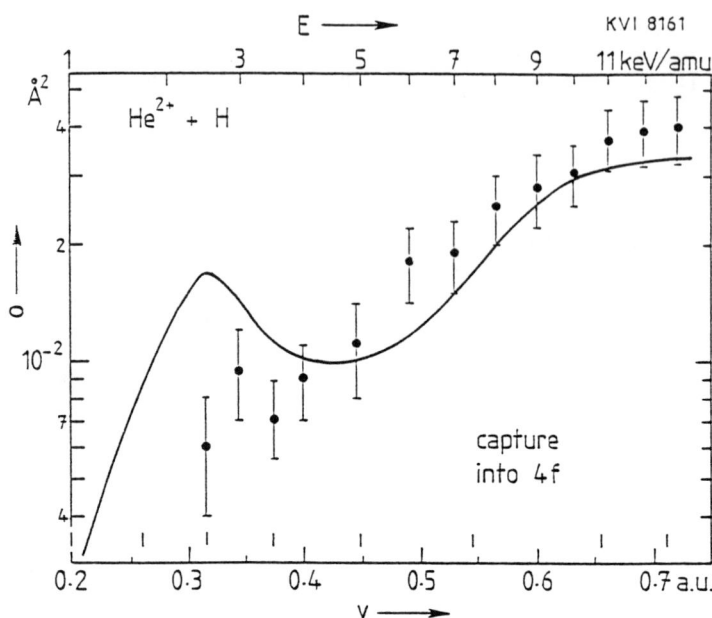

Fig. 5. Absolute electron capture cross sections for charge transfer into He^+ ($4f$) in collisions of He^{2+} on atomic hydrogen. Theoretical curve: extended AO calculations by Fritsch[9].

more than two orders of magnitude smaller than the dominant one for capture into $2p$ the theoretical calculations seem to describe the charge exchange processes to a fair extent.

CONCLUSION

It has been shown that the dominant charge transfer into C^{5+*} ($n=4$) and O^{7+*} ($n=5$) in collisions of C^{6+} and O^{8+} with atomic hydrogen is well described by the most extensive expansion methods, MO[4], AO[5,9] and AO-MO[6]. However, only the MO calculations and recent extended AO calculations[9] show good agreement over the whole velocity range covered by the experiment for capture in the first n level above the dominant one, respectively $n=5$ for carbon and $n=6$ for oxygen. The discrepancies between experiment and theory for the non-dominantly populated high-n states ($n=7$ and $n=8$ in CVI) indicate a necessity to improve the description of these minor electron capture processes.

In case of He^{2+}- H collisions we have shown that it is possible to unravel the contributions from the various degenerate $4l$-states by means of measuring the HeII ($4 \rightarrow 3$) emission profiles along the ion beam axis. In this way information relevant for fusion plasma diagnostics is obtained.

ACKNOWLEDGEMENT

The authors gratefully acknowledge the excellent technical support of J. Sijbring and J. Eilander. We would like to thank Dr. W. Fritsch (Hahn Meitner Institute, Berlin) for communicating his theoretical results. This work is part of the research program of the Stichting voor Fundamenteel Onderzoek der Materie (FOM) with financial support by the Stichting voor Nederlands Wetenschappelijk Onderzoek (NWO). The He^{2+} research is also supported financially by the Joint European Torus Undertaking under contract JP7/9006.

REFERENCES

1. R.K. Janev and H. Winter, Phys. Rep. 117, 265 (1985).
2. R.J. Fonck, D.S. Darrow and K.P. Jähnig, Phys. Rev. A29, 3288 (1984).
3. A. Boileau, M. von Hellermann, L.D. Harton, J. Spence and H.P. Summers, Plasma Phys. Controlled Fusion 31, 779 (1989).
4. T.A. Green, E.J. Shipsey and J.C. Browne, Phys. Rev. A25, 1364 (1982).
5. W. Fritsch and C.D. Lin, Phys. Rev. A29, 3039 (1984).
6. M. Kimura and C.D. Lin, Phys. Rev. A32, 1357 (1985).
7. W. Fritsch, Phys. Rev. A30, 3324 (1984).
8. H. Ryufuku, JAERI-M-82-031 (1982).
9. W. Fritsch, J. Physique Coll. 50, 87 (1989) and private communication.
10. F.W. Meyer, A.M. Howald, C.C. Havener and R.A. Phaneuf, Phys. Rev. A32, 3310 (1985).
11. R. Hoekstra, D. Ćirič, F.J. de Heer and R. Morgenstern, Phys. Lett. A124, 73 (1987).
12. R. Hoekstra, D. Ćirič, A.N. Zinoviev, Yu.S. Gordeev, F.J. de Heer and R. Morgenstern, Z. Physik D8, 57 (1988).
13. R. Hoekstra, D. Ćiriç, F.J. de Heer and R. Morgenstern, Phys. Scr. T28, 81 (1989).
14. D. Ćirič, D. Dijkkamp, E. Vlieg and F.J. de Heer, J. Phys. B: At. Mol. Phys. 18, 4745 (1985).
15. D. Dijkkamp, D. Ćirič, E. Vlieg, A. de Boer and F.J. de Heer, J. Phys. B: At. Mol. Phys. 18, 4762 (1985).
16. F. Aumayr, M. Fehringer and H. Winter, J. Phys. B: At. Mol. Phys. 17, 4201 (1984).
17. D. Dijkkamp, D. Ćirič and F.J. de Heer, Phys. Rev. Lett. 54, 1004 (1985).
18. R. Hoekstra, F.J. de Heer, R. Morgenstern and W. Fritsch, to be published.

EFFECTS OF ELECTRIC FIELDS ON AUTOIONIZATION AND IMPLICATIONS FOR DIELECTRONIC RECOMBINATION

R.R. Jones and T.F. Gallagher
Physics Department, University of Virginia, Charlottesville, VA 22901

ABSTRACT

The autoionization rates of Ba $6p_{1/2}nk$ Stark states have been measured in static and 9 GHz microwave fields using a multistep laser excitation technique. The experiments have been performed for a range of field values between zero field and the $1/16n^4$ classical ionization limit. The results of all the experiments can be reproduced quite well by a linear Stark effect model. The connection between these experiments and the actual situation in a plasma is discussed along with implications for dielectronic recombination.

INTRODUCTION

The effect of electric fields on autoionization has been a topic of study in atomic physics for over a decade.[1-3] Studies have been motivated by purely academic curiosity as well as practical applications. These practical applications lie primarily in the understanding of dielectronic recombination, a stabilization process which is responsible for cooling in plasmas. Here we present the results of a dynamic field experiment which has both academic and practical interest.

We begin with a brief discussion of the dielectronic recombination process and its relationship to autoionization. Next we present a standard model[2] which has been used to explain the effects of static fields on the autoionization process. A set of previous experiments which have verified the predictions of this model will also be discussed. We then present the practical motivation for an experiment to study the effects of dynamic fields on autoionization, followed by the results of this experiment. We conclude with the proposal of a future experiment which would allow us to examine autoionization rates in fields more closely resembling those produced in a plasma.

THE DIELECTRONIC RECOMBINATION PROCESS

The dielectronic recombination process can be written schematically as

$$A^{+(Z+1)}(\rho) + e^- \rightarrow A^{+Z}(\rho',n\ell) \rightarrow A^{+Z}(\rho'',n'\ell') + \phi$$

A free electron scatters from an ion, $A^{+(Z+1)}$, of charge, $+(Z+1)$, and electron configuration ρ. After the scattering process, the electron cloud of the ion is promoted to a more energetic orbital ρ' and the electron is captured into an $n\ell$ state. The autoionizing state thus formed may then radiatively stabilize, by emission of a photon ϕ, to a bound state of the ion A^{+Z}.

Obviously, the dielectronic recombination (DR) process is precisely the inverse of autoionization. Therefore, if we are to thoroughly understand DR we must have a full understanding of autoionization. Also, any effects pertaining to the autoionization process can be directly applied to DR.

Since all the autoionization experiments we will discuss here have been performed with Ba $6pn\ell$ states, Fig. 1 shows the DR process in which a Ba$^+$ ion captures a free electron to form a 6pnℓ autoionizing state. The final step of the DR process occurs when the Ba⁺ 6p core electron radiatively decays to form a Ba 6snℓ bound Rydberg state. For the 6pnℓ states studied, the radiative decay rate of the nℓ electron ($R_R < 200$ kHz) can be neglected compared to the decay rate of the 6p electron ($R_R \simeq 25$ MHz).[4] Thus, the radiative stabilization rates of the 6pnℓ states is taken to be 25 MHz for all 6pnℓ levels.

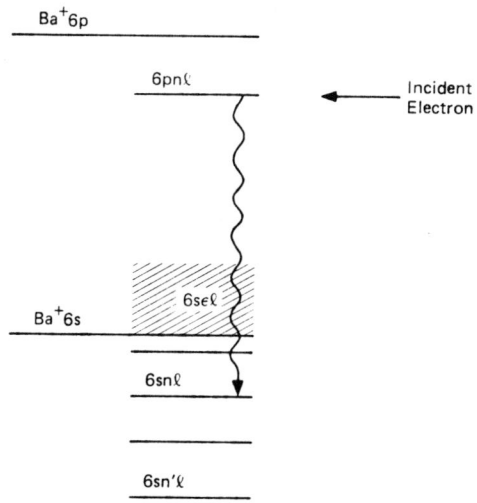

Fig. 1. Schematic of the dielectronic recombination process occuring via capture into a Ba 6pnℓ autoionizing level.

In zero external field the dielectonic recombination rate for electron capture into a given nℓ Rydberg state can be written as

$$A_{DR} = \beta \frac{A_A(n\ell)A_R(n\ell)}{A_A(n l) + A_R(n\ell)} \quad (1)$$

where $A_A(n\ell)$ and $A_R(n\ell)$ are the autoionization and radiative decay rates for the captured autoionizing state and β is a constant. The ratio given by Eq. 1 is simply the rate of capture into a given autoionizing state multiplied by the probability of radiative stabilization of this state.

By examination of Eq. 1 we observe that for $A_A > A_R$, the DR rate is equal to the radiative decay rate. Thus for the Ba 6pnℓ states we study, the DR rate is equal to the radiative decay rate of the Ba⁺ 6pnℓ level as long as this condition holds. On the other hand, if $A_A < A_R$ the DR rate is equal to the autoionization rate. For the Ba 6pnℓ states, $A_A(n\ell) < A_R(n\ell)$ for large n and large ℓ. The autoionization rates scale as n^{-3} due to the normalization of the bound state wave function. The dependence of $A_A(n\ell)$ on ℓ is a bit more complicated and has a strong dependence on the final state of the autoionization process.[5]

In Fig. 2 we show the scaled total decay rate of Ba 6p12ℓ states as a function of ℓ. The scaled total decay rate is defined as the total decay rate, $A_T = A_A + A_R$, multiplied by a factor of n^3 to remove the n dependence from the autoionization rates. Fig. 2 clearly shows the significant decrease in the autoionization rates with increasing ℓ. The constant total decay rate shown for $\ell \geq 8$ is simply the spontaneous decay rate of the Ba$^+$ 6p level multiplied by n^3. Thus, from Fig. 2 it is apparent that the DR rate for capture into a 6pnℓ state states is constant for $\ell \leq 8$ and then drops rapidly for $\ell > 8$. When considering the total DR cross section in zero field we may neglect contributions from states with $\ell \geq 9$.

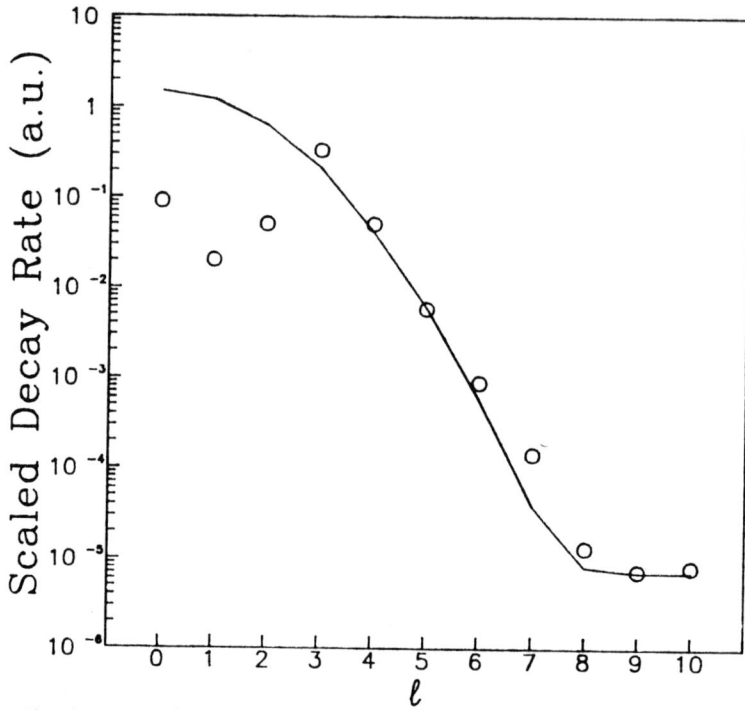

Fig. 2. Plot of the scaled total decay rates for the Ba $6p_{1/2}12\ell$ states vs. ℓ.

ELECTRIC FIELD EFFECTS

Jacobs et al.[2] were the first to discuss the effects of electric fields on the DR cross section. The thesis of their paper was that the plasma environment has a significant effect on ionization and recombination of impurities in a plasma. Specifically, they treated the rapid electron interactions as collisions and the relatively slow ion motion as a static electric field.

The presence of the electric field has two primary effects on DR. First, Rydberg states of ions are further ionized by static electric fields of $F = (Z^3)/(16n^4)$. Thus, states of high n which are placed in the field induced continuum cannot contribute to the total DR cross section. Secondly, the electric field mixes the allowed angular momentum states of a given principal quantum number n to form Stark states. In hydrogen the Stark mixing is uniform and all states in the manifold have equal admixtures of high and low ℓ states. In non-hydrogenic systems this is not the case due to the non-zero quantum defects of the low ℓ states. However, for the moment, let us assume the hydrogenic case.

Since all the Stark states of a given n manifold have the same admixture of low and high ℓ states, each Stark state will have the same autoionization rate. This common rate is simply the average autoionization rate of the zero field ℓ states. Thus,

$$A_A(nk) = \sum_\ell \frac{A_A(n\ell)}{n} \qquad (2).$$

If we explicitly write the n dependence of the zero field autoionization rates $A_A(n\ell) = A_\ell n^{-3}$, we have

$$A_A(nk) = \sum_\ell A_\ell n^{-4} \qquad (3).$$

Thus, the presence of a static electric field should show autoionization rates which scale as n^{-4}.

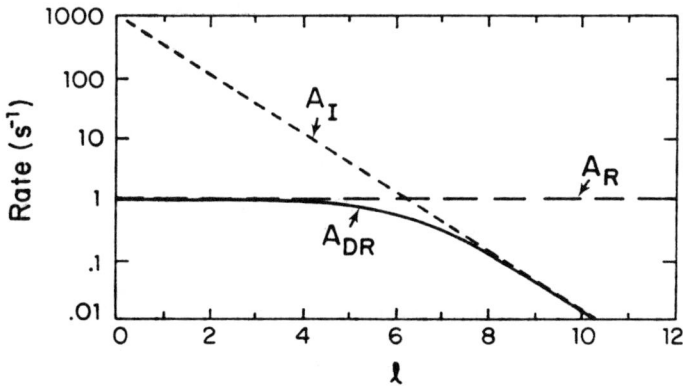

Fig. 3. The dielectronic recombination rates for a typical atom plotted vs. ℓ. Also shown are the autoionization and radiative stabilization rates.

The effect of the electric field on the total DR cross section is clear after examination of Fig. 3. The zero field low ℓ states can share much of their rates without effecting their DR rate. But the high ℓ states which do not contribute to the DR cross section are converted by the field to Stark states which do contribute. Therefore, we observe a constant DR rate for all Stark states rather than for just a few low ℓ levels.

As previously mentioned, this model is strictly valid for hydrogen only. For atoms with non-zero quantum defects, the Stark effect is only linear over a restricted range of electric fields. Fig. 4 shows a typical Stark map for several n manifolds in a typical atom. For low fields, the lowest ℓ states are not well mixed with the inner manifold states. Also, for fields greater than $F \simeq 1/(3n^5)$, the Inglis-Teller limit, the Stark states from adjacent manifolds have avoided level crossings which do not appear in hydrogen. Thus the static field linear Stark effect model is only strictly valid for fields greater than those necessary to mix the lowest ℓ states and also for fields less than $1/(3n^5)$ where the states from the n and n + 1 manifolds overlap.

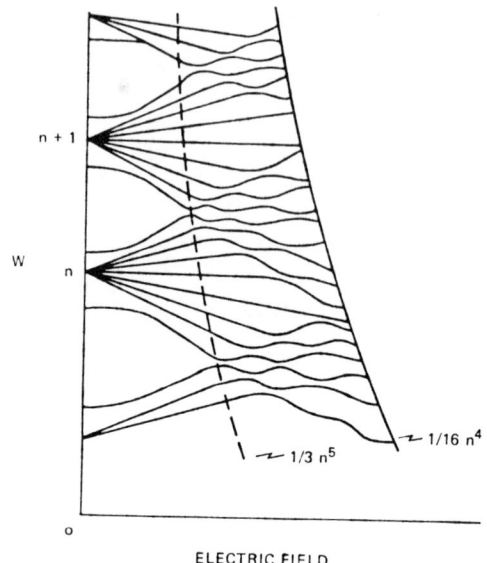

Fig. 4. Typical Stark map for a non-hydrogenic atom showing the quadratic Stark effect for low fields and avoided level crossings for fields $> 1/3n^5$.

Several experiments have been performed which verify the predictions of the linear Stark model. Safinya et al.[3] have measured the autoionizing rates of Ba 6pnk states in static fields where the linear Stark picture is valid. They found that all the Stark states for a given manifold had identical autoionization rates. They also determined that these rates are independent of the electric field amplitude within the linear Stark regime discussed above.

Dunn and his collaborators[6] observed enhancement of DR cross section in the presence of static fields. Their crossed beam experiments showed conclusively that electric fields of $\simeq 10$ V/cm could increase the total DR cross section by a factor of 5 or more. Thus, it seems that the static linear Stark picture is valid at least within certain field limits. But can we use the model in situations which do not conform to the field restrictions?

In an attempt to answer this question Jaffe et al.[7] measured the autoionization rates of Ba 6pnk Stark states for a range of n states and mixing field amplitudes. They observed effectively constant autoionization rates for all the Stark states in a given manifold for fields ranging from $1/(3n^5)$ to the static field ionization limit at $1/(16n^4)$. Recall that in this region the Stark effect is non-linear due to the many avoided level crossings where Stark states from adjacent n manifolds interact. Furthermore, the group found the n scaling of the autoionization rates to be $A_A(nk) = 0.6\ n^{-4}$ (a.u.) in excellent agreement with Eq. 3 which predicts $A_A(nk) = 0.53\ n^{-4}$ (a.u.). Thus, it appears that for static fields the linear Stark model is valid as long as the fields present are greater than those necessary to thoroughly mix the low ℓ states into the manifold.

DYNAMIC FIELD EXPERIMENT

Now, recall that the original discussion of electric field effects on DR and autoionization was initiated in order to describe the actual situation in a plasma. However, in reality, the plasma microfields due to ion motion are by no means static. The thermal ion motion has a characteristic frequency of $v_T \rho^{1/3}$ where v_T is the average thermal ion velocity and ρ is the ion density in the plasma. For a dense plasma of $\rho \simeq 10^{15}$ cm^{-3} and $T \simeq 10^5$ K the ion microfields have a characteristic frequency of $\simeq 100$ GHz. Although this is a slow change compared to the rapid autoionization rates of the low ℓ and low n states, it is very rapid compared to the high ℓ and high n rates which can be < 1 MHz. Is it reasonable to assume that these states are mixed by a static field?

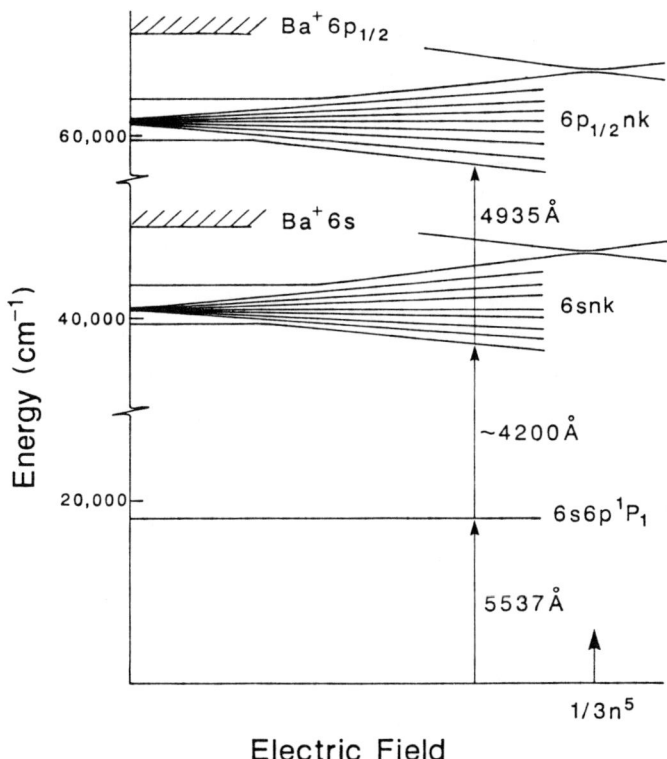

Fig. 5. Schematic of the laser excitation scheme used to excite 6pnk states in the presence of a microwave field.

In an attempt to answer this question we have performed an experiment in which we measured the autoionization rates of Ba 6pnk states in dynamic fields. Specifically, we used a multistep laser excitation of these autoionizing states in the presence of 9 GHz microwaves.[8] The laser excitation scheme is shown in Fig. 5. First we use two pulsed tunable dye lasers to excite ground state Ba atoms to a 6pnk state via a 6s6p intermediate state in the presence of a microwave field. The final autoionizing state is populated using a third dye laser which drives the core electron from a 6s to a 6p level. This transition is essentially an ion transition with the outer nk electron remaining as a spectator.[9,10]

Fig. 6 shows a diagram of the laser, microwaves, and detection equipment. The first two dye lasers are pumped using the second and third harmonics of a Nd:YAG laser respectively. The third laser is pumped by a XeCl excimer laser which is triggered by the Nd:YAG laser after a variable delay. The three laser beams interact with a beam of Ba atoms in a vacuum chamber as shown in Fig. 5.

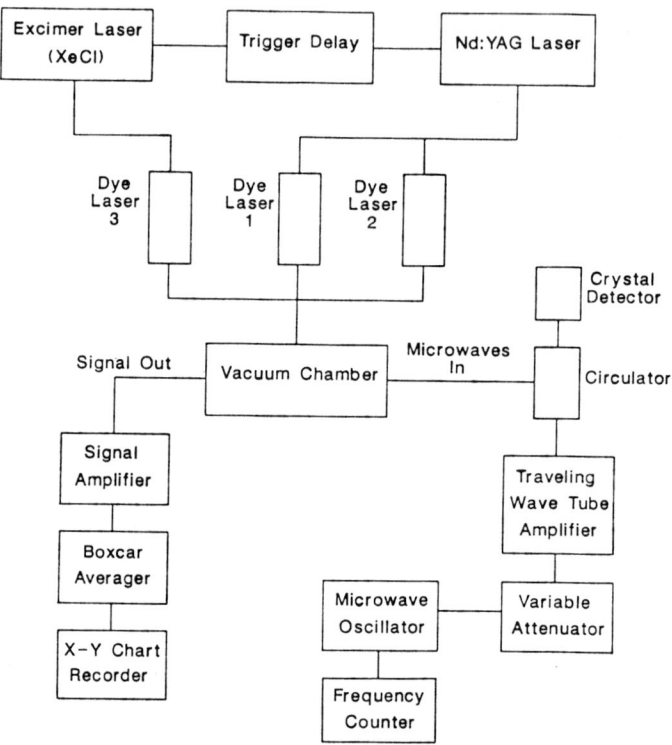

Fig. 6. Diagram of the experimental apparatus showing the laser, microwave, and detection equipment.

A detailed diagram of the interaction region is shown in Fig. 7. The microwave cavity has a resonant TE_{109} mode at 9.25 GHz and is made of a piece of X-band wave guide with two end flanges. The three laser beams and the atomic beam enter the cavity anti-parallel to each other through two small holes in the walls of the cavity. After the Ba atoms have autoionized, a small voltage is applied to the septum in the cavity. This voltage pushes the Ba^+ ions or electrons through a small hole toward a microchannel plate detector located above the cavity.

Fig. 8 shows a static field Stark map of the n = 18 manifold in Ba. We may consider the microwaves as a quasistatic field in which the Stark splittings change as the microwave field amplitude oscillates between its minimum and maximum values. Note the large zero field quantum defects of the levels with $\ell < 4$. The Ba atom presents a good test of the low field limit of the linear Stark model since the d and f states do not become part of the manifold until near the $1/3n^5$ Inglis-Teller limit.

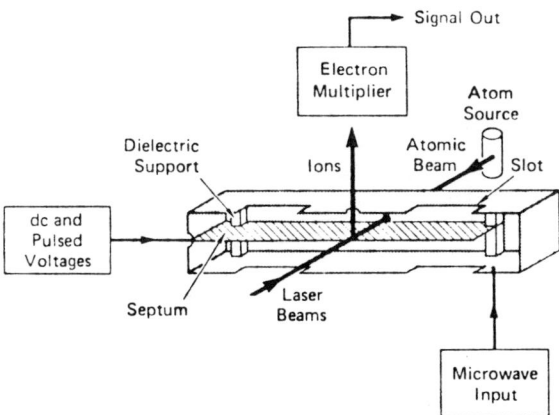

Fig. 7. Detailed view of the laser/atomic beam interaction region.

In order to determine if we are thoroughly mixing the low and high ℓ states we observed the cross section for excitation of the 6pnd state vs. frequency as a function of microwave field amplitude. First we excited the 6snd state in the presence of the microwave field and then removed the microwaves from the cavity non-adiabatically. Thus, the Stark states in the microwave field were projected onto the zero field ℓ states. If the initial Rydberg state we excited had any high ℓ character, then a fraction of the population was left in the zero field high ℓ states when the microwaves were removed. Next we scanned the frequency of the third laser near the Ba^+ 6s - 6p transition. The profiles we observed reflect the character of the Rydberg states left in the interaction region after the microwaves were removed. Low ℓ states are characterized by relatively broad profiles and peak positions which differ from the ion transition line due to non-zero quantum defects. High ℓ states are characterized by sharp resonances very near the ion line due to their long lifetimes and extremely small quantum defects. By observing frequency scans of the third laser for various field amplitudes we determined the appropriate fields at which the d states are mixed with the higher ℓ manifold states.

Fig. 9 shows frequency scans of the third laser at various microwave field amplitudes and with the third laser firing either before or after the microwaves were removed from the cavity. Obviously, the zero field d states do not mix with the higher ℓ states for microwave fields less than half of the Inglis-Teller limit. Furthermore, these states do not become true Stark states in fields less than 70% of the microwave ionization limit.

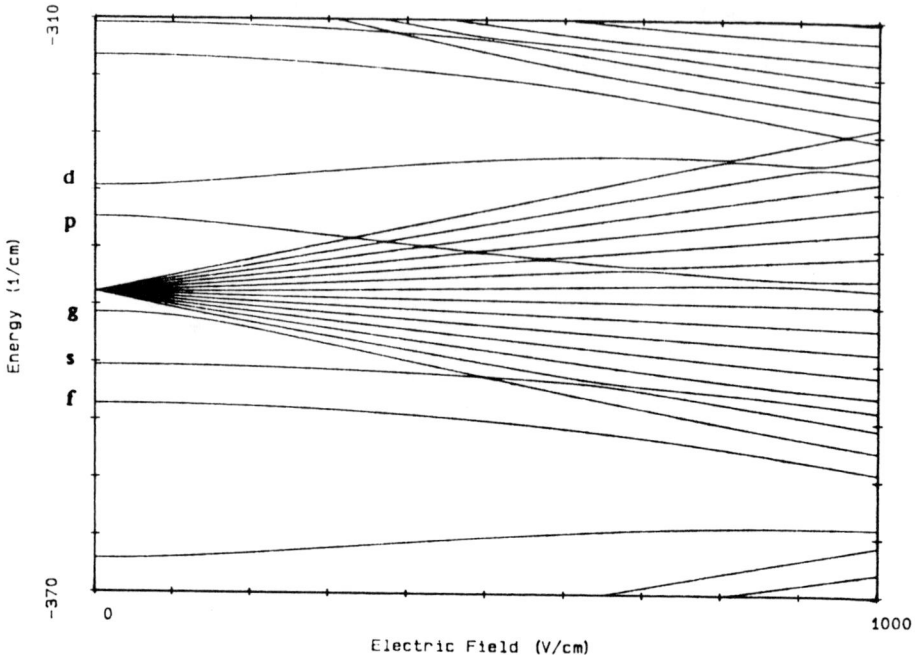

Fig. 8. Stark map of the 6p18k manifold as a function of electric field. Note the nonlinearity of the Stark effect for fields less than the Inglis-Teller limit.

Next we excited a series of Stark states in microwave fields ranging from 70% to 100% of the microwave ionization limit. The widths of the resonances we observe when scanning the third laser frequency over the 6pnk level positions is directly proportional to the autoionization rate of these states. The widths of the profiles are plotted vs. n^{-4} in Fig. 10. The linear dependence is obvious and demonstrates that the linear static Stark picture can be used to describe the situation in dynamic fields as well. If we think of the microwaves as a quasistatic field as discussed above, we expect to see the same result as in a static field. More formally, the atomic system in the presence of microwaves can be correctly described by Floquet theory.[11] Although we will not go into the mathematical complexities here, the results of the Floquet treatment yield the same results as the more intuitive quasistatic picture.[8]

The fitted line in Fig. 10 predicts autoionization rates which scale as $A_A \simeq 0.25$ n^{-4} (a.u.). However, Eq. 3 predicts a scaling of $A_A \simeq 0.53\ n^{-4}$ (a.u.). We believe that the discrepancy can be attributed to incomplete mixing of the Ba 6pnf states with the rest of the manifold. Fig. 8 clearly shows that the f state energies do not deviate significantly from their zero field positions until the Inglis-Teller microwave ionization threshold. If we ignore the f state composition in the Stark states then Eq. 3 predicts $A_A \simeq 0.22\ n^{-4}$ (a.u.) in excellent agreement with our result. This adjustment of Eq. 3 may seem somewhat ad hoc but is a necessary correction since Eq. 3 assumes that we have used field values in excess of those necessary to mix the low ℓ levels into the manifold.

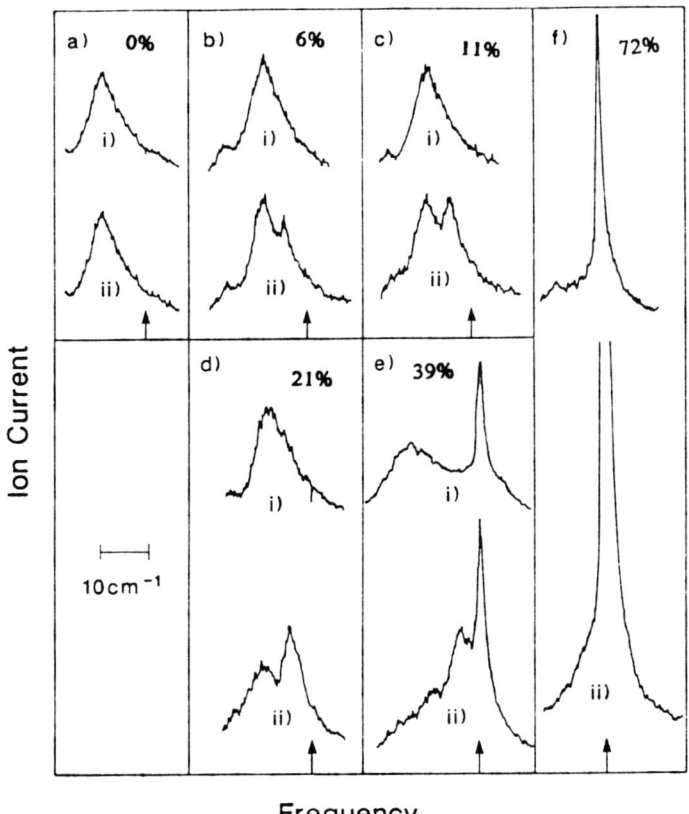

Fig. 9. Series of frequency scans of the third laser showing mixing of the 6p18d state with the higher angular momentum states. i) shows spectra taken with the third laser firing after the microwaves were removed from the cavity and ii) shows spectra generated with the third laser firing in the microwave field. Field amplitudes are shown as percentages of the Inglis-Teller limit, $F = 1/3n^5$.

CONCLUSIONS

We can now be confident of the correctness of the linear Stark picture for describing the effects of electric fields on autoionization and DR. Experiments have shown that the model works not only in the static linear Stark effect regime, but also in field regimes where the Stark effect is non-linear. Furthermore, the model also accurately predicts the redistribution of autoionization rates throughout the available states in dynamic fields.

Once again we must consider the application of the linear Stark model to plasma microfields. These microfields differ in one important aspect from the electric fields used in the experiments discussed here. In our experiments the laser and microwave fields have parallel polarizations which makes m (the projection of angular momentum on the z axis) a conserved quantity. This is not the case in a plasma. Due to the random motion of the ions in the plasma, there is no well defined field polarization and m is not conserved.

The effect of the random field polarization can be extremely important to the DR cross section. Since high m values are only allowed for high ℓ states, high m states do not contribute to the total DR cross section in linearly polarized fields. The mixed m and ℓ levels provide $2n^2$ accessible states per principal quantum number so that the autoionization rates should scale as n^{-5}. Furthermore, if the m states are mixed along with the ℓ states then all the states for a given principal quantum number will contribute to DR. Thus, the random field direction should enhance the total DR cross section by an even greater amount than a simple linearly polarized field.

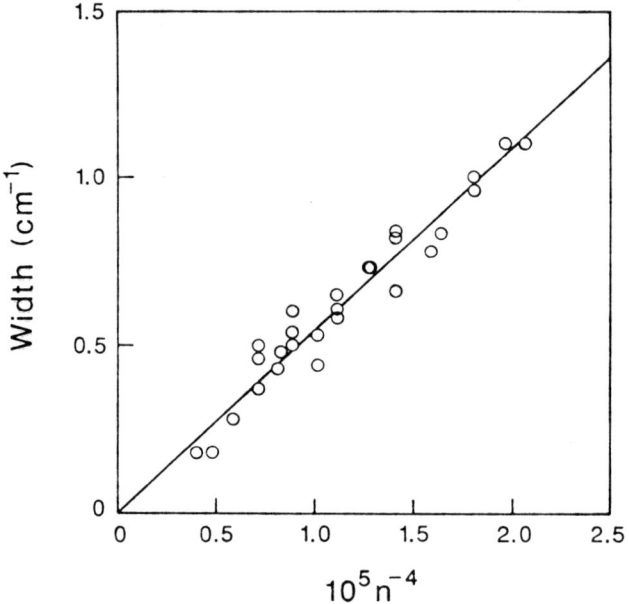

Fig. 10. Experimental widths of 6pnk autoionizing states plotted vs. $n.^{-4}$ Also shown is the best linear fit to the data which has a slope of .25 a.u.

ACKNOWLEDGEMENTS

This work has been supported by the U.S. Department of Energy, Office of Basic Energy Sciences, Chemical Sciences Division.

REFERENCES

1. W.R.S. Garton, W.H. Parkinson, and E.M. Reeves, Proc. Phys. Soc. London **80**, 860 (1962).
2. V.L. Jacobs, J. Davis, and P.C. Kepple, Phys. Rev. Lett. **37**, 1390 (1976).

3. K.A. Safinya, J.F. Delpech, and T.F. Gallagher, Phys. Rev. A 22, 1062 (1980).
4. A. Lindgard and S.E. Nielsen, At. Data Nucl. Data Tables 19, 613 (1977).
5. R.R. Jones and T.F. Gallagher, Phys. Rev. A 38, 2846 (1988).
6. G.H. Dunn, Electronic and Atomic Collisions, D.C. Lorents, W.E. Meyerhof, and J.R. Peterson eds. (1986).
7. S.M. Jaffe, R. Kachru, N.H. Tran, H.B. van Linden van den Heuvell, and T.F. Gallagher, Phys. Rev. A 30, 1828 (1984).
8. R.R. Jones and T.F. Gallagher, Phys. Rev. A 39, 4583 (1989).
9. W.E. Cooke, T.F. Gallagher, S.A. Edelstein, and R.M. Hill, Phys. Rev. Lett. 40, 178 (1978).
10. N.H. Tran, P. Pillet, R. Kachru, and T.F. Gallagher, Phys. Rev. A 29, 2640 (1984).
11. S.H. Autler and C.H. Townes, Phys. Rev. 100, 703 (1955).

Databases

THE ALADDIN ATOMIC PHYSICS DATABASE SYSTEM

Russell A. Hulse
Princeton University, Plasma Physics Laboratory, Princeton, NJ 08543

ABSTRACT

ALADDIN is an atomic physics database system which has been developed in order to provide a broadly-based standard medium for the exchange and management of atomic data. ALADDIN consists of a data format definition together with supporting software for both interactive searches as well as for access to the data by plasma modeling and other codes.

The ALADDIN system is designed to offer maximum flexibility in the choice of data representations and labeling schemes, so as to support a wide range of atomic physics data types and allow natural evolution and modification of the database as needs change. Associated dictionary files are included in the ALADDIN system for data documentation. The importance of supporting the widest possible user community was also central to the ALADDIN design, leading to the use of straightforward text files with concatentated data entries for the file structure, and the adoption of strict FORTRAN 77 code for the supporting software. This will allow ready access to the ALADDIN system on the widest range of scientific computers, and easy interfacing with FORTRAN modeling codes, user developed atomic physics codes and databases, etc. This supporting software consists of the ALADDIN interactive searching and data display code, together with the ALPACK subroutine package which provides ALADDIN datafile searching and data retrieval capabilities to user's codes.

ALADDIN has been adopted as the standard international atomic physics data exchange format for magnetic confinement fusion applications by the International Atomic Energy Agency (IAEA). Entry of critically evaluated atomic data sets into ALADDIN format is to be coordinated by the IAEA Atomic and Molecular Data Unit, which will also coordinate long-term development and distribution of updated software and documentation. The increasingly widespread adoption of the ALADDIN data format can be expected to greatly facilitate access to atomic data both within and outside of this original fusion application area.

INTRODUCTION

The basic motivation for development of the atomic physics data system to be described here was the need to facilitate the exchange and use of atomic and molecular data across a wide range of atomic physics researchers and those in other fields who are users of such data for various applications. The goal was thus to provide a basic, common language for atomic and molecular data exchange, creation of databases, and accessing atomic data by user applications. ALADDIN[1] (A Labeled Atomic Data Interface) was developed at the Princeton Plasma Physics Laboratory to address these needs, originally within the context of the atomic and molecular data needs of the magnetic confinement fusion energy (MFE) research community. The invited papers presented at the "Atomic Processes for Fusion" sessions at this conference illustrate some of the many applications of atomic data in MFE research. ALADDIN was adopted May 1988 by the International Atomic Energy Agency (IAEA) Atomic and Molecular Data Unit as a standard data format for MFE applications, and ALADDIN-based atomic data exchange is now becoming broadly established within the MFE community. Extensive data sets are now in place, and others are under continuing development as the user group expands and existing applications codes and databases are adapted to ALADDIN. ALADDIN is, however, potentially applicable in a similar role for use in many other research areas, and could be developed as a basis for mutually beneficial sharing and exchange of atomic, molecular, surface physics, other data amongst various application areas. Hopefully, this presentation will serve to encourage such application of the ALADDIN system to atomic physics related research activites outside of the fusion community.

In short, the ALADDIN objective is to provide a basic, broadly-based standard medium for the exchange and management of atomic data, which therefore must be straightforward to use, highly flexible, and widely transportable.

WHAT EXACTLY IS ALADDIN ?

The ALADDIN system comprises several related components, consisting of a data format, supporting software, specific data labeling schemes, and finally, the data itself. Users can select and adapt various elements of the system as best suits their own needs.

The most fundamental ALADDIN component is a defined but extremely flexible data format for entering and labeling atomic data as entries in ASCII (text) data files.

Next, the ALADDIN system includes specific supporting FORTRAN-77 software to facilitate user access to ALADDIN data. The ALADDIN code provides interactive data search and access from a user's terminal, while the ALPACK subroutine package provides a "black box" subroutine interface to ALADDIN data files from a user's FORTRAN codes.

Any given practical implementation of ALADDIN relys on specific choices of ALADDIN data labeling schemes appropriate for these applications. These labeling schemes can combine standardized systems established by the IAEA and other data centers along with a user's own schemes to describe their application-specific data.

Finally, the ALADDIN system includes specific data available either directly in ALADDIN-based datafile compilations or from other databases and codes which can read and write ALADDIN format files for data exchange. Associated ALADDIN dictionary files provide documentation and user information via keyword reference from the data files.

A schematic overview of how the ALADDIN system can serve to communicate atomic data between various sources and applications is illustrated in Fig. 1.

BASIC TECHNICAL CONSIDERATIONS IN THE ALADDIN DESIGN

The basic ALADDIN data storage format employs ASCII data files consisting of concatenated, independent entries. Such sequential "text" data files are by far the most readily transmitted and read across all computer systems, using media such as floppy disks, tape, computer networks and electronic mail. Also, the user's favorite text editor can serve as a basic and familiar way of searching, reading, and editing data entries. These data files can also be simply sorted by data type and printed out, allowing easy publication and use of the data even by those without convenient access to computers. Each ALADDIN data entry contains searchable attribute labels, together with the data itself.

Following the same emphasis on portability, primary supporting ALADDIN software is written in strict FORTRAN 77. This is, in practice, still the principal programming language for scientific applications, and as such, it is widely available on a full range of systems from PC's to Crays. Conveniently, the FORTRAN 77

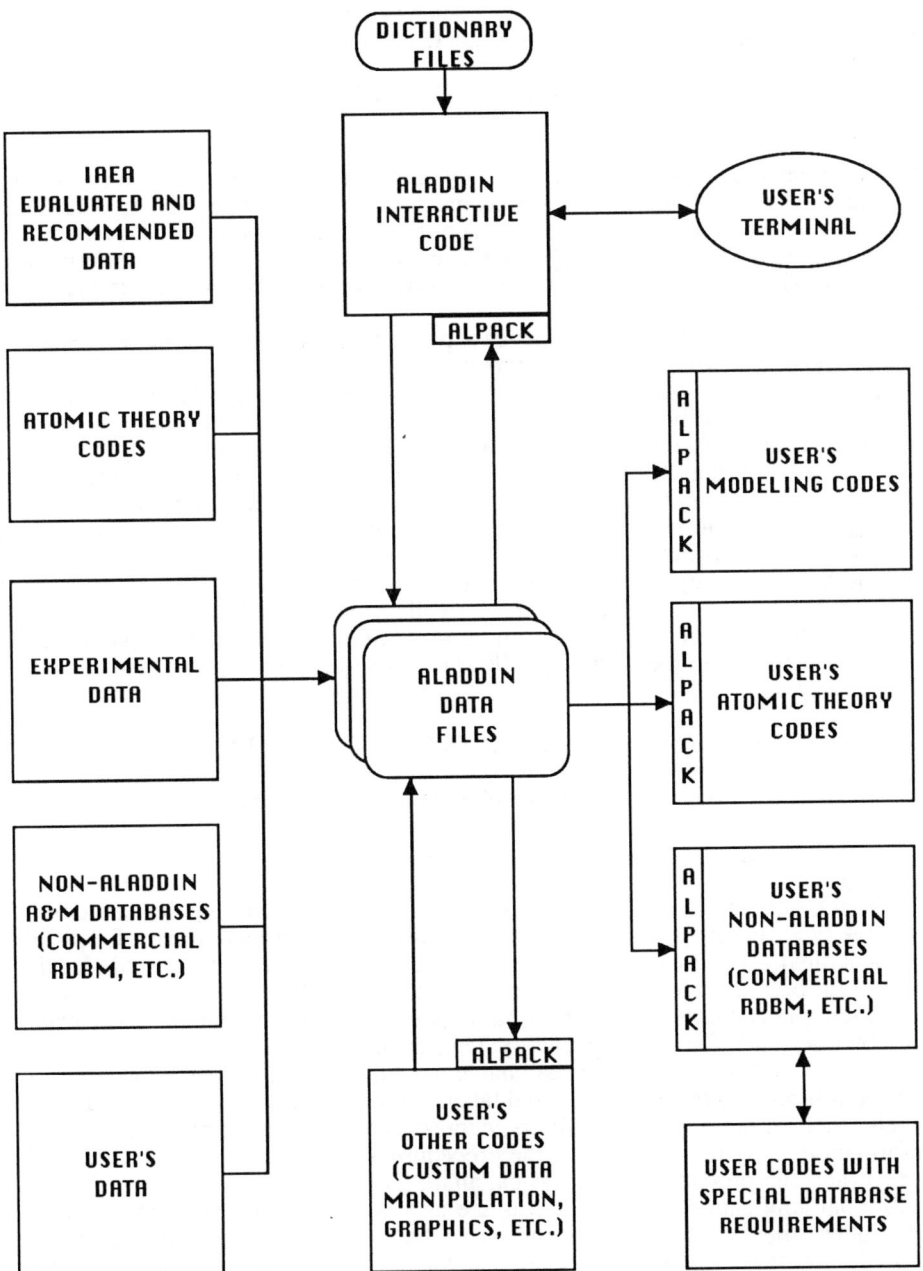

Fig. 1

standard also offers some support for character string manipulation, which simplifies ALADDIN's label-based searching and data handling approach.

DATA FILES vs. SUBROUTINES

An important point incorporated in the ALADDIN design is the observation that information in data files is much more easily manipulated than information coded into software. ALADDIN therefore keeps as much information as possible in data files, with the software kept as universal and data-independent as possible. However, the structure of languages such as FORTRAN require that some procedural information associated with interpreting the data must be contained in lines of code compiled into each program that uses the data.

This procedural information can be seperated into two types. First, we have the data file read, data search, and data store procedures. The ALADDIN approach is to design a general data labeling and search retrieval method, insensitive to data types being read. This software procedure receives only key searching information, and returns data identifying information together with the data itself into generalized buffer arrays. Secondly, we have the procedure(s) necessary for returning data values at specific points from the stored data representation. Here, a subroutine library is provided to interpret each data representation (by using the mathematical form of the data fitting equation, for example), with the subroutine to be called selected based on a keyword label associated with the data itself.

To summarize, ALADDIN uses "self-identifying" data containing keyword references to associated procedures and documentation as much as possible. This philosophy is akin to what would be called an "object-oriented" approach in computer terminology.

NECESSARY COMPONENTS OF THE DATA ENTRIES

Atomic data to be stored can be looked at as a function of the form

$$F_{ijk...}(x, y ... | a, b, c ...)$$

where (i, j, k ...) are a set of (discrete) selection variables describing the data (atomic process type, element, charge state, etc.) and the functional form of F. This mathematical form uses one or more supplied fitting coefficients (or tabular data points) (a, b, c ...), which are used to return the desired data value (cross section, rate coefficient) corresponding to input of the (typically continuous) independent variables (x, y ...) (for example, energy, temperature).

ALADDIN represents the searchable attributes (i,j,k...) of each data entry by lists of arbitrary character string-based labels in the entry header. These "searchable labels" include the physics identifiers, data source and revision number, and other information, as well as identifying the procedure to call to read this entry's specific format and the procedure to call to calculate the data function F. An important additional application for these labels are as keywords referring to entries in the "dictionary files" associated with the ALADDIN data files. These dictionary files provide whatever documentation is necessary to fully support the data entries themselves. Distinct from the information contained in these searchable labels for each entry are the "coefficients" (a, b, c...), defined by ALADDIN as non-searchable information which is to be passed as input to the function evaluation routine.

HIERARCHICAL AND BOOLEAN SEARCHABLE LABELS

As just described, the searchable labels serve to identify each ALADDIN entry. These labels are treated by ALADDIN as a list of arbitrary character strings, and can be placed in both hierarchical and boolean search structures to provide flexibility in designing specific labeling schemes.

Hierarchical labels are defined as a sequence of labels where the order of appearance is significant. Much information (especially physics labelling) naturally has this form. For example, a simple hierarchical label sequence for spectral line excitation rate coefficient data might be set up as follows:

$$\text{SLEXR} \quad 26 \quad 24 \quad 255.2$$

where these labels describe the data type, element ($Z = 26$), charge state (XXIV), and wavelength (255.2 Å), respectively.

Boolean labels are used to represent independent attributes whose entry order does not have a fixed sequence, and which are tested individually in a search using the Boolean logical operators .AND. and .NOT.. Typical boolean label information would be the data source, fitting accuracy, a dictionary file keyword reference to documentation, etc. ALADDIN uses two reserved boolean label types for special functions. An "access label" with a $ prefix flags special procedures to read the coefficient field data, while the "evaluation label" denoted by a # prefix specifies the data fitting function and associated procedure call.

ALADDIN DATA FORMAT SPECIFICATION

Figure 2 illustrates the structure and components of an ALADDIN entry. The first label of any entry is the access label. Access labels always begin with the reserved special symbol $, with an access label consisting of a $ symbol alone denoting a standard single-precision floating point data list in the coefficient field. Following the access label is the list of ordered hierarchical labels, which are simply character strings delimited by intervening blanks. The hierarchical label field is terminated by the & symbol, whereupon follows the similarly blank-delimited boolean label field. The first boolean label is always taken to be the access ($) label, followed by the (optional) other entries in this boolean label field. The final boolean label is always the evaluation label beginning with a # symbol, which also serves to delimit the end of the entire searchable label header. The header field may extend across many lines, as necessary.

On the next line following the end of the searchable label header may appear one or more optional comment lines, beginning with the ! character. These may contain any desired text, but extensive or repetitive documentation is best provided in the dictionary files and referred to by keyword boolean labels rather than placed in these embedded comment lines.

On the next line following the searchable label header and optional comment lines begins the coefficient field. This field again may extend across an arbitrary number of lines, and contains the actual "data" in the entry. These coefficient field lines are initially handled by ALADDIN only as character strings, to provide complete freedom in the choice of data structures. As mentioned above, the default structure later assumed for this field is blank-delimited single-precision floating point numbers.

Figure 3 shows a simple test data file in ALADDIN format, including spectral line excitation data represented using the form due to Mewe and a simple function represented using the linearly interpolated TAB1D tabular data form[3]. The actual

ALADDIN DATA FORMAT

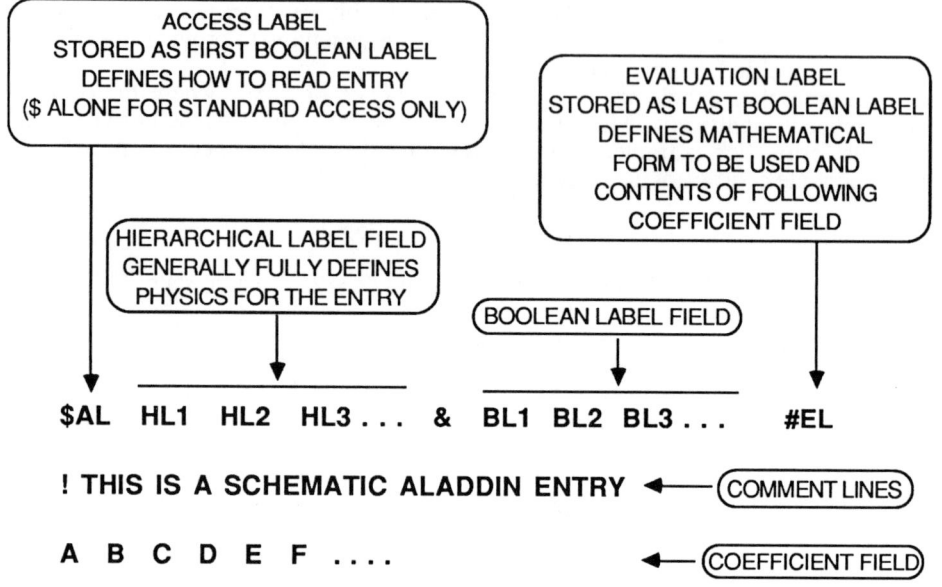

Fig. 2

SIMPLE ALADDIN DATA FILE EXAMPLE

```
THIS IS A TEST DATA FILE FOR ALADDIN, WITH #MEWE AND #TAB1D DATA
$ SLEXR  8  3   703.36  & RAH/PPPL #MEWE
703.36  0.0   0.18    1.0    0.6    0.0    0.0     0.28
$ SLEXR  8  4   790.36  & RAH/PPPL #MEWE
790.36  0.0   0.15    1.0    0.6    0.0    0.0     0.28
$ SLEXR  8  8    18.97  & RAH/PPPL #MEWE
18.97   0.0   0.4162  1.0    0.04   0.21  -0.04    0.28
$ SLEXR  8  8   102.5   & RAH/PPPL #MEWE
! THIS IS A COMMENT LINE IN THE FOURTH ENTRY
0.0    774.37  0.0791  0.1100  0.27   0.08   0.0     0.28
$ SLEXR 26 24   192.1   & FEXXIV RAH/PPPL #MEWE
192.1   0.0   0.0478  1.0     0.6848  0.9652 -0.4783  0.28
$ SLEXR 26 24   255.2   & FEXXIV RAH/PPPL #MEWE
255.2   0.0   0.0177  1.0     0.6648  0.9652 -0.4783  0.28
$ TEST  #TAB1D
0.0 0.0  100.0 100.0  200.0 100.0
```

Fig. 3

ALADDIN data file also allows for a comment section at the beginning of the file, as shown, before the first data entry denoted by occurence of the $ character. This figure hopefully makes clear how simple, in practice, it is to make up an ALADDIN labeling scheme and start using data in ALADDIN form. For further discussion, and standardized labeling schemes, please refer to the IAEA ALADDIN publications[2,3].

ALADDIN SUPPORTING SOFTWARE

ALADDIN supporting software is intended to provide the functionality essential to allow new users to rapidly incorporate ALADDIN into their work. At the present, this consists of a set of FORTRAN 77 routines, conceptually divided into two functional groups: the ALADDIN interactive code, and the ALPACK subroutine package.

The ALADDIN interactive code provides a basic interactive data searching, display, and manipulation capability from the user's terminal. It is a command-based program, which can be installed on a wide variety of hardware, requiring only a true FORTRAN-77 compatible compiler. The user specifies the hierarchical and boolean labels to be searched for in a given ALADDIN data file, and the ALADDIN interactive program will sequentially search through the data file until a matching data entry is found. The search strings include "wild cards" and other constructs to provide a flexible searching capability. Once found, data entries can be displayed to the terminal, and the data function (fitting function) can be evaluated at specified points (energies, temperatures, etc.) with the resulting data values (cross sections, rate coefficients, etc.) displayed to the screen or written to a file. On-line access to documentation describing not only the ALADDIN data files but also other ALADDIN nomenclature, fitting forms, references to the literature, etc., can also be accessed on-line from the ALADDIN interactive program by asking for keyword searches in associated ALADDIN dictionary files. A list of the present ALADDIN interactive code command set indicates the code's basic, straightforward approach:

H, {C/R}	HELP (SHOW THIS COMMAND SUMMARY)
{LABEL}?	QUERY THE ALADDIN DICTIONARY ABOUT {LABEL}
SL, L	SEARCH LABELS (DEFINE NEW SEARCH LABELS)
S	SEARCH FOR NEXT MATCHING ENTRY
G##	GO TO ENTRY AT SPECIFIED SEQUENCE NUMBER
N	NEXT (GO TO NEXT SEQUENTIAL ENTRY NUMBER)
D{FEL!CS*}	DISPLAY TO TERMINAL ITEM(S) SPECIFIED BY SUFFIX CHARACTER(S): Files, Entry, Labels, !comments, Coefficients, Search labels, *all display fields
WR	WRITE CURRENT ENTRY TO OUTPUT ALADDIN FILE
EV	EVALUATE FITTED DATA POINTS FOR CURRENT ENTRY
R	REWIND INPUT ALADDIN DATA FILE
F	FILE (SELECT INPUT ALADDIN DATA FILE)
OF	OUTPUT FILE (SELECT WR COMMAND OUTPUT FILE)
EF	EV FILE (SELECT EV COMMAND OUTPUT DATA FILE)
QF	QUERY FILE (SELECT QUERY DICTIONARY FILE)
EX, EN	EXIT OR END ALADDIN

The ALPACK subroutine package provides a simple interface to ALADDIN data files for user's applications codes, as shown in Fig. 4. ALPACK reproduces the same basic search capability as the on-line ALADDIN interactive code, but now via subroutine calls from the user's code. Successive data entries are read in using ALREAD. The desired hierarchical and boolean search label sequences (again, with possible wild card constructs) are passed to the ALCOMP subroutine, which compares these with the entry header, flagging successful matches. The label and coefficient field information from the ALADDIN entry are then available in COMMON blocks for use by the calling code. The subroutine ALRECF provides conversion of standard (default) coefficient field data from character strings to real (floating point) values which are then passed to the appropriate evaluation subroutine.

It is not possible or desirable to present here a comprehensive user's manual describing the entire evolving ALADDIN system in detail. For this purpose, the reader is encouraged to consult the references at the end of this paper, and contact the IAEA Atomic and Molecular Data Unit, Vienna, Austria, which will be coordinating ALADDIN development and distribution of ALADDIN software and data files.

ROLE OF THE INTERNATIONAL ATOMIC ENERGY AGENCY (IAEA)

ALADDIN was adopted as the standard IAEA atomic and molecular data format for MFE applications at Consultant's meeting of the IAEA Atomic and Molecular Data Unit[2] in May, 1988. Continued IAEA coordination of ALADDIN development is planned so as to ensure the organized, effective growth of the ALADDIN system.

This coordinating role will occur at several levels. First, the IAEA Atomic and Molecular Data Unit will continue to provide international coordination of atomic physics efforts towards providing comprehensive, critically evaluated ALADDIN data sets, particularly for MFE applications. Already, a wide range of international MFE atomic data centers have adopted the ALADDIN system as a standardized format for communication of their atomic physics data. The IAEA also will act to provide standardized ALADDIN labeling schemes for a wide range of atomic, molecular, and surface physics data, as well as standardized ALADDIN software development, documentation and distribution[3].

SUMMARY

ALADDIN is designed to provide a basic, broadly-based standard system for the exchange and management of atomic data. To meet this goal of a universal data format, highest priority has been placed on a system which is straightforward to use, highly flexible, and widely transportable.

The ALADDIN system consists of several elements. Most fundamentally, ALADDIN defines a basic data format, based on ASCII data files, which allows a wide variety of data types to be flexibly accomodated within a common structure based on self-describing data entries, associated subroutine libraries and "dictionary" documentation files. Supporting software provides access to ALADDIN files both interactively and from user's codes.

ALADDIN has been adopted as the standard international atomic data format by the IAEA Atomic and Molecular Data Unit for MFE applications. The IAEA is providing continued coordination of ALADDIN labeling conventions, system development and the provision of recommended data. ALADDIN-based atomic physics data compilations and data exchange are now becoming broadly established in the MFE community, and can be profitably extended for use in other applications.

New ALADDIN users and contributors are strongly encouraged!

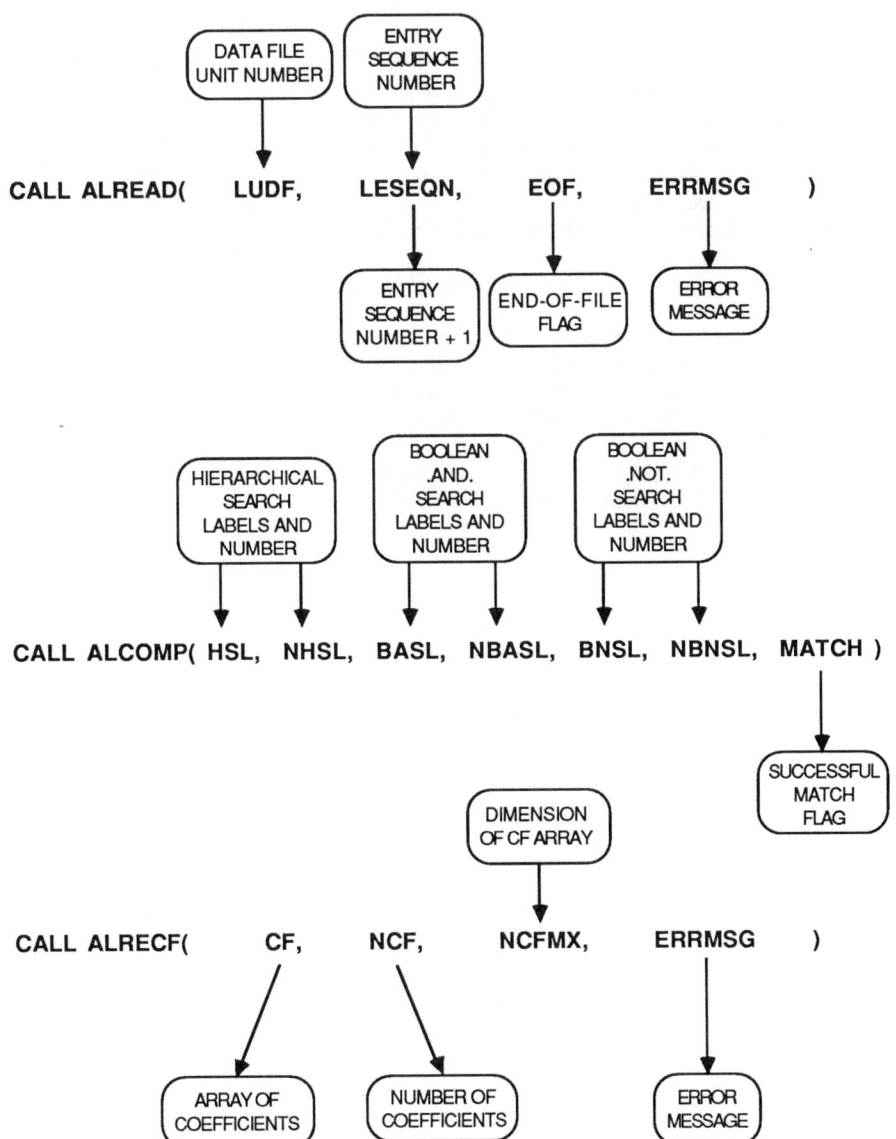

Fig. 4

ACKNOWLEDGEMENTS

The rationale for the ALADDIN system arises from the need to deal with the large body of diverse atomic physics data required for MFE and other research programs. The existence and continued development of this body of data is, of course, due to the efforts of many theoretical and experimental atomic physicists. It is to be hoped that, in return for their efforts, ALADDIN will provide them with the satisfaction of even more widespread and fruitful application of the results of their research. The efforts of R. Janev and J. Smith at the IAEA have been and continue to be crucial to realizing the international coordination of MFE atomic physics data base development in general and ALADDIN in particular. Particular thanks for discussions, encouragement, and ALADDIN beta testing go to R. Phaneuf at ORNL.

This work was supported by U.S. DoE contract # DE-AC02-76-CHO-3073.

REFERENCES

(1) R.A. Hulse, "ALADDIN - A Labeled Atomic Data Interface for Fusion Applications", to be submitted for publication. (see also presentation in reference 2)

(2) "IAEA Consultant's Meeting on the Atomic Database and Fusion Applications Interface", May 9-13, 1988, Vienna, Austria, IAEA Summary Report INDC(NDS)-211/GA, Sept. 1988.

(3) "ALADDIN Manual, Version 1.0", IAEA Atomic and Molecular Data Unit report IAEA-NDS-AM-17, Vienna, Austria, June 1989.

THE OPACITY PROJECT: PROGRESS AND METHODS

Yan Yu
Department of Astronomy, University of Illinois
1011 West Springfield Avenue, Urbana, Illinois 61801

ABSTRACT

Large amounts of accurate atomic data have been calculated in a collaborative effort referred to as the Opacity Project. The computational methods and various physical processes of interest are discussed here, with reference to the opacity spectrum and some preliminary results.

1. INTRODUCTION

An outline of the scope of the Opacity Project and the theoretical methods used is given in the first two of a series of papers entitled "Atomic data for opacity calculations"[1,2]. Aspects of the plasma effects on the internal atomic partition function are discussed in a companion series of papers entitled "The equation of state for stellar envelopes"[3,4,5].

The present project is concerned with the calculation of stellar envelope opacities in the temperature-density domain $3.5 \leq \log T \leq 7$, $-12 \leq \log \rho \leq -2$, where the major absorption processes are bound-bound and bound-free transitions by atomic ions. So far, energy levels and radiative data for the first 10 isoelectronic sequences, H–Ne, as well as atoms and ions of other astrophysically abundant elements, have been calculated, and work is well under way to complete the Fe ion data. The close-coupling approximation of electron-ion collision theory is used in the entire atomic data calculations, which provides a unified treatment and ensures consistent accuracy for both the bound states and the free electron states.

2. SUMMARY OF FORMULAE

In the close-coupling approximation[6], the wave functions Ψ of an $(N+1)$-electron system are expanded in terms of the N-electron target wave functions ϕ_i and the orbital functions θ_i of the colliding electron,

$$\Psi = A \sum_i \phi_i \theta_i, \tag{2.1}$$

where A is the anti-symmetrization operator and where the summation is over all channels of a total L, S and parity combination. Bound-type wave functions of the $(N+1)$-electron system are usually included in (2.1) to improve convergence when the summation is truncated to a finite number of terms.

The close-coupling equations are solved in an inner region of configuration space, $r \leq a$, using the R-matrix method[7,8]. The value of a is chosen such that the short range effects can be neglected for $r \geq a$.

In the R-matrix formulation, a set of basis functions ψ are defined in terms of discrete zeroth-order functions ϕ satisfying some fixed boundary condition at $r = a$, thus

$$\psi = C\phi, \tag{2.2}$$

such that ψ diagonalizes the $(N+1)$-electron Hamiltonian,

$$C\left(\phi|\mathbf{H}|\phi\right)_\mathrm{I} C = \mathbf{E}, \tag{2.3}$$

where $\mathbf{E}$ is a diagonal matrix with elements E_k and the notation $(\cdots)_\mathrm{I}$ indicates matrix elements evaluated in the inner region. The complete $(N+1)$-electron wave function at energy E is then expended in terms of the basis functions,

$$\Psi(E) = A(E)\psi, \tag{2.4}$$

where the column vector $A(E)$ contains the simple energy dependence $(E_k - E)^{-1}$ and incorporates the matching to the desired asymptotic forms through the radial wave functions evaluated at $r = a$.

Radiative matrix elements constructed in the inner region using wave functions given by (2.4) reduce to linear combinations of the elements of the dipole matrix between basis states in the initial and final symmetries, which need be calculated only once.

In the outer region exchange between the colliding electron and the target electrons can be neglected. The radial wave functions of the outer electron are given by

$$(h + \mathbf{v})\mathbf{F} = \epsilon \mathbf{F}, \tag{2.5}$$

where $\mathbf{v}$ is the multipole potential matrix, the column vector h has elements (in z-scaled notation)

$$h_i = -\frac{d^2}{dr^2} + \frac{l_i(l_i+1)}{r^2} - \frac{2}{r}, \tag{2.6}$$

and the channel energy ϵ has elements

$$\epsilon_i = E - E_i. \tag{2.7}$$

In (2.7) E is the total energy, E_i the energy of the parent state to which the channel is attached. A channel i is open if $\epsilon_i \geq 0$, or closed if $\epsilon_i < 0$. The effective quantum number ν_i of a closed channel is given by

$$\epsilon_i = -1/\nu_i^2. \tag{2.8}$$

We define a critical value, $\nu_{\max}$, for the effective quantum number such that the channel i is said to be weakly closed if $\nu_i > \nu_{\max}$, or strongly closed if $\nu_i < \nu_{\max}$.

Solutions of (2.5) are linear combinations of functions $\mathbf{S}$ and $\mathbf{C}$ for open channels, with

$$\mathbf{S}(r) \underset{r\to\infty}{\sim} s(r), \quad \mathbf{C}(r) \underset{r\to\infty}{\sim} c(r), \tag{2.9}$$

and $s(r)$, $c(r)$ as defined in ref. 9, or Θ for closed channels, with

$$\Theta_{ij} \underset{r\to\infty}{\sim} 0. \tag{2.10}$$

When all channels are closed, we put

$$\mathbf{F} = \Theta \mathbf{X}, \tag{2.11}$$

and the condition for matching of bound state solutions at $r = a$ is

$$\mathbf{F}' = \Theta' \mathbf{X}. \tag{2.12}$$

A method for solving (2.11) and (2.12) to determine the bound states is given in ref. 10.
When some channels are open or weakly closed, we put

$$\mathbf{F}(K) = \begin{pmatrix} \mathbf{S} + \mathbf{C}K \\ \Theta \mathbf{X} \end{pmatrix}, \qquad (2.13)$$

where K and $\mathbf{X}$ have partitioning

$$K = \begin{pmatrix} K_{oo} & K_{ow} \\ K_{wo} & K_{ww} \end{pmatrix}, \qquad (2.14)$$

$$\mathbf{X} = (\mathbf{X}_{so} \quad \mathbf{X}_{sw}) \qquad (2.15)$$

with the subscripts o for open, w for weakly closed, s for strongly closed.

The normalization (2.13) is used in the final state wave functions in constructing the real dipole matrix $D(K)$, which has partitioning

$$D(K) = \begin{pmatrix} D_o(K) \\ D_w(K) \end{pmatrix}. \qquad (2.16)$$

It can be shown (see ref. 11) that the dipole matrix for photoionization is

$$D^+(\mathbf{S}) = D_o^+(\chi) - \chi_{ow}[\chi_{ww} - \exp(-2\pi i \nu)]^{-1} D_w^+(\chi), \qquad (2.17)$$

where the χ-normalization is related to the K-normalization thus

$$\chi = (1 + iK)(1 - iK)^{-1}, \qquad (2.18)$$

$$D^+(\chi) = -i(1 - iK)^{-1} D(K). \qquad (2.19)$$

The oscillator strength for a bound-bound transition from state a to state b is

$$f(b, a) = \frac{\Delta E S(b, a)}{3g} \qquad (2.20)$$

and the cross section for photoionization is

$$\sigma = \frac{4\pi^2 \alpha a_0^2 \Delta E}{3g} S. \qquad (2.21)$$

In (2.20) and (2.21) ΔE is the energy difference of the initial and final states in Rydbergs, g is the statistical weight of the initial state, and the (generalized) line strengths are

$$S(b, a) = |D(b, a)|^2, \qquad (2.22)$$

$$S = |D^+(\mathbf{S})|^2. \qquad (2.23)$$

In (2.23) a summation over all contributing open channels is implied for cross section for photoionization into a specific final state; furthermore, a summation over all final states is implied for the total cross section, which is the relevant quantity for opacity calculations.

The weakly closed channels, when present, give rise to a series of narrow resonances characterized by the term $\exp(-2\pi i \nu)$ in (2.17). Following Gailitis, we define the average generalized line strength in a unit interval of effective quantum number thus

$$\langle S \rangle = \int_{\nu_0 - \frac{1}{2}}^{\nu_0 + \frac{1}{2}} S(\nu) d\nu. \qquad (2.24)$$

By assuming that χ remains constant over the small energy range corresponding to the unit interval in effective quantum number (a good approximation except, perhaps, in the vicinity of perturber resonances due to the presence of strongly closed channels), it can be shown[9] that the above integral for the total generalized line strength reduces to the simple form

$$\langle S \rangle = \sum_{i=o+w} |D_i^+(\chi)|^2. \tag{2.25}$$

Since the weakly closed channels become open and χ reduces to S at the onset of the next threshold for photoionization, equation (2.25) implied that the Gailitis averaged cross section below a threshold joins smoothly to the cross section above the threshold.

The technique of Gailitis averaging can also be applied to the region just below the first ionization threshold. By taking the number of open channels to be zero, the expression (2.21) with (2.25) gives the mean photoabsorption cross section for the Rydberg series of lines converging to the ionization threshold[11,13].

3. SAMPLE CALCULATIONS

We use C II as an illustrative case. The features discussed in this section are also found in most other systems.

The target for the close-coupling expansion consists of the states $1s^2 2s^2$ ^{1}S, $2s2p$ ^{3}P°, ^{1}P°, and $2p^2$ ^{3}P, ^{1}D, ^{1}S of C III.

3.1 Photoionization spectra including Gailitis average

Figure 1 shows a series of partial cross section spectra for photoionization from bound states in the ^{2}P° symmetry to the ^{2}S final symmetry. Resonance structures converging to each threshold were resolved up to $\nu_{max} = 10$, and Gailitis averaging was used to join the cross section smoothly across the threshold. In some instances (such as the region below the ^{3}P threshold) perturber resonances belonging to strongly closed channels mix with the weakly closed resonances, and the former retain their structure in the Gailitis average due to the large variation in the χ matrix. The structure just below the first threshold is a reflection of the existence of a high-lying perturber state, $2s2p(^3P°)3p$, in the final state symmetry.

3.2 The PEC resonance

An especially prominent feature in almost all the photoionization spectra produced in the Opacity Project is the family of resonances due to photoexcitation of the core (PEC), where an $(N+1)$-electron atom or ion absorbs a photon and the N-electron core makes a dipole transition to a higher state, leaving the total system in the continuum. This is the reverse process of dielectronic recombination.

As an example, the C II photoionization is dominated by the PEC process

$$2s^2(^1S)nl + h\nu \rightarrow 2s2p(^1P°)nl \rightarrow 2s^2(^1S) + e \tag{3.1}$$

in the vicinity of the energy for the core transition, $E(^1S \rightarrow {}^1P°)$. Figure 2 shows the spectra of the total cross section for photoionization from the ^{2}D bound states of C II, which contain a PEC resonance series slightly below 1 Rydberg in photon energy. It will be seen in the following section that the PEC resonances (3.1) also enter the opacity spectrum prominently.

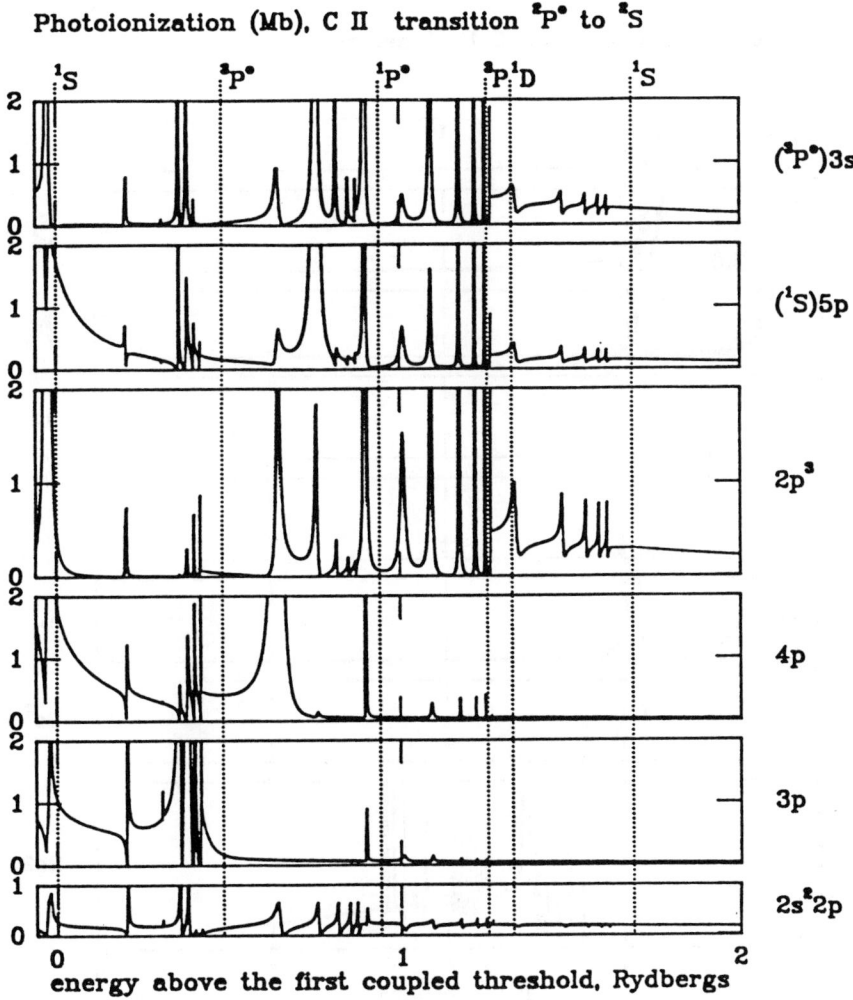

Fig. 1 Partial cross section (Mb) vs. electron energy (Rydberg) for C II photoionization from bound states in the $^2P^o$ series to the ^2S final state. The dotted vertical lines mark the threshold positions.

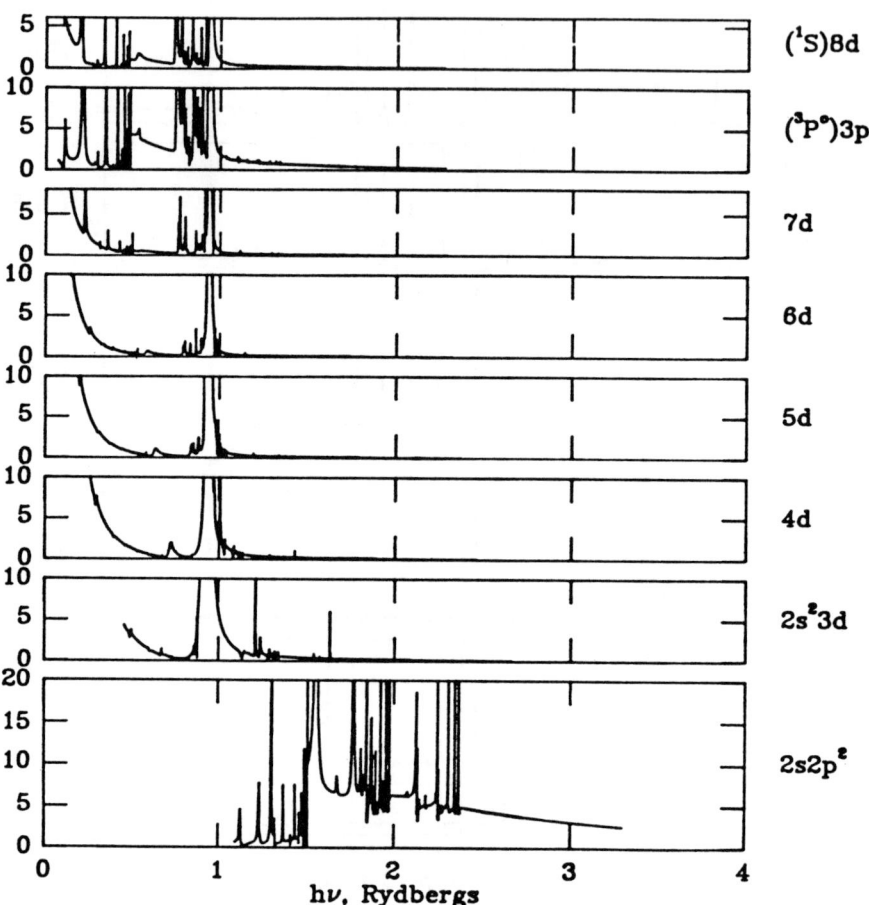

Fig. 2 Total cross section (Mb) vs. photon energy (Rydberg) for C II photoionization from bound states in the ^{2}D series.

4. THE OPACITY

The contribution of a bound-bound spectral line to the monochromatic opacity is

$$\kappa_\nu(a \to b) = \frac{\pi e^2}{mc} N_a f(b,a) \phi_\nu, \tag{4.1}$$

where the profile factor ϕ_ν is normalized to

$$\int \phi_\nu d\nu = 1. \tag{4.2}$$

The contribution due to photoionization is

$$\kappa_\nu = N_a \sigma_\nu. \tag{4.3}$$

In (4.1) and (4.3) N_a is the population of the initial level. Other processes contributing to the opacity are free-free and electron scattering. Except for the latter component, the monochromatic opacity is corrected for stimulated emission thus

$$\kappa'_\nu = \kappa_\nu [1 - \exp(1 - h\nu/kT)]. \tag{4.4}$$

The Rosseland mean opacity κ_R is given by

$$\frac{1}{\kappa_R} = \frac{\int_0^\infty \frac{1}{\kappa'} g(\nu) d\nu}{\int_0^\infty g(\nu) d\nu} \tag{4.5}$$

with the weighting function

$$g(u) = u^4 e^{-u} (1 - e^{-u})^{-2} \tag{4.6}$$

and

$$u = h\nu/kT. \tag{4.7}$$

Figure 3 shows a section of the monochromatic opacity spectrum for pure carbon at $T = 10^{4.5}$ K, $\rho = 10^{-8}$ g/cm^3, with the weighting function (4.6) drawn to an offset scale (upper plot). The ionization fractions are as follows:

C II 0.9×10^{-3}
C III 0.6034
C IV 0.3955
C V 0.2×10^{-3}

Several points in the spectrum are of interest:

(1),(2),(3) these are the photoionization edges of, respectively, the ground state 2s^2 ^{1}S of C III, the first excited state 2s2p ^{3}P$^\circ$ of C III, and the ground state 2s^{2}2p ^{2}P$^\circ$ of C II. Note that absorption by the low-lying states of the abundant ions takes place in the tail of the weighting function;

(4) the strong line at this frequency is the transition 2s^2 ^{1}S $\to$ 2s2p ^{1}P$^\circ$ in C III. The broad feature in the corresponding continuum is due to the family of PEC resonances (3.1).

By assessing the order-of-magnitude variation in the weighting function and the opacity spectrum it can be concluded that the Rosseland mean is sensitive to contributions from a wide range of excited states of the dominant ions, as well as from less abundant ions whose radiative excitation and ionization energies are near the peak of the weighting function at a given temperature. (It is important to note that the Rosseland mean opacity is a harmonic mean, and is therefore most sensitive to regions of low absorption.) The total Rosseland mean is 7.45 cm^2/g.

80 The Opacity Project

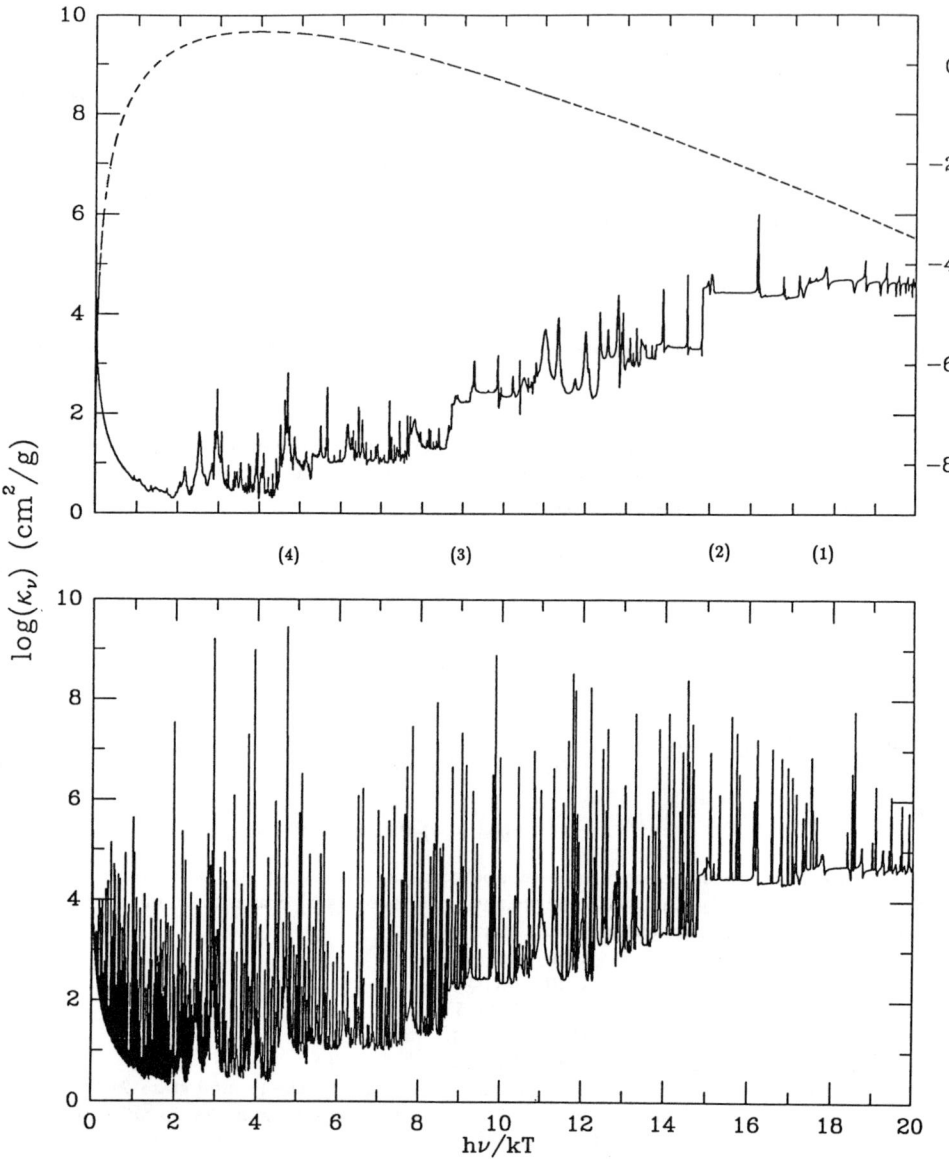

Fig. 3 Spectrum of the monochromatic opacity for carbon at $\log T = 4.5$, $\log \rho = -8$. Upper: contribution from the free-free and the bound-free radiative processes; lower: the total. The Rosseland mean weighting function is drawn (dash) to the logarithmic scale at right.

When the contributions from C II, III and IV are in turn set to zero, the Rosseland mean drops to 63%, 25% and 99% of the total value respectively. Though the Rosseland integral (4.5) is not additive, this numerical experimentation gives an indication of the relative importance of the absorption processes of each ion. In particular, it is seen that an ion of less than 0.1% in fractional abundance may contribute significantly to the Rosseland mean opacity.

Collaborators in the Opacity Project are:
Queen's University of Belfast. K.A. Berrington, P.G. Burke, V.M. Burke, W.B. Eissner, A. Hibbert, A.E. Kingston, D.J. Lennon.
University of London. J.A. Fernley, G. Peach, H.E. Saraph, M.J. Seaton, K.T. Taylor.
Paris Observatory. M. LeDourneuf, Vo Ky Lan, C. Zeippen.
Nice Observatory. J.A. Tully.
Munich University Observatory. K. Butler.
IBM Venezuela. C. Mendoza.
University of Illinois. D. Mihalas, Yan Yu.
Ohio State University. A.K. Pradhan.
JILA. D.G. Hummer.
University of Florida. C.F. Hooper.
NIST, Gaithersburg. L. Woltz.

Work supported in part by a grant from NSF, AST 85-19209.

REFERENCES

1. M.J. Seaton, *J. Phys. B: At. Mol. Phys.* **20**, 6363 (1987).
2. K.A. Berrington, P.G. Burke, K. Butler, M.J. Seaton, P.J. Storey, K.T. Taylor, Yu Yan, *J. Phys. B: At. Mol. Phys.* **20**, 6397 (1987).
3. D.G. Hummer, D. Mihalas, *Ap. J.* **331**, 794 (1988).
4. D. Mihalas, W. Däppen, D.G. Hummer, *Ap. J.* **331**, 815 (1988).
5. W. Däppen, D. Mihalas, D.G. Hummer, *Ap. J.* **332**, 261 (1988).
6. P.G. Burke, M.J. Seaton, *Meth. Comput. Phys.* **10**, 1 (1971).
7. P.G. Burke, W.D. Robb, *J. Phys. B: At. Mol. Phys.* **4**, 153 (1971).
8. P.G. Burke, W.D. Robb, *Adv. Atom. Molec. Phys.* **11**, 143 (1975).
9. M.J. Seaton, *Rep. Prog. Phys.* **46**, 167 (1983).
10. M.J. Seaton, *J. Phys. B: At. Mol. Phys.* **18**, 211 (1985).
11. Yu Yan, M.J. Seaton, *J. Phys. B: At. Mol. Phys.* **20**, 6409 (1987).
12. M. Gailitis, *Sov. Phys.-JETP* **17**, 1328 (1963).
13. J. Dubau, M.J. Seaton, *J. Phys. B: At. Mol. Phys.* **17**, 381 (1984).

UPDATE ON THE OPAL OPACITY CODE

F. J. Rogers, C. A. Iglesias and B. G. Wilson
Lawrence Livermore National Laboratory
Livermore, CA 94550

ABSTRACT

Persisting discrepancies between theory and observation in a number of astrophysical properties has led to the conjecture that opacity databases may be inaccurate. The OPAL opacity code has been developed to address this question. The physical basis of OPAL removes several of the approximations present in past calculations. For example, it utilizes a much larger and more detailed set of atomic data than was used to construct the Los Alamos Astrophysical Library. This data is generated online, in LS or intermediate coupling, from prefitted analytic effective potentials and is of similar quality as single configuration, relativistic, self-consistent-field calculations. The OPAL code has been used to calculate opacities for the solar core and for Cepheid variable stars. In both cases, significant increases in the opacity compared to the Los Alamos Astrophysical Library were found.

INTRODUCTION

Attempts to understand the structure and observed properties of stars require a detailed knowledge of the equation of state and radiative opacity of stellar compositions. This involves a combination of statistical mechanics, atomic physics, and line broadening theory which results in a substantial computational effort. In the past, most astrophysical opacity data has been provided by groups working at Los Alamos National Laboratory.[1,2,3]

Improved observational capabilities in recent years have led to a number of discrepancies between theoretical prediction and experimental measurement in a wide range of astrophysical observables. The most well known of these discrepancies is the so called "solar neutrino problem" where the measured solar neutrino flux is only about one-third of that predicted.[4] Some other discrepancies involve period ratios of Cepheid variable stars[5] and lithium depletion in the Hyades cluster.[6] Theoretical treatments of these observables all require a detailed knowledge of the radiative opacity and have led to a plea from the astrophysics community for reexamination of opacity calculations. The strongest evidence for this need has been presented by Simon[7] and later by Andreasen[8] who showed that a factor of 2-3 increase in the opacity could explain the mass anomalies in classical Cepheid variables.

The fundamental importance of understanding these observed properties has stimulated several new efforts[9-12] to calculate the equation of state and opacity for stellar conditions. Thus far, new results applicable to Cepheid variable stars have appeared that support Simon's supposition.[11,12] These reported increases in opacity are mostly due to improved atomic physics. In particular, the rich line spectrum from $\Delta n=0$ transitions and configuration term splitting. These effects are missing in the Los Alamos codes,[1-3] due to their hydrogenic approximation to the oscillator strengths and almost complete neglect of the term structure.

An opacity code consists of three major pieces: 1) equation of state, 2) atomic physics, and 3) photoabsorption algorithms. The basic philosophy we use to construct each of these parts, as well as some recent calculations are described in the following sections.

EQUATION OF STATE AND OCCUPATION NUMBERS

We use the so-called "physical picture" approach starting from the grand canonical ensemble of a system of electrons and nuclei interacting through the Coulomb interaction.[13] The effect of plasma environment on the internal states is obtained directly from the statistical mechanical analysis. Our approach is based on the fact that many particle correlations are highly classical and, consequently, it is restricted to regions where the de Broglie wavelength is less than the plasma screening length. We have, however, developed methods for systematically introducing quantum mechanics into the many-particle correlations.

Our first step is to develop the logarithm of the grand partition function into an activity expansion. Following, the standard method, the activity is eliminated in favor of the density to obtain a virial expansion. Since each viral coefficient is divergent for the Coulomb potential, it is necessary to diagrammatically redevelop the virial expansion to eliminate the divergences and thus obtain a plasma density expansion. In this process each virial coefficient is replaced by (screened) virial coefficients for the Debye potential, but with some low order diagrams subtracted out. These screened objects are referred to as Abe nodal functions. Although the Abe nodal functions involve the weak coupling Debye-Hückel potential, the plasma density expansion recovers both the weak and strong coupling limits.[13]

Astrophysical plasmas are, in general, only partially ionized and cannot easily be treated by expansions in the density. In the region of partial-ionization activity expansions are much more amenable to analysis. Consequently the next step in our theoretical development is to convert the plasma density expansion into a plasma activity expansion. This could also be accomplished directly through a diagrammatic resummation of the initial activity series but it proves to be much easier to follow the procedure just described. The cluster coefficients that arise in the plasma activity expansion are for a Debye-like potential with some low order diagrams subtracted out, analogous to the Abe nodal functions. The plasma activity expansion thus obtained views the system in terms of electrons and nuclei and is not properly ordered for partially ionized regions.

To overcome this shortcoming, the final step in our equation of state theory is to define activities for composite particles. These are built up from products of the electron and nucleus quantities. A renormalization of the electron-nucleus activity expansion leads to a new series in terms of an augmented set of activities. It is important to note that the composite particle activities we actually work with in calculating the equation of state are effective activities, since the Boltzmann factors are split into bound and continuous parts according to the Planck-Larkin prescription.[14-15] Correspondingly, the occupation numbers that arise are effective occupation numbers.[16] There are very good analytical reasons for doing this as described in the cited literature. Nevertheless, our opacity calculations require the total state occupation numbers and these are obtained from a supplementary calculation after the equation of state calculation has been carried out.[16]

Some important results of the analysis just described are: 1) the number of bound states is determined by the screened potential; 2) bound states are unscreened except near the plasmas continuum; 3) scattering states are screened; and 4) most of the effect of high lying states appears in the Coulomb corrections to the free energy, i.e., the Debye-Hückel correction at low density. The bound states appearing in the virial and cluster coefficients, which were screened in the initial steps of developing the partially-ionized plasma activity expansion, become unscreened in the final renormalization, i.e., these screening effects now appear in the correlational terms associated with the appearance of composite particles.

ATOMIC PHYSICS

The spectrum of complex ions contains millions of lines that can potentially contribute to the opacity so that, for the sake of computational simplicity, it seems plausible to replace groups of closely spaced lines by a single representative feature,[12] e.g., an unresolved transition array[17] (UTA). However, at the low density usually associated with envelopes of stars the individual lines may be too narrow to merge and make this a reliable procedure. At the other extreme one can construct a detailed database using state of the art atomic physics codes. Fortunately, astrophysical applications almost always require the Rosseland mean opacity which is relatively insensitive to small displacements of line positions. Because of this, we have chosen to use prefitted analytic effective potentials as input to the Dirac equation to obtain configurationally averaged energies on-line.[18] Others have had considerable success with similar approaches.[19-20] We split the configurational averages into term energies by using standard perturbation methods in either LS or intermediate coupling.

The parametric potentials we use are chosen so that they incorporate electronic shell structure. It was found that one Yukawa term per shell gives good results and, consequently, our potential was chosen to have the following form[18]

$$V = -\frac{2}{r}\left[(Z-v) + \sum_{n=1}^{n^*} N_n e^{-\alpha_n r}\right] \quad (1)$$

(in Rydbergs), where

$$v = \sum_{n=1}^{n^*} N_n \quad (2)$$

is the number of electrons for the parent ion, N_n the number of electrons in the shell with principal quantum number n, n* the maximum value of n for the parent configuration, and α_n the screening parameter for electrons in shell n.

The screening parameters, α_n, for shell n were determined by interactively solving a spin-averaged Dirac equation and matching the experimental configurational averaged ionization energies for ground state parent configurations. The parameter values were then fitted along isoelectronic sequences to simple four parameter functions of the following form

$$\alpha_n = (\xi_n + 1) \sum_{j=0}^{3} \frac{a_j(v_n)}{\xi_n^j} \quad (3)$$

where

$$\nu_n = \sum_{i=1}^{n} N_n \tag{4}$$

is the total electronic occupation up through shell n and

$$\xi_n = Z - \nu_n \tag{5}$$

is the net charge at the nth shell for the parent-configuration ion. The potential V is also used to calculate the configurational average energies for all excited states of the running electron, so that a large amount of data is generated from a small set of fit coefficients. The parametric potential approach produces energy levels, oscillator strengths, and photoionization cross sections comparable in accuracy to single configuration, self-consistent-field calculations with relativistic corrections.[18]

In addition to all the states connected with the ground state parent configuration, opacity calculations involve a large number of states corresponding to excited parent configurations. These are also obtained from potentials of the analytic form given by Eq. (1). To accomplish this the OPAL code first calculates all states of the running electron in the ground state potential and then uses < r > calculated from the resulting wavefunctions to rescale V to account for promoted parent electrons.

PHOTOABSORPTION

In OPAL, the state occupation numbers are obtained from a renormalized activity expansion of the grand canonical ensemble as described in Section II. This approach avoids the ad hoc cutoff procedures necessary in Saha-Boltzmann calculations. The method also provides a systematic procedure for including plasma effects in the photon absorption coefficients. We do not follow the ideal gas mixing procedure described by Huebner et al.[3] for combining the various photon absorption coefficients from the different elements. Instead, coupled equations for the full mixture are solved at each density and temperature point.

Photoionization processes are considered individually for every subshell (subshells are defined by principal and orbital angular momentum quantum numbers, n and l) in every electron configuration of the various ion stages. The calculation is done in the central field approximation for electrons with n greater than 4. For n less than 5, we now include the configuration term splitting in the LS angular momentum coupling scheme.

The bound-bound transitions are calculated for every subshell in each configuration of the various ion stages explicitly. The absorption line energies and oscillator strengths are computed in the LS coupling scheme for transition electrons with both initial and final n less than 5. For jumping electrons with initial n greater than 4, configuration term splitting is neglected and transitions are computed for the nl subshell only. A recent modification is that for electrons with initial n less than 5 but final n greater than 4, the lower configuration term splitting is included in the LS coupling scheme. Transitions to states with $n > 10$ are not explicitly included, but instead are approximately treated by extending the photoionization cross sections to lower photon energies.

Three additional approximations are made with regards to the angular momentum coupling. Due to memory and computer time constraints the number of lines in a given configuration-to-configuration transition array is restricted to 10,000. At present, if the actual number of lines is greater, then the term splitting is ignored and only a single average line is considered. For similar reasons the term splitting is neglected for special lines with energies greater than 13.5 kT (kT = temperature in energy units). At these higher photon energies the Rosseland mean weighting function is less than 1% of the peak value and the details of the bound-bound transitions should not affect the Rosseland mean opacities. Finally, excited spectator electrons with n greater than 5 are included in the calculation of the configuration-averaged energies and wavefunctions, but are not included in the angular momentum coupling. This approximation reduces, in principle, the number of spectral lines in the calculation. However, these electrons usually belong to configurations of low abundance. Furthermore, they reside in large orbits which do not couple strongly to the actual jumping electron. Interestingly, due to the size constraint above, this last approximation can actually increase the number of lines considered in the calculation for ions with partially filled M-shell configurations.

Except for some special cases, line shapes are assumed to be Voigts with the Gaussian widths given by Doppler broadening and the Lorentz width given by electron impact formulas from Dimitrijevic and Konjevic.[21] Their electron impact formulas require radial dipole integrals which are obtained from our parametric potential approach. The exceptions include ion Stark broadening. The Lyman and Balmer series of hydrogenic ions include fine structure and linear Stark effects due to ion electric microfields using subroutines generously provided by Lee.[22] Similar subroutines are used for the Lyman series of helium-like ions and transitions out of the $n = 2$ level in lithium-like ions. These subroutines have been extended in order to treat neutral atoms. For one-electron atoms and ions we have modified the method by Griem[23] in order to treat transitions from levels with n greater than 2. The modifications involve improved ion microfield distributions[24] and electron impact widths.[22]

COMPARISON WITH EXPERIMENT

The best way to validate our calculations is by comparison with high quality photoabsorption experiments on LTE plasmas. This is not yet possible but progress is being made.[25] There are however a number of good hydrogen emissivity experiments covering a wide density range.[26-29]

We have previously shown[30] good agreement with the relatively low density measurements of Wiese et al.[26] However these experiments are not a sensitive test for equation of state or occupation numbers, but rather they are an excellent test for Stark broadening of hydrogenic spectral lines as originally intended by the authors. The reason is that the Saha-type equation of state formulations with principle quantum number cutoff above the Inglis-Teller limit will reproduce reasonably well the experimental data.[31] The difficulty is in reproducing the spectrum near threshold where a careful theory (not presently available) would need to Stark mix many states with different quantum numbers plus the continuum. What has been done with some success are phenomenological methods[31-32] that mimic the line broadening of overlapping lines. These methods are in effect "smoothing procedures" that conserve the oscillator strengths. Just as important, they restrict the line radiation to the region near threshold, while standard line broadening theories redistribute the oscillator strength to regions far from line center leaving a spectral window

In Fig. 1 we compare the OPAL code with the hydrogen emissivity experiment of Radtke et al.[28] for the case T=22200 K and free electron density n_e=8.4x10^{17}/cm^3, which is an order of magnitude denser than Wiese et al.[26] A small silicon contamination is present in the emission but Radtke et al estimate its effect on the observed intensity to be small except at the position of strong line features. The H_α line is cutoff by the planckian and is used to estimate the temperatures. The agreement of our calculation with the experimental results is good, although we note that our lines are slightly too broad and that somewhat better agreement with the observed emissivity could be obtained with a few percent decrease in n_e. This would also improve the agreement in the valley between H_β and H_γ. The good comparison of the OPAL code with the both the Wiese et al. and the Radtke et al. experiments shows that our hydrogenic line broadening theory is accurate over essentially the entire density range that distinct line radiation can be observed, i.e. no line features at all should be observed at a few times higher electron density.

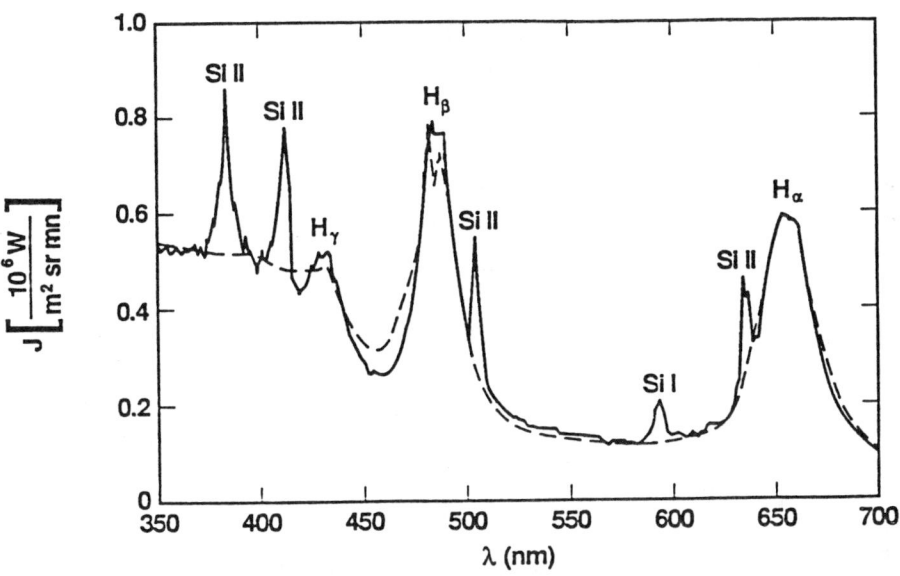

Fig. 1. Comparison of OPAL (dashed line) with the emissivity measurements of Radtke et al. (solid line).

RESULTS

Calculations with the OPAL code have so far been limited to condition occurring in the solar core[30] and in Cepheid variables.[33] In both cases significant increases in the calculated opacity were found in certain temperature density regions.

Solar Core Apacities

The Rosseland mean opacities are compared with the results from the Los Alamos Astrophysical Library published in Bahcall et al.[34] The comparisons are presented in Table 1. The assumed solar mixture (labelled Ross-Aller'76) was obtained from Table IV of Ref. 34 and the Los Alamos Rosseland means from their Table VI. We see no significant differences until the matter temperature drops below 7×10^6 K. As the temperature continues to drop, the opacity difference increases to approximately 18% at 1×10^6 K. Such increases are significant to Helioseismological data.[35]

The opacity is also sensitive to uncertainties in the element mixture. For comparison, we did our calculations assuming the same temperatures and densities but assumed a more recent Aller[36] element abundance (labelled Aller'86). This mixture is richer in both neon and iron. In Table 1 one can see that near the solar center where iron is important to the opacity there is a few percent increase in the Rosseland mean due to changes in the mixture when using the OPAL code. Similarly, near 3×10^6 K neon is important and there is an 8% increase.

TABLE 1. Comparison of OPAL and Los Alamos[34] Rosseland mean opacities (cm^2/g) where X=0.35 and Z=0.0179)

T (10^6K)	ρ (g/cm^3)	Ross-Aller'76		Aller'86
		Ref. 34	OPAL	OPAL
15.7	135.0	1.18	1.17	1.21
12.8	73.4	1.34	1.33	1.37
11.3	50.5	1.45	1.44	1.47
10.0	35.0	1.61	1.58	1.60
7.0	12.0	2.54	2.53	2.55
4.5	3.19	5.85	6.23	6.39
3.0	0.945	13.6	14.6	15.8
1.8	0.204	31.0	35.4	36.5
1.0	0.035	49.9	59.3	57.3

The differences between the results of the two codes are mostly due to the improved atomic physics package in OPAL. We have done some comparisons of the occupation numbers, but found small differences for the most abundant states. There are also differences in the treatment of line broadening since the Lorentz widths are not the same in the two codes. The subject of line broadening, in particular line wings, remains an open question in opacity calculations. The problem of line broadening of spectral lines in multi-electron ions has just begun to be explored.

Cepheid Variable Opacities

The Rosseland mean opacities from the OPAL calculations for the King IVa mixtures are plotted in Fig. 2. The results are compared to those used by Simon[7] in his Cepheid model studies, based on the Los Alamos opacity calculations. We found a significant enhancement of the opacity at temperatures around 2.7×10^5 K. It is interesting to apply to the Simon opacities the enhancement suggested by Andreasen.[8] The result is also shown in Fig. 2. As can be seen, the OPAL and Andreason plots are similar and it should prove interesting to try the new OPAL results in a Cepheid model study.

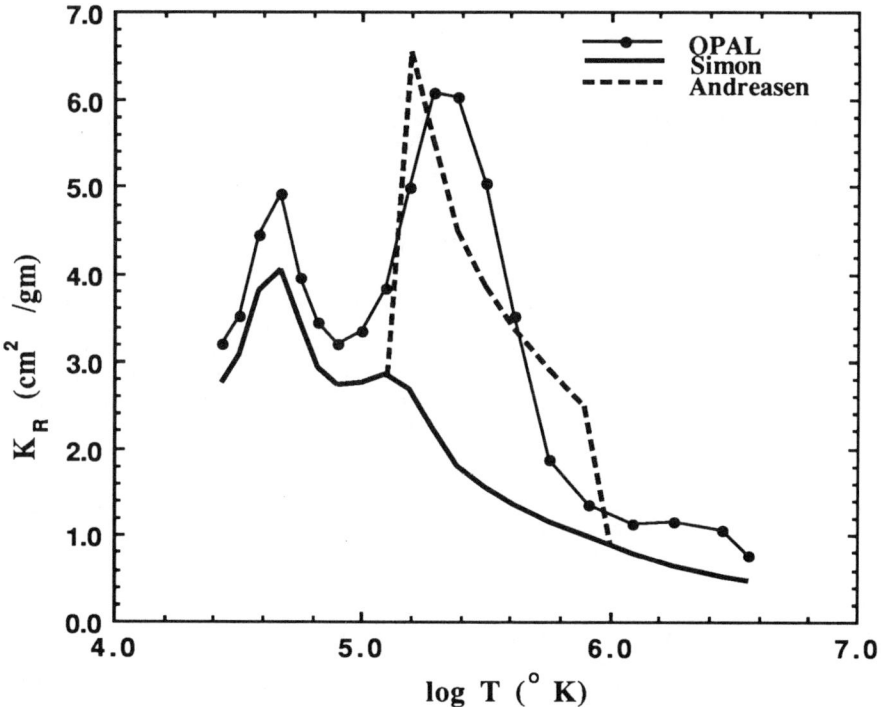

Fig. 2 Rosseland mean opacities along a Cepheid variable track. Densities corresponding to the dots in the OPAL results are given in Ref. 33.

Most of the increase over the Los Alamos results can be traced to the improved atomic physics in OPAL. In particular, the hydrogenic oscillator strengths used by Los Alamos in their bound-bound transitions are a poor approximation for complex ions since they do not predict any $\Delta n=0$ transitions (transitions within the same principal quantum number) which, in fact, can produce strong spectral features. In addition, the configuration term splitting due to angular momentum coupling, which is almost completely absent in the Los Alamos codes,[1-3] replaces single, narrow spectral lines (these contribute little to the Rosseland mean opacity) with many weaker lines spread out over a significant energy range. These weak lines tend to fill in transparency windows raising the Rosseland mean opacity.

In order to determine possible variations due to line widths, we recomputed the opacity at T=20.992 eV, $\rho=2.1831 \times 10^6$ g/cm^3. For this check, the Lorentz widths in OPAL were replaced by the Los Alamos hydrogenic-scaled, electron impact formula.[37] This resulted in a 6% reduction from the OPAL Rosseland mean opacities plotted in Fig. 2. Although the small decrease suggests that our enhancement of the opacity over Los Alamos is not due to line width issues, there still remain questions concerning line broadening (see for example Ref. 11).

The increased opacity in Fig. 2 is significantly larger than our earlier results.[11] The larger enhancement is in part due to the King IVa mixture which not only has more elements than the mixture in the earlier work, but these have higher atomic number. It is for these more complex ions that the Los Alamos atomic physics approximations are least suited. A substantial increase comes from more extensive treatment of the bound-bound transitions. The number of lines due to LS coupling is now larger due, in part, to more efficient use of the computer. Also, the new feature where the configuration term structure for initial levels is included for all transition electrons with n less than 5 regardless of final n, greatly increases the number of spectral lines.

CONCLUDING REMARKS

Improved opacity calculations have confirmed that opacities in standard astrophysical databases significantly underestimate the opacity in important density, temperature zones. Preliminary studies indicate that these increases will lead to improved agreement between theory and observation. The main source of the increases, so far reported, can be traced to improved atomic physics. Presently these "astrophysics experiments" remain the best test of opacity codes but it is anticipated that useful laboratory measurements will become available in the near future.

Acknowledgments: This work was performed under the auspices of the U.S. Dept. of Energy by LLNL under Contract # W-7405-Eng-48.

REFERENCES

1. A. N. Cox and J. N. Stewart, Ap. J. Suppl. **19**, 246 (1970).
2. A. N. Cox and T. E. Tabor, Ap. J. Suppl. **31**, 271 (1976).
3. W. F. Huebner, A. L. Merts, N. H. Magee, and M. F. Argo, Los Alamos Scientific Report LA-6760-M (1977).
4. J. N. Bachall and R. K. Ulrich, Rev. Mod. Phys. **60**, 297 (1988).
5. A. N. Cox, Ann. Rev. Astr. Ap. **18**, 15 (1980).
6. P. H. Bodenheimer, Ap. J. **142**, 451 (1965); G. S. Stringfellow, F. J. Swenson, and J. Faulkner, Bull. A.A.S. **19**, 1020 (1987).
7. N. R. Simon, Ap. J. (Letters) **260**, L87 (1982).
8. G. K. Andresen, Astron. Astrophys. **201**, 72 (1988).
9. Y. Yu, this issue; M. J. Seaton, *Spectral Line Shapes* 4, ed. R. J. Eaton (A Deepak, Hampton, 1987) p. 583.
10. A. Merts, this issue.
11. C. A. Iglesias, F. J. Rogers, and B. G. Wilson, Ap. J. (Letters) **322**, L-45 (1987).
12. B. F. Rozsnyai, Ap. J. **341**, 414 (1989).
13. F. J. Rogers, Phys. Rev. A23, 1008 (1981) and references therein .
14. C. Pisani and B. H. J. McKellor, Phys. Rev. A40, 6597 (1989); D. Bolle', Phys. Rev. A36, 3259 (1987).
15. F. J. Rogers, Phys Rev. A19, 375 (1979).
16. F. J. Rogers, Ap. J. **310**, 723 (1986).
17. C. Bauche-Arnoult, J. Bauche and M. Klapisch, Adv. At. Mol. Phys. 23, 131; and references therein; S. D. Bloom and Goldberg, Phys. Rev. A34, 2865 (1986); Phys. Rev. A36, 3152 (1987);

18. F. J. Rogers, B. G. Wilson, and C. A. Iglesias, Phys. Rev. **A38**, 5007 (1988).
19. M. Klapisch, Comput. Phys. Commun. **2,** 239 (1971).
20. P. S. Ganas and A. E. S. Green, J. Chem. Phys. **73,** 3891 (1980).
21. M. S. Dimitrijevic and N. Konjevic, J. Quant. Spectrosc. Rad. Transf. **24,** 451 (1980); Astron. Astrophys. **163,** 297 (1986); Astron. Astrophys. **172,** 345 (1987).
22. R. W. Lee, J. Quant. Spectrosc. Rad. Transf. **40,** 561 (1988).
23. H. R. Griem Ap. J. **132,** 883 (1987).
24. C. A. Iglesias, H. E. DeWitt, J. L. Lebowitz, D. MacGowan, and W. B. Hubbard, Phys. Rev. **A31,** 1968 (1985).
25. S. J. Davidson, J. M. Foster, C. C. Smith, K. A Warburton, S. R. Rose, Appl. Phys. Lett . **52,** 847 (1988)
26. W. L. Wiese, D. E. Kelleher and D. R. Paquette Phys. Rev. **A6** , 1132 (1972)
27. K. Behringer, Z. Phys. **246,** 333 (1971)
28. R. Radtke, K. Gunther, and R. Spanke, Contrib. to Plasma Phys. **26,** 151 (1986).
29. E. A. Ershov-Paulov, L. E. Krat'ko, N. I. Chubrik, V. D. Shimanovitch, Contrib. Plasma Phys. **29,** 299 (1989).
30. C. A. Iglesias and F. J. Rogers, IUA Colloquium 121, *Inside the Sun Conference,* Versailles, France, May 22-26 (1989).
31. W. Dappen, L. Andreson, and D. Mihalas, Ap. J. **319,** 1954 (1987).
32. L. D' yachkov, G. Kobzev, and P. Pankratov, J. Phys. **B21,** 1939 (1988), and references therein.
33. C. A. Iglesias, F. J. Rogers, and B. G. Wilson, Ap. J. (to be published)
34. J. Bahcall, W. Huebner, S. Lubaw, P. Parker, and R. Ulrich, Rev. Mod. Phys. **54,** 767 (1982).
35. S. G. Korzennik, and R. K. Ulrich, Ap. J. **339,** 114 (1989).
36. R. Allen, *Spectroscopy of Astrophysical Plasmas,* Ed. A. Dalgarno and D. Laymer (Cambridge University Press, Cambridge, 1986).
37. N. H. Magee, private communication (1984).

Low-Density Plasmas

A DISCUSSION OF SOME LESS WELL ACCOUNTED FOR ATOMIC PROCESSES RESPONSIBLE FOR XUV EMISSION FROM MAGNETICALLY CONFINED FUSION PLASMAS

Michael Finkenthal
Department of Physics and Astronomy
Johns Hopkins University, Baltimore, MD 21218*

ABSTRACT

With the advent of new technologies enabling high-near normal incidence reflectivities in the soft x-ray range, new types of plasma spectroscopy diagnostics such as for instance, short time scale fluctuation measurements based on line emission will become feasible both at the edge and in the center of magnetically confined fusion plasmas. Also, in those cases in which time resolved local electron temperature and density estimates are needed, the classical spectroscopic method of line intensity ratios is used. Such measurements require an accurate knowledge of the atomic processes determining the spectral line intensities. The present paper discusses some less accounted for processes responsible for level populations, such as the effect of resonances on the electron impact excitation cross sections, inner shell ionization of M and L-shell charge states, and excitation-autoionization. It is shown that these processes may have very important effects on the XUV radiative patterns of high temperature plasmas.

INTRODUCTION

The spectral line intensity in the high temperature - low density, magnetically confined fusion plasmas is determined by the charge state distribution under given electron temperature and density conditions, the collisional excitation (de-excitation) and radiative decay rates from excited states. Usually, ionization and recombination by electron impact and electron and proton collisions together with electric dipole transition probabilities are the main ingredients used in collisional-radiative models producing the temperature and density dependence of individual lines and line ratios. (Comprehensive reviews on the subject in general, and the emission patterns of ionized atoms in tokamaks in particular, can be found in De Michelis and Mattioli[1] and Isler[2], respectively). Charge exchange with neutral hydrogen or deuterium, the typical working gases in fusion experiments, and dielectronic recombination will affect the ionization balance and, by shifting the fractional abundance distributions, the line intensities.

The present review presents and discusses observations in various magnetically confined fusion plasmas (mainly tokamaks), related to the effect of resonances on electron impact excitation cross sections, inner-shell ionization and excitation-autoionization on the line intensities of N, M and L-shell ions of atoms ranging from Z=6 to Z=66. Also, it will be shown that in some cases, in which the best available atomic data is used in modeling, the discrepancies between the measured and predicted line intensities remain unexplained. In cases in which only relative changes in the electron temperature or

*On sabbatical leave from Racah Institute of Physics, Hebrew University, Jerusalem, Israel.

density are measured (such a case will be mentioned in one of the next sections), the accuracy of the prediction may be less critical an issue; however, in most cases an absolute temperature and/or density measurement is required. (Remember that in astrophysics the line emission is the only way to measure these parameters). The next section discusses the effect of the resonances on line ratios of low Z ions of laboratory and astrophysical interest; some unsolved problems related to the Be I-like carbon to neon are presented and discussed also. The effect of inner-shell ionization-excitation on the singlet/triplet transitions of Be I-like Ti^{18+} and the effect of excitation-autoionization on the emission of Ga I-like rare earth ions emission will also be discussed.

THE EFFECT OF RESONANCE STRUCTURES ON ELECTRON EXCITATION RATES

Although this problem has been discussed in detail in a previous work of ours[3], it is appropriate to mention it here as an introduction to this discussion on the effects of less accounted for processes on spectral line ratios. Figure 1 summarizes the comparison between the results of two collisional-radiative models and experimental data obtained for Si III from a tokamak (TEXT at Fusion Research Center, University of Texas, Austin) and a magnetic mirror device (TMX-U at Lawrence Livermore National Laboratory). The two models assumed the lines of interest being emitted at the Si III ion ionization equilibrium temperature (although as one can see from the figure higher and lower temperatures have also been

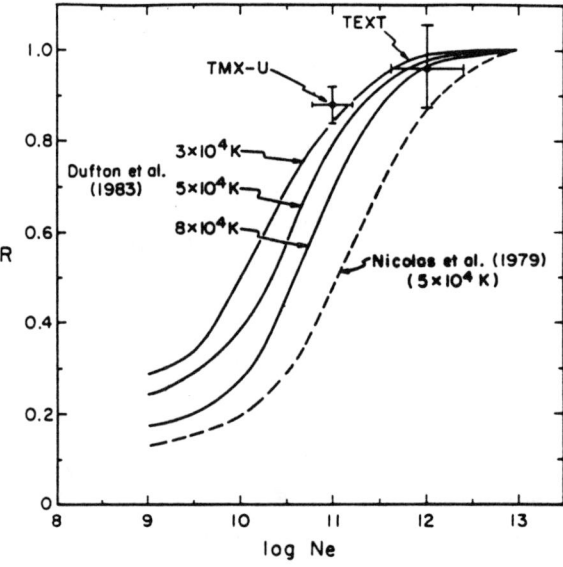

Fig. 1 Effect of resonances on the line ratio R (see text) of Si III; a comparison between theory and tokamak (TEXT) and mirror (TMX-U) data.

considered). The levels included in the two models were the same; the only significant difference was that in the electron impact excitation rates used to predict the line ratio R, Nicolas et al.[4] used the distorted wave approximation (DWA) method, whereas Dufton et

$$R = \frac{I(3p^{2\,3}P_2 \rightarrow 3s3p^3P_2)}{I(3s3p^1P_1 \rightarrow 3s^{2\,1}S_0)}$$

al.[5], used close coupling - R matrices calculations[6]. Strong resonances near the threshold enhance the electron impact cross sections of the $3s^2\,^1S_0 \rightarrow 3s3p\,^3P_1$ transition. Since the triplet level has a low radiative decay rate to the ground, it contributes significantly to the population of the $3p^2\,^3P$ levels, by collisional excitation. Therefore the accurate calculation of the $3s^2\,^1S \rightarrow 3s3p\,^3P$ collision rate is the key to an accurate density sensitivity estimate, as the comparison with the experiment shows. In the range in which this line ratio is density sensitive, the discrepancy between the two models, would lead to an order of magnitude difference in the electron density estimate.

After the experimental confimation of the importance of the resonances and their effect on the density sensitive line ratios for low Z, Mg I-like ions, an extensive experimental effort has been undertaken on the TEXT tokamak, to study the Be I-like emission of low and intermediate Z ions Z=6 to 26 [7,8]. Similarly to the previously discussed Si III case, an analysis of the solar C III and O V emission by Dufton et al.[9] has pointed out the noticeable effect of resonances on the electron impact cross section of the $2s^2\,^1S \rightarrow 2s2p\,^3P_1$ transition. Here too, the ratio
will be density sensitive.

$$R = \frac{(2p^2\,^3P_2 \rightarrow 2s2p\,^3P_2)}{I(2s2p\,^1P_1 \rightarrow 2s^{2\,1}S_0)}$$

However, as Z increases, the Be I-like ions exist and emit in the plasma at temperatures higher than those corresponding to the energy gap between the ground and the triplet 2s2p excited state. As we move away from the threshold, the DWA calculations should in principle be as good as the closed-coupling ones. This is the problem we have investigated, and reported in detail in references 7 and 8.

The elements of interest, from carbon (Z=6) to nickel (Z=28), have been injected into the tokamak plasma by a laser blow-off technique[10] (gases have been puffed in through fast valves in controlled amounts). Radial scans of the lines of interest have been performed and radial profiles of the emissivity have been established for each charge state by Abel inversion. The electron temperature and density profiles have been determined independently, by non-spectroscopic means such as Langmuir probes, Thompson scattering of a ruby laser light and microwave interferometry. The comparison of the experimental data with the predictions of the collisional-radiative models using both R-matrix and DWA calculations for electron impact excitation rates, has confirmed that for $\Delta n=0$ transitions above $Z \approx 15$ the calculated line ratios are very similar, and both agree (within the errors of the experiment and uncertainties in the theoretical estimates) with the experimental data.

The next step was to study $\Delta n=1$ transition line intensities, where a temperature sensitivity appears while comparing them with $\Delta n=0$ lines. The detailed discussion of the results can be found in reference 11; discrepancies between theory and experiment have

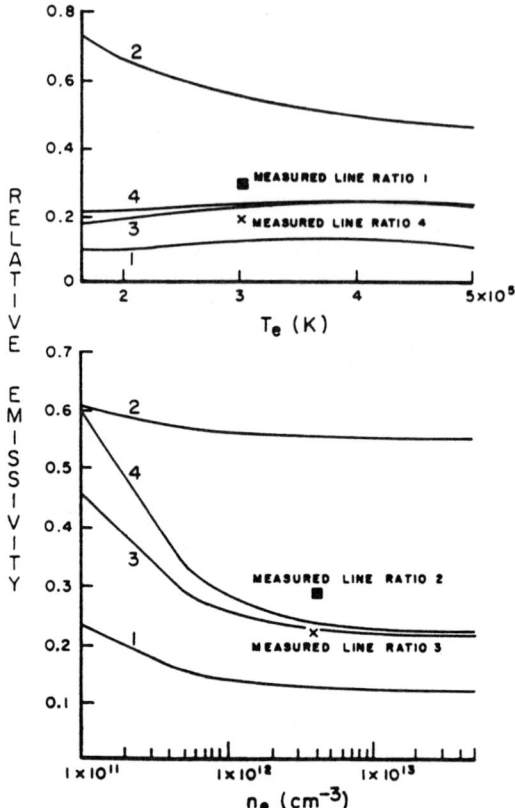

Fig. 2 Computed O V line intensities relative to the 2s2p ^{3}P-2s3d ^{3}D line intensity as a function of electron density and temperature. The numbered lines correspond to the following transitions: (1) 2s^2 ^{1}S-2s3p ^{1}P, (2) 2s2p ^{3}P-2s3s ^{3}S, (3) 2s2p ^{1}P-2s3d ^{1}D, (4) 2s2p ^{1}P-2s3s ^{1}S. The upper figure is for an electron density of 4×10^{12}cm^{-3}. The lower figure is for an electron temperature of 3×10^5 K. The measured line ratios 3 and 4 were in good agreement with the computed values, but the measured line ratios 1 and 2 differ by a factor of 2 from the computed values.

been found for some line ratios, while the same theory was in agreement with other ratios. A summary of these results is shown in figure 2. As an example of a discrepancy, the ratio R,

$$R = \frac{I(2s3s\ ^3S \rightarrow 2s2p\ ^3P; \lambda 215\text{Å})}{I(2s3d\ ^3D \rightarrow 2s2p\ ^3P; \lambda 193\text{Å})}$$

has been measured to be twice larger than predicted[11]. Since the 2s3s ^{3}S level is mainly populated by cascades from 2s3p ^{3}P(~63%) and to a lesser extent by collisions from the ground and 2s2p ^{3}P (~32%), one may assume that the collision rates into the mentioned excited states are overestimated. A collisional radiative model prediction of T. Kato using the most recent, and to our knowledge the best available data produced by the group at Queen's University in Belfast, does not resolve the discrepancy[12]. Figure 3 presents the results of these calculations, for the above mentioned ratio, at two densities, representing the scrape-off and approximately the central density of a tokamak plasma. It can be seen that the experimental value is far below the predicted curves over a very large range of temperatures. Moreover, following a previous work in which the effect of inner shell ionization on the emission lines of O IV has been studied[13], T. Kato included the possible effect of a direct transfer of population from the ground state (2s^2 2p) of O IV into the excited state 2s2p of O V. As one can see from figure 3, the effect is negligible under 100 eV. In the TEXT experiments we have found the OV lines to be emitted around 30 eV. The conclusion seems to be that either there are processes not accounted for in the models used, or some of the collisional excitation rates involved are not accurate. More discrepancies have been found for other line ratios of oxygen, fluorine and neon isoelectronic with BeI (see detailed discussion in reference 11).

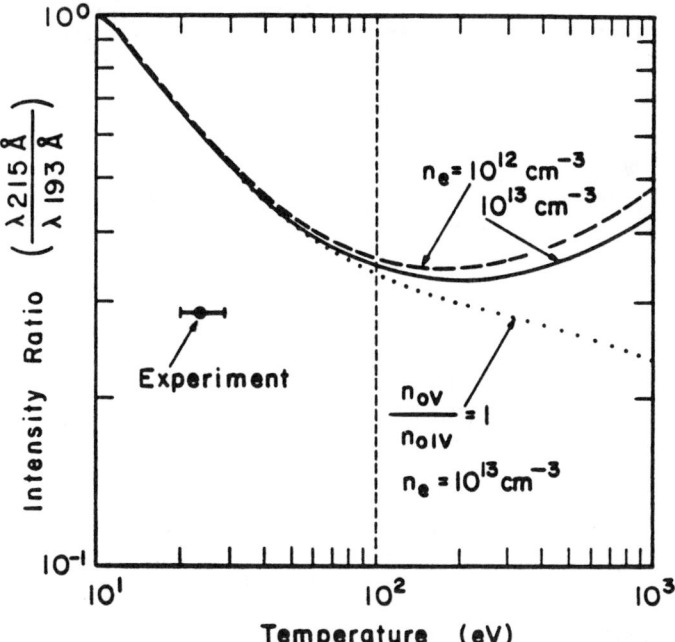

Fig. 3 A comparison between experiment and a collisional-radiative model calculation for the OV, 2s2p-2s3s/2s2p-2s3d line ratio. R-matrix calculations have been used to produce the electron impact collision strength; a case in which inner shell ionization of OIV has been considered is shown too (work performed by T. Kato in Nagoya).

As already mentioned in the introduction, in some fusion relevant experiments, relative changes in the electron density and temperature are of interest. In that case, in spite of the above discussed problems, the BeI-like low Z ion emission can be used to determine local changes of n_e and T_e. The results of an experiment performed on the TEXT tokamak, in collaboration with the TEXT group and a group from General Atomic (San Diego), related to the effect of the ergodic magnetic limiter[14] (EML) on the tokamak plasma will serve as an example. The EML creates a randomization of the magnetic field at the plasma edge; in between the nested magnetic surfaces and this random field, magnetic islands are formed. As a result, poloidal changes in the electron density and temperature profiles correlated with the magnetic island structure are expected. Since at the radial position of the islands, O V is the dominant charge state, ratios of $\Delta n=1$ lines (in the 200 Å range) and $\Delta n=0$ lines (at 600-700 Å), can be used as temperature diagnostics. On the other hand, the ratio $\lambda 761 Å/\lambda 630 Å$ is electron density sensitive around $n_e \approx 10^{12} cm^{-3}$.[7] Two photometrically calibrated spectrometers, a 1 meter grazing incidence instrument for the lower wavelength range and a normal incidence for the EUV, have been used, together with a glancing angle scanning mirror which enabled poloidal scan.[15] Figure 4 shows clear poloidal changes in the electron temperature (they are correlated with EML produced island chains where m=6 at 23 cm and m=7 at 25 cm[16]). Notice that only percent changes of T_e are given, although the quite "safe" $\lambda 193/\lambda 630 Å$ line ratio was used; the reason for that is that beside the atomic physics problems, experimental difficulties in the photometric calibration over large spectral ranges may introduce significant uncertainties in the line ratio measurements.

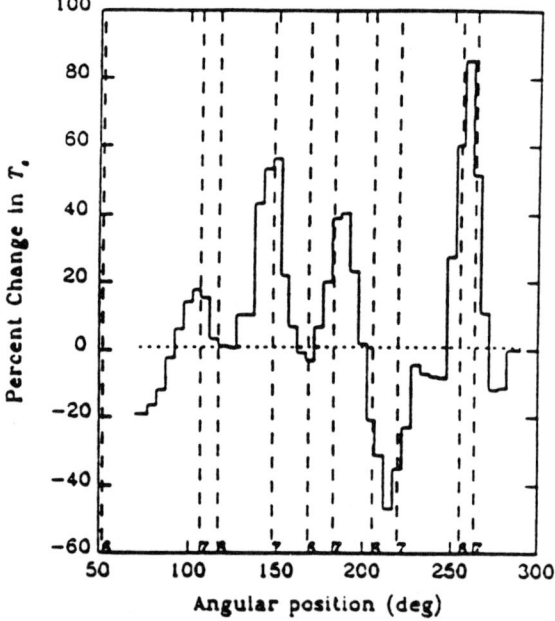

Fig. 4 Spectroscopically determined percentage change in T_e at 25 cm during 2.5 kA EML.

THE EFFECT OF THE INNER-SHELL IONIZATION (ISI) ON THE LINE EMISSION

In a series of experiments performed on two different tokamaks, TFR-600 (Fontenay-aux-Roses, France) and PLT(Princeton Plasma Physics Laboratory), we have measured unexpectedly high line intensity ratios between the $3s^2\ ^1S_0 - 3s3p\ ^3P_1$ and $3s^2\ ^1S_0 - 3s3p\ ^1P_1$ transitions of Mg I like ScX (Z=21) to Mo XXXI (Z=42).[17,18] Collisional-radiative models used to interpret the relatively high intensities of the intercombination lines, failed to point out any "hidden" cause (effect of cascades from high n levels, large errors in the rate coefficients used in modeling, etc.). However, when the inner-shell ionization of the Al I-like ions was assumed, i.e. a 3s electron is ejected from the ground state $3s^23p$, we found that the $3s3p\ ^3P$ level populations were significantly enhanced at the electron temperature where these ions emitted in the tokamak plasmas[19].

As mentioned, the effect of the inner shell ionization on O IV was observed in the JIPPT-II U tokamak in Nagoya[13]; also, Kato *et al.* solved some earlier reported discrepancies in the He I like, Ti XXI spectra from PLT, by including the inner-shell ionization of Li I like ions in the modeling of the He I-like line emission[20].

A very interesting observation related probably to ISI, which I will present and only briefly discuss here comes again from the JIPPT-II-U tokamak in Nagoya. The experimental data has been obtained by K. Sato and co-workers during ion-cyclotron radio-frequency (ICRF) heating experiments. The ICRF pulse is on for about 60 ms; as a result, the central electron temperature increases from a sub keV value to approximately 1200 eV. Figure 5 presents, the experimental line ratio,

$$R = \frac{I(2s2p\ ^3P_1 \to 2s^2)}{I(2s2p\ ^1P_1 \to 2s^2)}$$

of Be I-like, Ti XIX, before and during the ICRF pulse (titanium is an intrinsic impurity in JIPPT-II-U). The ratio is about 0.035 in the ohmically heated phase, in agreement with our previous observations on the TEXT tokamak[8]. This measured value is higher by a factor of two than that predicted by the model discussed in reference 8. By including the effect of ISI from the ground state $2s^22p$ of the BI-like ions (in a similar way to that above mentioned for Al I/Mg I charge states), one obtains a predicted ratio of 0.03 in good agreement with the experiment (this result, together with the following ones will be presented and extensively discussed in a paper by Sato, Bhatia and the author, in preparation). However, the very interesting point is the large increase in the ratio R, as the ICRF heating is turned on; the change is by a factor of seven. A first attempt to explain this increase was made by adding highly excited levels, 2s4l (l=s,p,d) to the collisional-radiative model of A.K. Bhatia; as the temperature increased, one would expect them to become significantly populated. The calculations showed that even by increasing the electron temperature to 3 keV, the ratio R remains practically unchanged. The next step was to include the effect of ISI. The $2s^2\ 2p \to 2s2p$ ionization rates were calculated using the formulas given by Arnaud and Rottenflug,[21] based on measured and quantum mechanical calculations using Coulomb-Born and DW approximations. At 600 eV (the temperature of maximum fractional abundance for Ti XIX at ionization equilibrium), the ISI rate thus obtained, 7.8×10^{12} cm^3/sec is smaller than, the electron impact excitation rate for $2s^2 \to 2s2p\ ^3P_1$ transition; at 1200 eV, the ISI and the direct excitation rate into the $^3P_1$ level are of the same order. It is clear therefore, that in

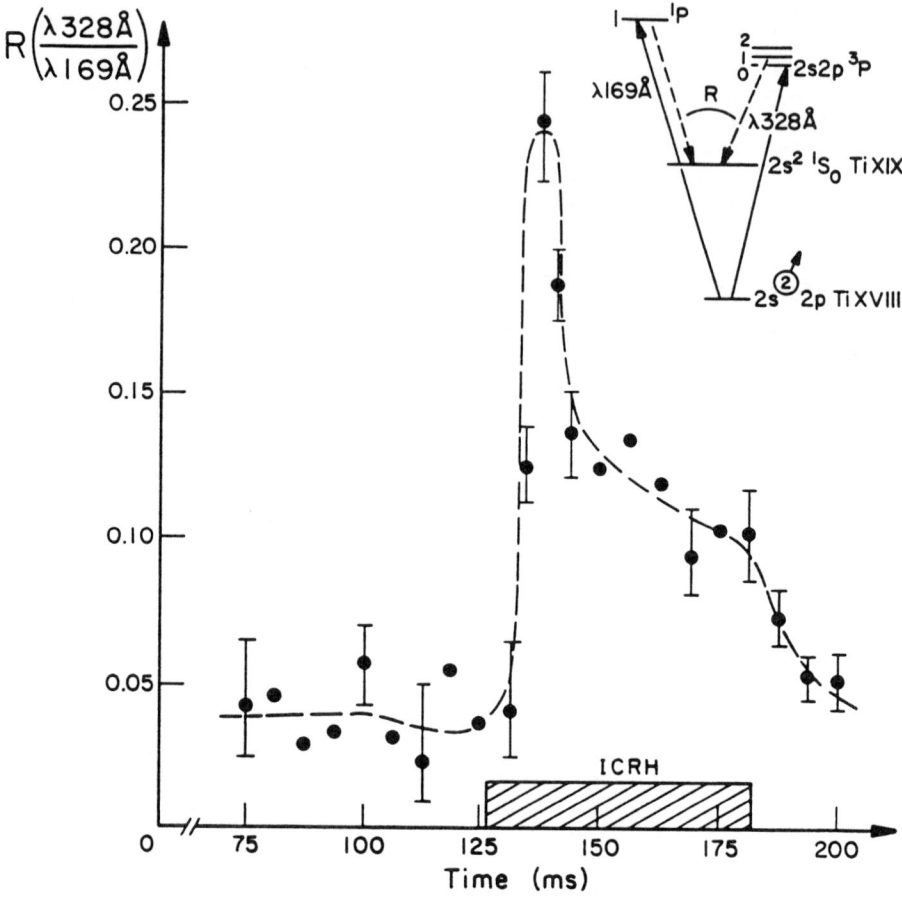

Fig. 5 The time history of the Ti XIX, intercombination/singlet line ratio during ICRF heating on the JIPPT-II-U tokamqak in Nagoya (observation of K. Sato).

a higher temperature plasma the ISI process will increase the ratio R; however the increase is not large enough to explain the experimental result. We had therefore to assume a fractional abundance, $f = n(BI)/n(BeI)$, of the order of five to ten to match the experimental result. Such a situation would occur if the rapid heating would affect mainly the outer half of the plasma. As mentioned, work is in progress on both the atomic physics and the plasma physics implication of these results.

THE EFFECT OF EXCITATION-AUTOIONIZATION (EA) PROCESSES ON LINE EMISSION

Finally, this section discusses briefly the effect of the inner-shell excitation into

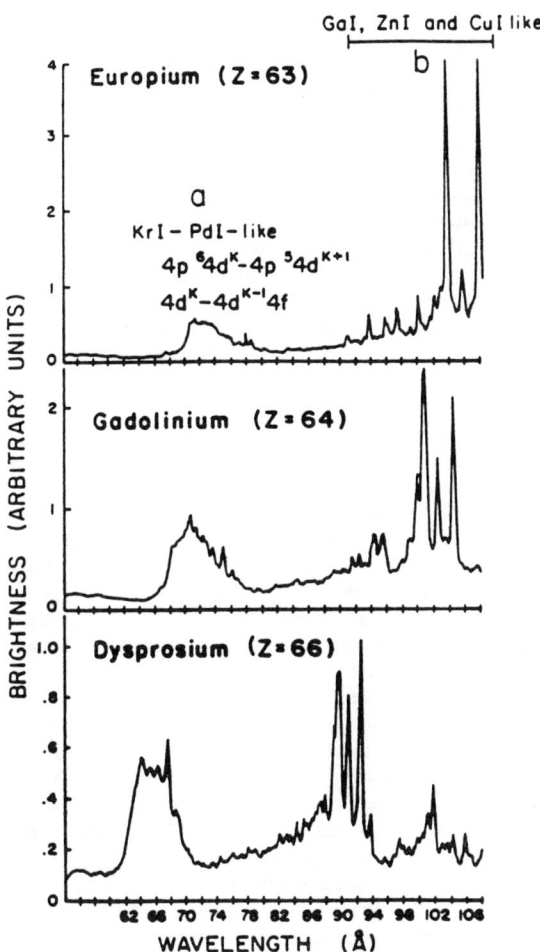

Fig. 6 The soft X-ray spectrum of three rare earth atoms, Eu, Gd and Dy emitted from the tokamak plasma.

autoionion states on GaI-like heavy ions (Z>50) line intensities. The effect of the EA process on the ionization equilibrium calculations has been discussed in connection with astrophysical plasmas as early as 1969.[22] Since, theoretical work on the evaluation of the contribution of the EA process to the total ionization rates by electron impact has been performed mainly for the relatively simple isoelectronic sequences, NaI,[23] CuI[24] and very recently ArI.[25] The experimental work was confined to low and intermediate Z elements.[26] EA rates have been measured for atoms with Z≤10 but experimental evidence for the direct effect of this process on soft x-ray line intensities of highly ionized heavy ions has been observed, to the best of our knowledge, for the first time in the TEXT tokamak produced spectra of highly ionized xenon and rare earths ions. These results, discussed in detail in a work submitted for publication by Mandelbaum et al.[27], are resumed in the following.

The Xe (Z=54) gas was puffed by a fast valve and prasaeodymium (Z=59), europium (Z=63), gadolinium (Z=64) and dysprosium (Z=66) were introduced into the tokamak by the already mentioned laser blow-off method.[10] A limited range of the soft x-ray emission of Eu, Gd and Dy is shown in figure 6. At the longer wavelength side one observes three prominent lines representing Zn I, Ga I and Cu I-like (in order of increasing wavelength). The Zn I and Cu I-like lines are the $4s^2\,^1S_0$-$4s4p\,^1P_1$ and $4s\,^2S_{1/2}$-$4p\,^2P_{3/2}$ transitions respectively; the Ga I-like line originates from a $4s^24p$-$4s4p^2$ transition. One can see that the relative intensity of the GaI-like line compared with both the Zn I and Cu I like is increasing with increasing Z. In the Xe and Pr spectra, the Ga I-like lines are barely measurable; the same has been observed on Xe spectra recorded from the TFR 600 tokamak.[28]

The interpretation of these results is as follows: in the described TEXT experiments, the three charge states under discussion emit near, or from the hot center of the tokamak plasma,[29] at electron temperatures around, and higher than 1 keV. At these temperatures, the electron impact excitation rates from the GaI-like ground state, $3d^{10}4s^24p\,^2P$, into the $3d^94s^24p4f$ excited state are high. For atoms with Z<65, almost all the levels of this excited state are above the ionization energy limit; thus ions excited into these levels have high autoionizing rates. Collisional-radiative calculations performed in the 500-1000 eV range, in which the autoionizing process has been included had shown that for Pr XXIX ionization, the EA process is more important than the direct (ground to ground) ionization process. Although the quantitative analysis is to be found in reference 27, I will mention here that at 600 eV (the ionization equilibrium temperature of this ion) the EA/direct ionization rate ratio is about four. As Z increases, the $3d^94s^24p\,4f$ levels are moving below the ionization energy limit; for Dy XXVI this same ratio, again at the ionization equilibrium temperature (in this case 1000 eV), is only 0.3.

There are several implications to these observations: the first and the most obvious is that one must include the EA process whenever fractional abundance calculations are performed, especially in transient plasmas. Second, the ratios of the GaI/ZnI and GaI/CuI-like lines are sensitive diagnostics for local increases in the electron tempeature. Also, in conjunction with known electron temperature profiles, one can use these ratios as indicators for enhanced inward transport of heavy impurities in the magnetically confined plasmas.

In conclusion, atomic processes as those above mentioned may significantly affect the low density - high temperature laboratory and astrophysical line intensities used as plasma diagnostics. Therefore, collisional-radiative models used to predict electron temperature and density dependences should incorporate them.

ACKNOWLEDGEMENTS

I would like to acknowledge the contributions of Dr. A.K. Bhatia in the modeling of the ISI effects on BeI-like ions, Dr. T. Kato for her calculations of the OV $\Delta n=1$ line ratios, Dr. K. Sato who produced the experimental data related to the titanium emission during ICRF heating on JIPPT-II-U tokamak. Also collaborators at Johns Hopkins University, H.W. Moos, S. Lippmann (now at GA, San Diego), L.K. Huang, A. Zwicker, P. Mandelbaum at the Hebrew University, the TEXT group at the Fusion Research Center, Austin and W. Goldstein and A. Osterheld at Lawrence Livermore National Laboratory, are acknowledged for contributions to various experimental and theoretical parts of the mentioned works. This work is supported by the U.S. Department of Energy grant # DE-FG02-85ER53214.

REFERENCES

1. C. De Michelis and M. Mattioli, Nuclear Fusion **21**, 677 (1981)
2. R. Isler, Nuclear Fusion **24**, 1599 (1984).
3. T.L. Yu, M. Finkenthal and H. W. Moos, Astrophys.J. **305**, 880 (1986).
4. K.R. Nicolas, J.D.F. Bartoe, G.E. Brueckner and M.E. Van Hoosier, Astrophys.J. **233**, 741 (1979).
5. P.L. Dufton, A. Hibbert, A.E. Kingston and G.A. Doschek, Astrophys.J. **274**, 420 (1983).
6. P.G. Burke and W.D. Robb, Adv.Atomic Molec. Phys. **11**, 143 (1975).
7. M. Finkenthal et al., Astrophys. J. **313**, 920 (1987).
8. L.K. Huang, S. Lippmann, T.L. Yu, B.C. Stratton, H.W. Moos, M. Finkenthal, W.L. Hodge, W.L. Rowan, B. Richards, P. Phillips and A. K. Bhatia, Phys.Rev.A **35**, 2919 (1987).
9. P.L. Dufton, K.A. Berrington, P.G. Burke and A.E. Kingstone, Astron. Ap. **62**, 111 (1978).
10. E. Marmar, J. Cecchi and S. Cohen, Rev.Sci.Instr. **46**, 1169 (1975).
11. L.K. Huang, S. Lippmann, B.C. Stratton, H.W. Moos and M. Finkenthal, Phys.Rev. A **37**, 3927 (1988).
12. T. Kato, private communication
13. T. Kato, K. Masai and K. Sato, Phys. Lett. **108A**, (1985).
14. W. Fenneberg and G.H. Wolf, Nuclear Fusion **21**, 669 (1981).
15. S. Lippmann, M. Finkenthal and H.W. Moos, Physica Scripta, in print, March 1990.
16. S.C. McCool et al., Nuclear Fusion 29, 547 (1989) and Nuclear Fusion **30**, 167 (1990).
17. M. Finkenthal, R.E. Bell, H.W. Moos and TFR group, Phys.Lett. **88A**, 165 (1982).
18. M. Finkenthal, E. Hinnov, S. Cohen and S. Suckewer, Phys. Lett. **91A**, 184 (1982).
19. M. Finkenthal, B.C. Stratton, H.W. Moos, A. Bar Shalom and M. Klapisch, Phys. Lett. **108A**, 71 (1985).
20. T. Kato, S. Morita and K. Masai, Phys.Rev.A **36**, 795 (1987).
21. M. Arnaud and R. Rottenflug, Astron.Astrophys.Suppl.Ser. **60**, 425 (1985).
22. J.W. Allen and A.K. Dupree, Astrophys.J. **155**, 27 (1969).
23. R.D. Cowan and J.B. Mann, Astrophys. J. **232**, 940 (1979).
24. M.H. Chen and D.L. Moores, Phys. Rev. A, **41**, 550 (1989).
25. V.L. Jacobs, P.L. Hagelstein, M.H. Chen, R. Minner and J.F. Seely, Phys.Rev.A **41**, 1041 (1990).
26. D.C. Griffin, M.S. Pindzola and C. Bottcher, Phys.Rev.A **36**, 3642 (1986) and references

therein.
27. P. Mandelbaum et al., submitted to Phys.Rev.A.
28. C. Breton et al., Physica Scripta **37**, 33 (1988).
29. M. Finkenthal et al., J. Appl. Phys. **59**, 3644 (1986).

LASER OPTOGALVANIC AND FLUORESCENCE SPECTROSCOPY IN GLOW DISCHARGE PLASMAS

J. E. Lawler and E. A. Den Hartog
Department of Physics, University of Wisconsin,
Madison, Wisconsin 53706

ABSTRACT

Diagnostic techniques based on laser optogalvanic spectroscopy and laser induced fluorescence are providing spatial and temporal maps of key quantities in glow discharge plasmas. Such maps are particularly useful in studying the cathode region. This region is of fundamental interest because of the failure of the local field approximation, and of practical interest because of widespread use of glow discharges in plasma processing and other areas. Laser techniques are used to map space charge electric fields, charged particle densities, average energies of charged particles, flux densities of charged particles, excited atom densities, and other key quantities. These maps provide valuable insights into the dominant physical processes in the cathode region and provide a stringent test of numerical models. A more quantitative understanding of the cathode region is emerging from laser studies and from advanced modeling efforts.

INTRODUCTION

Weakly ionized (glow) discharge plasmas are widely used in materials processing, as sources of coherent and incoherent light, and in pulsed power devices. Although these important plasmas have been studied for more than 50 years, major questions about some aspects of the plasmas have been difficult to answer. The modeling of the cathode region of typical cold cathode glow discharge plasmas is particularly challenging. The local field (or hydrodynamic equilibrium) approximation and other traditional approximations used in modeling glow discharge plasmas fail in the cathode region.[1] The cathode region is of particular importance because most of the voltage drop and power dissipation can occur in this region, and because it determines the flux and energy of ions at the surface. Laser optogalvanic and fluorescence spectroscopy has proven to be quite valuable in studying the cathode region because of the excellent sensitivity, spatial resolution, and temporal resolution provided by these noninvasive techniques. Accurate absolute maps of various key quantities provide insight and enable one to test models of the cathode region.

The experiments described herein were performed in a clean, stable helium (He) dc discharge between plane parallel aluminum electrodes. Figure 1 is a schematic of the apparatus. The discharge pressure was fixed at 3.5 Torr and the electrode separation was fixed at 0.62 cm. Experiments were carried out over a range of discharge current densities from 0.190 mA/cm^2 to 1.50

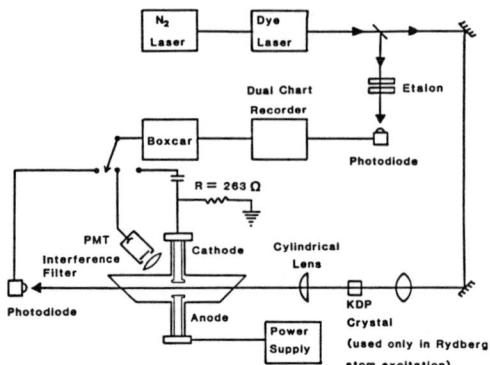

Fig. 1. Schematic of the experimental apparatus showing three detection methods: optogalvanic, fluorescence, and absorption detection.

mA/cm^2. The low current density corresponds to a near normal cathode fall voltage of 173V and the high current density corresponds to a highly abnormal cathode fall voltage of 600V. The simple geometry and clean He gas were chosen for direct comparison to microscopic models of the cathode region. The diagnostics developed in this work are broadly applicable to a variety of dc and rf discharge plasmas.

The cathode region is often divided into cathode fall and negative glow regions. The cathode fall region has a large, rapidly changing electric field and a low density of high energy "beam" electrons. The negative glow region can be described as an "overshoot" region next to the cathode fall. It has very small electric fields and a high density of low energy electrons. Ionization in the negative glow is provided by a low density of high energy beam electrons injected from the cathode fall region.

Accurate maps of the electric field in the cathode fall region are essential for determining the current balance at the cathode surface and other key parameters of the cathode region. In 1984 Doughty, Salih and Lawler developed a technique for mapping electric fields based on optogalvanic detection of Rydberg atoms.[2,3] Fragile Rydberg atoms are an ideal discharge probe. They exhibit large, linear Stark effects which can be accurately ($\sim 1\%$) measured, and easily interpreted using nearly hydrogenic single-electron calculations. Optogalvanic methods make it possible to detect the Rydberg atoms which do not survive long enough to fluoresce in a typical discharge. Laser techniques have a number of important advantages over the traditional electron beam deflection technique.[4] Laser techniques are useful at higher pressures and discharge current densities. It is easier to use a spectroscopic technique in a high purity discharge system. A laser technique is truly nonperturbing when performed with a nanosecond pulsed laser because the field is measured when the atoms or molecules absorb the laser light, and any perturbation to the discharge fields occurs on a longer time scale. Lasers have the potential for making measurements with a few microns spatial resolution and nanosecond temporal resolution. A technique based on Rydberg spectroscopy has broad applicability because all atoms and molecules have Rydberg levels. It also has a wide dynamic range because one can choose an appropriate principal quantum number to provide required sensitivity to the electric field. A much more detailed discussion of electric field measurements using optogalvanic detection of Rydberg atoms has been published.[5]

Figure 2 shows the electric field maps for five different current densities. The fields are accurate on average to ∼1%. This accuracy is verified by integrating the fields across the cathode fall to determine a potential difference within ∼1% of the discharge voltage measured using a digital voltmeter. A slight negative curvature is visible in Fig. 2 but the field magnitude decreases almost linearly with distance from the cathode.

These field maps were analyzed by Doughty, Den Hartog, and Lawler to determine the current balance or ratio of ion to electron current at the cathode surface.[6] According to Poisson's equation the spatial gradient of the field equation is the net space charge density. The net space charge density is essentially the ion charge density because of a very low electron density in the high field cathode fall region. Ions, unlike the electrons, obey the local field or hydrodynamic equilibrium approximation. This means that the ratio of electric field to gas density (E/N or reduced field) can be used to determine the average ion velocity in most of

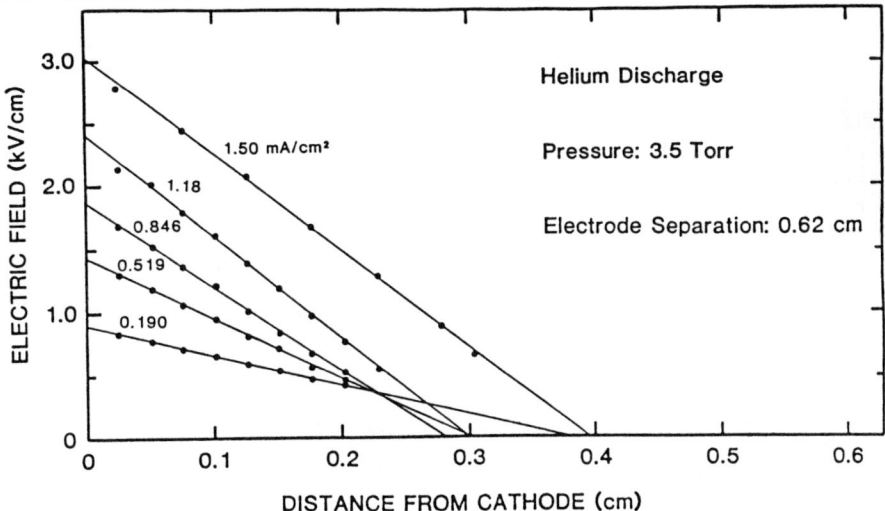

Fig. 2. Electric field as a function of distance from the cathode for five current densities, all at 3.50 Torr. The lines are linear-least-square fits to the data. The anode corresponds to the right-hand side of the figure.

the cathode fall region. The short equilibration distance of the ions is largely due to the symmetric charge exchange reaction,

$$He^+ (fast) + He (slow) \rightarrow He (fast) + He^+ (slow), \qquad (1)$$

which has a large cross section with only weak energy dependence. Charge exchange effectively stops the ion every collision and wipes out all memory of its history. The cathode fall region has a thickness of 50 to 100 mean-free-paths for symmetric-charge-exchange. This intuitive argument is found to be true even when effects due to large field gradients and a distributed ion source are rigorously included.[7] It is important to note that the local field approximation, which is the central approximation of gaseous electronics, fails to describe electrons in the cathode fall region.[1] Modeling the electrons in a realistic fashion is one of the most difficult challenges in modeling the cathode region.

The final requirement for determining the current balance is a measurement of the gas density in the cathode fall. The discharge tube pressure in these experiments was fixed at 3.5 Torr. In the cathode fall region of an abnormal discharge gas heating is often significant. The symmetric charge exchange reaction which produces a short equilibration distance also serves to convert most of the electrical power dissipated in the fall directly into heavy particle (atomic) motion and into heat. Doughty et al. used the Doppler width of a carefully chosen He transition to determine the gas temperature and gas density versus discharge power.[6]

The field maps of Fig. 2 and the gas density measurements based on Doppler widths were used by Helm's to determine the average ion velocity at the cathode. Helm's precise He^+ ion mobilities were used for the lower currents where the E/N is less than 1500 Td.[8] The equilibrium drift velocity $\sqrt{2eE/(m_+ \pi \sigma N)}$, where m_+ is the ion mass and e the unit charge was used for higher currents. The symmetric charge exchange cross section, σ, was taken from Sinha, Lin and Bardsley.[9] These theoretical cross sections agreed with Helm's experimentally derived cross sections.

The product of the ion charge density and average ion velocity at the cathode is the ion current density J_+^o. Figure 3 is a plot of the current balance $J_+^o/(J_D - J_+^o)$ versus the discharge current density J_D. It is interesting to note that the current balance, or ratio of ion to electron current, at the cathode is ~ 3.3 and is essentially independent of total discharge current.

The current balance is related to the coefficient, γ, for electron emission by ion bombardment of the cathode. Two effects are included in the current balance which are not included in γ^{-1} as measured by shining an ion beam on the cathode in vacuum. Some electrons emitted from the cathode are backscattered from gas atoms in the discharge, and are not included in the current balance which relates the net ion and electron currents at the cathode. Metastable atoms and possibly UV or VUV photons play a major role in electron emission from a cold cathode, and their effect is included in the current balance.

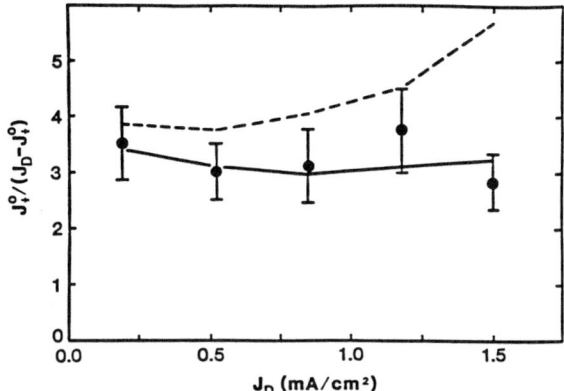

Fig. 3. Ratio of ion current to electron current at the cathode surface $J_+^0/(J_D-J_+^0)$ as a function of discharge current density. The points are the empirically derived values; the lines are from Monte Carlo simulations [Ref. 5].

Laser techniques have also proven quite useful in studying the negative glow region. Metastable atom densities throughout the cathode fall and negative glow region were mapped by Den Hartog et al.[5] using laser techniques. The maps of the 2^3S and 2^1S He metastables show an unexpected suppression of the 2^1S metastable density in the negative glow due to the spin exchange reaction

$$He(2^1S) + e^- \rightarrow He(2^3S) + e^- + 0.79V.$$

Some of the metastable density maps are shown in Fig. 4. This reaction had been studied previously and was known to have a large rate constant for low energy electrons.[10] This reaction will proceed in the forward (exothermic) direction only until the relative singlet and triplet metastable densities come into Boltzmann equilibrium with the low energy electrons in the negative glow. Thus, the empirical ratio of singlet to triplet metastable density provides an upper limit of ~0.25 eV on the average electron energy in the negative glow. A detailed model of the metastable kinetics and transport in the cathode fall and negative glow regions has been used to extract more information, specifically an effective rate for the spin exchange reaction.[5] This effective rate is proportional to the product of the electron density and an electron energy dependent rate constant. It provides a relation between the electron density and average energy in the negative glow. Although we often refer to the average electron energy in terms of an electron temperature, one must remember that the electrons in the negative glow have a distinctly non- Maxwellian distribution function due to a low density of high energy beam electrons injected from the cathode fall region.

Additional information on the negative glow electrons was gained from studies of collisional coupling of the populations of low Rydberg levels. These experiments were performed by Den Hartog

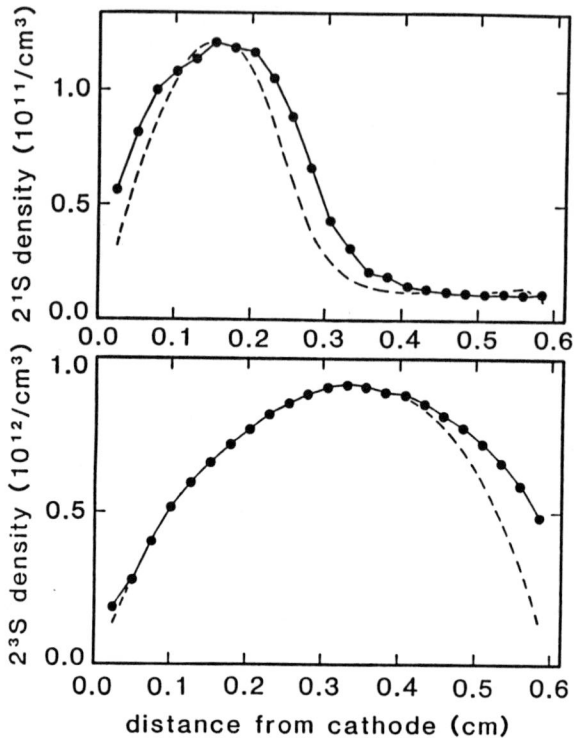

Fig. 4. 2^1S and 2^3S metastable densities vs distance from the cathode. The points are experimental measurements. The dashed curves are calculated densities using the fitting process described in Ref. 5.

et al.[11] Atoms in low Rydberg levels, unlike those in the higher levels used for field measurements, do survive sufficiently long to fluoresce in a typical glow discharge plasma. In these experiments the 5^3P and 6^3P levels were populated by a short laser pulse and fluorescence from the 5^3D, 6^3D, 7^3D, and 8^3D levels was observed. Populations of levels with the same principal quantum number "n" but different electronic orbital angular momentum "ℓ" are strongly coupled by neutral atom collisions. The coupling of populations in levels with different principal quantum numbers is much weaker, but is enhanced by the high density of low energy negative glow electrons as shown by the data in Fig. 5. Laser induced fluorescence data of this type is analyzed using a standard rate equation approach. An effective coupling rate which is the product of the electron density and an electron energy dependent rate constant was determined from a detailed analysis of the laser induced fluorescence data. This information and similar information from the metastable spin conversion reaction is plotted in Fig. 6 for the 0.846 mA/cm^2 He discharge. The effective rates

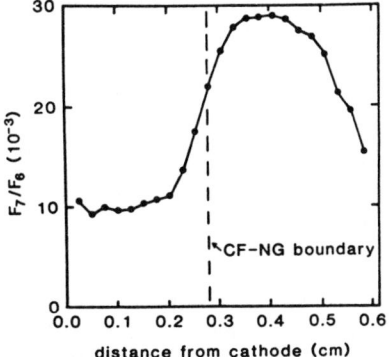

Fig. 5. Ratio of LIF signal from n = 7 to that from n = 6 vs distance from the cathode when n = 6 is populated with the laser.

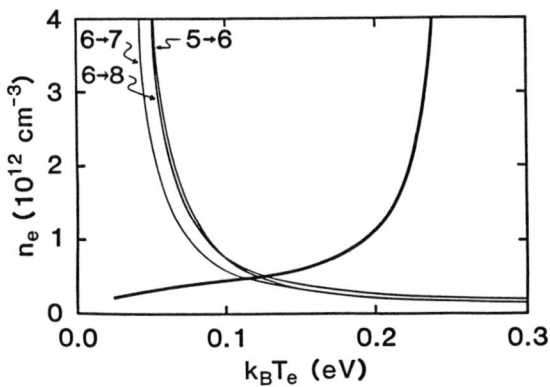

Fig. 6. Electron density vs electron temperature for the low-energy electrons in the negative glow. The three light lines arise from the Rydberg-atom diagnostic. The bold line arises from the metastable analysis.

are represented as curves in an electron density versus electron energy (temperature) plot. The curves for the inelastic or endothermic reactions of electrons and low Rydberg levels are all similar to each other but are quite different from the curve for the superelastic or exothermic spin exchange reaction of electrons on He metastables. The intersection of the curves determines the density and temperature of the low energy negative glow electrons.

A vigorous modeling effort was run in parallel with the experiments described in the preceeding paragraphs. The detailed spatial maps of electric field and the measurements of the current balance in clean He discharges have provided a unique opportunity for testing models of the cathode region. The first model calculations to be compared were Monte Carlo simulations of the electron avalanches in the cathode region.[5] These simulations use the empirical field maps and thus do not predict a self-consistent electric field. Monte Carlo codes based on the null collision algorithm developed by Boeuf and Marode are quite realistic because all of the known details of electron-atom collisions, including anisotropic scattering, are readily included.[12] The most interesting conclusion from comparing experiments and Monte Carlo simulations is that most of the ions hitting the cathode are produced by electron impact ionization in the high field cathode fall region. This means that high energy "beam" electrons carry nearly all of the discharge current across the cathode fall-negative glow boundary where the cathode field extrapolates to zero.

Very recently it has become possible to produce a fully self-consistent kinetic model of the cathode fall region. This model by Sommerer, Hitchon, and Lawler is based on the Convective Scheme.[13] The Convective Scheme is actually a Green's function method applied on a phase space mesh. It is a truly kinetic approach which solves Boltzmann's equation and avoids any assumption about the phase space distribution function. It thus preserves all of the rich detail found in Monte Carlo simulations, but is much faster than a Monte Carlo simulation. The much higher speed and efficiency is in part due to the fact that the Courant-Friedrichs-Lewy criterion, which is important in an explicit finite difference calculation, does not limit the time step in a Convective Scheme calculation. It is sometimes difficult in Monte Carlo simulations to achieve good statistics throughout a discharge plasma, and to deal with trapped particles. The Convective Scheme can easily deal with both problems. Figure 7 shows the electric field produced in a fully self-consistent kinetic calculation of the cathode fall region.[13]

The most recent work using the Convective Scheme was on He rf discharge plasmas.[14] The central or bulk region of a rf plasma is rather like the negative glow of a dc discharge plasma. This

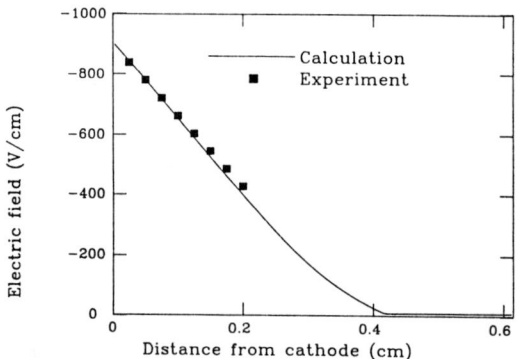

Fig. 7. Self-consistent electric field configuration as predicted by the convective scheme for the j_D = 0.190 mA cm^{-2} discharge of Ref. 5. Experimental field points are optogalvanic measurements from the same reference.

region has a low density of high energy electrons injected from the oscillating cathode sheaths, and has a high density of low energy electrons. Convective Scheme calculations were used to study the energy balance of the cold bulk electrons. The rf discharge calculations are believed to be the first fully self-consistent kinetic model of a complete (electrode-to-electrode) discharge plasma.

In conclusion there are good prospects for applying self-consistent kinetic models to a variety of important dc and rf discharge plasmas. The comparison of results from advanced laser diagnostics and fully self-consistent kinetic models will lead to a much more quantitative understanding of a variety of important dc and rf glow discharge plasmas.

REFERENCES

1. P. Segur, M. Yousfi, J. P. Boeuf, E. Marode, A. J. Davies, and J. G. Evans, Electrical Breakdown and Discharges in Gases, edited by E. E. Kunhardt and L. H. Luessen, NATO ASI Series B Vol. 89A (Plenum, New York, 1983) p.331.
2. D. K. Doughty and J. E. Lawler, Applied Physics Letter 45, 611 (1984).
3. D. K. Doughty, S. Salih, and J. E. Lawler, Physics Letter A103, 41 (1984).
4. R. Warren, Physical Review 98, 1650 (1955).
5. E. A. Den Hartog, D. A. Doughty, and J. E. Lawler, Physical Review A38, 2471 (1988).
6. D. A. Doughty, E. A. Den Hartog, and J. E. Lawler, Physical Review Letter 58, 2668 (1987).
7. J. E. Lawler, Physical Review A32, 2977 (1985).
8. H. Helm, Journal Physics B: Atomic Molecular Physics 10, 3683 (1977).
9. S. Sinha, S. L. Lin, and J. N. Bardsley, Journal Physics B: Atomic Molecular Physics 12, 1613 (1979)
10. A. V. Phelps, Phys. Rev. 99, 1307 (1955).
11. E. A. Den Hartog, T. R. O'Brian, and J. E. Lawler, Phys. Rev. Letts. 62, 1500 (1989).
12. J. P. Boeuf and E. Marode, Journal Physics D: Applied Physics 15, 2169 (1982).
13. T. J. Sommerer, W. N. G. Hitchon, and J. E. Lawler, Physical Review A 39, 6356 (1989).
14. T. J. Sommerer, W. N. G. Hitchon, and J. E. Lawler, Phys. Rev. Letts. 63, 2361 (1989).

A PLASMA SPECTROSCOPIC STUDY OF MOLECULAR HYDROGEN AND HYDROCARBONS IN A TOKAMAK; TECHNIQUES AND PROBLEMS

Takashi Fujimoto and WT-III group
Department of Engineering Science, Kyoto University, Kyoto, Japan
and
Department of Physics, Kyoto University, Kyoto, Japan

ABSTRACT

The cross sections for production of excited H atoms by electron collisions exhibit significantly different variations with the principal quantum number of the product atoms, depending on whether the production mechanism involves dissociative excitation of H_2, or direct excitation from ground-state H atoms. It is also known that, in a tokamak plasma, the excited state which constitutes the boundary between the corona-phase and the ladder-like excitation-ionization phase is located close to the states from which the lower members of Lyman or Balmer lines are emitted. On the basis of these premises we developed a method to determine the densities of atomic hydrogen, molecular hydrogen, and electrons from the observed Balmer or Lyman line intensities. We applied this method to the WT-III tokamak and found that, in the edge region of the main plasma, the H_2 density was forty times higher than the H atomic density. We discuss the possible use of a similar method for the treatment of a duterium plasma or a plasma containing hydrocarbons.

INTRODUCTION

The current standard method to determine the electron density in torus plasmas is to use Thomson scattering. However, in the outer region, where the electron density becomes low, this method becomes less reliable and an alternative method is required for density determination.

To understand the physics of the H-mode in tokamaks and the process of recycling of hydrogen (deuterium) in a torus plasma, it is important to know what hydrogen species are present in the outer region of these plasmas, whether atomic or molecular hydrogen, or even hydrogen bound in hydrocarbons. No experimental method for the study of this problem has been reported thus far.

In 1988 the authors proposed a method for the determination of the electron density from the observed intensity ratio of atomic hydrogen emission lines.[1] This method was based on the assumption that excited hydrogen atoms were produced from ground-state hydrogen atoms. We subsequently extended our method to the treatment of plasmas containing molecular as well as atomic hydrogen. The method now makes it possible to determine the densities of atomic and molecular hydrogen and electrons from measurements of intensities of atomic emission lines,[2] and we used it to determine these quantities in the WT-III tokamak plasma of Kyoto University.[3]

We presents a summary of this investigation and discuss a persisting difficulty arising from the uncertainties in the atomic and molecular excitation cross section.

THE COLLISIONAL-RADIATIVE MODEL

According to the standard collisional-radiative model of atoms and ions in a plasma, the population density of an excited atomic or ionic state is given by the sum of two terms: a term which is proportional to the density of the next ionization stage ions, and a term which is proportional to the density of the ground-state atoms or ions of the same ionization stage. We call the former component the recombining plasma component and the latter the ionizing plasma component.[4]

It may be easily shown that in a tokamak plasma the system of neutral hydrogen, protons and electrons constitutes a typical example of an ionizing plasma in which the second term predominates over the first term in the population density of all the excited states. If this plasma contains neutral hydrogen molecules as well as atomic hydrogen, the molecular hydrogen may produce excited hydrogen atoms through a dissociative excitation process induced by electron collisions. Accordingly, we must include in the population equation an additional term which is proportional to the molecular density.

In general, the population density of an excited state is determined by the equilibrium between the populating and the depopulating processes for this state under a particular set of plasma conditions. In the case of the ionizing plasma we can classify the excited states into two groups according to their characteristics. This classification is based on a comparison between the collisional and radiative processes and depends on the electron density and, to a lesser extent, on the electron temperature. For a plasma having a lower electron density than the critical density, or for excited states lying below the critical state, an excited state is in corona equilibrium. The critical state or the critical density is identical to that which has been known for some time as constituting an LTE criterion.[5] The dominant populating mechanism for these states is the direct excitation from the ground-state atoms or dissociative excitation from the ground-state molecules (by electron collisions), and their depopulation take place by radiative decay. Let p be the principal quantum number of the particular state in which we are interested. We note that the atomic excitation cross section is approximately proportional to p^{-3}, the p-dependence of the absorption oscillator strength of Lyman lines, and the molecular dissociative excitation cross section is proportional to p^{-6}. The radiative decay rate is, to a good approximation, proportional to $p^{-4.5}$. These p-dependences give the population-density distribution among the excited states. In the case in which atomic excitation is dominant, this distribution is approximately given by the expression $n(p)/g(p) \propto p^{-0.5}$, where $n(p)$ is the population density of state p and $g(p)$ is the statistical weight of this state, or $2p^2$, and in the case of molecular excitation we have $n(p)/g(p) \propto p^{-3.5}$. These substantially different p-dependences enable us to determine the densities of atoms and molecules from measurements of population densities of the excited atoms.

For a plasma with a higher electron density than the critical density, or for excited states lying above the critical state, the excited state becomes populated by a ladder-like excitation-ionization mechanism.[5] The dominant mechanism which determines the excited-state population is the stepwise excitation by electron collisions from one excited state to the adjacent higher state. Because the rate coefficient for the excitation $p \rightarrow (p+1)$ is approximately proportional to p^4, this gives rise to a population density distribution $n(p)/g(p) \propto p^{-6}$. This is substantially different from the distributions in a low-density plasma or for the lower-lying states. We thus have the possibility of determining the electron density from the same measurement of state population densities.

To apply this method to an actual plasma, we have constructed a collisional-radiative model, using the most reliable data for atomic and molecular cross sections. We assembled from literature a single set of cross section data for the excitation and ionization of atomic hydrogen, which, however, did not include cross sections for excitation from the atomic ground-state. We believe that this set is satisfactory and constitutes the best choice for the present. For the excitation cross section from the atomic ground state, we had a set of cross sections given by Johnson's standard semiempirical formula.[6] The result of a recent 15-state R-matrix calculation was also available for the excitation $1 \rightarrow 2$, 3, 4 and 5.[7] We found that this calculation produced a significantly different cross section than the original Johnson formula. Except for the complex resonance structure of the cross section there was good agreement for the excitation $1 \rightarrow 2$. For $1 \rightarrow 3$, the R-matrix calculation gave a larger cross section near the excitation threshold by a factor of 2, and for $1 \rightarrow 5$, this calculation gave cross section larger by about one order near the threshold. No information was available for the excitation of higher lying states. However, for the excitation $1 \rightarrow$ (very high state), we succeeded in estimating the threshold value of the cross section from the slope of the ionization cross section, which was well established experimentally. This method was based on the fact that various quantities continue smoothly from the discrete states across the ionization limit to the continuum states. The cross section thus obtained was about 20% larger than the original Johnson cross section.

For dissociative excitation from molecular hydrogen, many authors have reported the emission cross section for Lyman and Balmer lines. The cross section which we need, to which we shall refer as the production cross section, differs, however, from the emission cross section. The production cross section is the quantity which describes the production of excited atoms with any angular momentum quantum number. Knowing the ratio of the cross sections for production of the different-l levels with a common principal quantum number, which we call the branching ratio, we can translate the emission cross section into the production cross section. By making several corrections and approximations to the emission cross section data, and by estimating the branching ratio from experimental data,[8] we have succeeded in estimating the production cross section.

We constructed the collisional-radiative model by incorporating these atomic and molecular data. We calculated the excited-state populations under conditions relevant to the tokamak plasmas, and we converted these results into the emission intensities of the Lyman and Balmer lines.

We also included in our calculation the effect of proton collisions but, under the particular plasma conditions of interest to us, this effect was found to be minimal.

The WT-III tokamak has the major radius of 0.65m and the minor radius of 0.21m. The following experiment was carried out with the plasma in a Joule heating mode, with a toroidal magnetic field of 1.58T, plasma current of 68kA and loop voltage of 1.6V. The electron temperature and density in the central region of the plasma were determined from Thomson scattering, and they were 0.51keV and $1.7 \times 10^{19} m^{-3}$, respectively, at the plasma center. The electron temperature at the edge of the main plasma was determined by a Langmuir probe to be 15eV.

The plasma was observed with a visible-uv spectrometer from the direction of the major radius through a fused quartz window. The line of sight was scanned over the plasma by moving one of the mirrors of the light collecting optics. We registered the Balmer α, β and γ line intensities. On the shot-to-shot basis we scanned the plasma between the center and the edge by 1cm steps to obtain the chord-dependence of the emission intensityies of these lines. The Abel inversion was applied, and we obtained the radial dependence of the line intensities.

We analyzed the result by applying our collisional-radiative model. When we relied on the standard semiempirical formula to calculate the cross sections for excitation from the atomic ground-state, we obtained the following results: inside the main plasma the atomic hydrogen density was of the order of $10^{15} m^{-3}$ and at the edge of the plasma it increased to $1 \times 10^{16} m^{-3}$ at r=0.21m. Molecular hydrogen existed only in the outer region corresponding to $r \geqslant 0.18m$, and it reached the value $4 \times 10^{17} m^{-3}$ at r=0.21m. When we relied on the modified semiempirical formula based on the R-matrix calculation, the result was much the same for the inner region of the plasma. In the outer region, the density of the atoms decreased slightly and the molecular density was about $2 \times 10^{17} m^{-3}$ at r=0.21m.

We tentatively conclude that the result based on the original semiempirical formula is more reasonable. One of the reasons for our conclusion is that the decrease in the atomic density in the outer region is quite unlikely to occur. Reference 3 contains a graphical representation of these quantities.

So far we have neglected the effect of dissociative excitation from molecular ions which are expected to be present in the plasma and might contribute to the observed atomic line intensity. It is expected that the p-dependence of the respective rate coefficient should be similar to that of the cross section for excitation from the atomic ground-state ($\propto p^{-3}$), rather than of the cross section for dissociative excitation from neutral molecules ($\propto p^{-6}$). This is because the molecular ion, having one electron, lacks the crossings and avoided-crossings among various molecular potential curves, which exist in the case of neutral hydrogen. Therefore, the quantity which we have identified as the atomic hydrogen density would correspond to the sum of the atomic hydrogen density and the ionic molecular haydrogen density multiplied by a factpr. A distinction between these densities could not be accomplished in the present experiment.

If our plasma consistes of deuterium instead of hydrogen, we use the cross section data for deuterium. We expect that the cross section for excitation and ionization by electron collisions are almost exactly equal, and we found that the measured emission cross section for dissociative excitation was about 20% smaller than the corresponding cross section for all the excited states of hydrogen. We further expect that the branching ratio in this case is much the same as in the hydrogen case, and we obtained the production cross section which was about 20% smaller than for hydrogen. The present method may be applied to this plasma with a slight modification.

When the wall and limiter of our tokamak are made of graphite, we expect to have hydrocarbons as materials released from the wall, These hydrocarbons are expected to produce excited hydrogen atoms by dissociative excitation through electron collisions. The p-dependence of the emission cross section for light hydrocarbons is known to be similar to that for molecular hydrogen. However, the branching ratio is not known, and the preliminary result obtained by the authors of ref.8 indicates that the branching ratio for methane is quite different from that for molecular hydrogen. At present we do not have enough data to estimate the production cross section for hydrocarbons.

CONCLUSIONS

We have succeeded in developing a spectroscopic method to determine the densities of neutral hydrogen atoms, molecules, and electrons in an ionizing plasma. At present, our method is subject to an uncertainty which is due mainly to the uncertainties in the cross section for excitation from atomic hydrogen and the cross section for production from molecular hydrogen.

In order to establish the present method as a reliable diagnostic tool, we need reliable cross section data. In particular, we need i) the atomic cross section for excitation 1 → 3, 4 and 5, and ii) the dissociative excitation cross section for the production of the same upper state atoms. In order to establish the data set of the latter, we need a reliable experiment to determine the degree of the polarization of the emitted radiation and, depending on this information, the brancihing ratio.

REFERENCES

1. T. Fujimoto, S. Miyachi and K. Sawada, Nucl. Fusion 28, 1255 (1988).
2. T. Fujimoto, K. Sawada and K. Takahata, J. Appl. Phys. 66, 2315 (1989).
3. T. Fujimoto, K. Sawada, K. Takahata, K. Eriguchi, H. Suemitsu, K. Ishii, R. Okasaka, H. Tanaka, T. Maekawa, Y. Terumichi and S. Tanaka, Nucl. Fusion 29, 1519 (1989).
4. T. Fujimoto, J. Phys. Soc. Japan 47, 265 (1979).
5. T. Fujimoto, J. Phys. Soc. Japan 47, 273 (1979).
6. L.C. Johnson, Astrophys. J. 174, 227 (1972).
7. A. Pathak, A.E. Kingston, and K.A. Berrington, (private communication).
8. T. Ogawa, M. Taniguchi and K. Nakashima, Proceedings of the Fifteenth International Conference on the Physics of Electronic and Atomic Collisions, Brighton, United Kingdom, 1987, edited by J. Geddes et al., Abstracts of Contributed Papers (ICPEAC, Brighton, United Kingdom, 1987), p.339.

ATOMIC PROCESSES AND SPECTROSCOPIC TECHNIQUES APPLIED TO FUSION PLASMA DIAGNOSTICS

R. J. Fonck
Department of Nuclear Engineering and Engineering Physics
University of Wisconsin, Madison, WI 53706

ABSTRACT

A wide variety of spectroscopic techniques exploiting several different atomic processes are used to study the properties of high temperature magnetically confined plasmas. The methods of choice in present experiments use either charge exchange recombination reactions between plasma ions and light neutral atoms injected via high energy neutral beams or fluorescence of beam-injected neutrals excited via collisions with the plasma constituent species. While these techniques allow accurate studies of plasma equilibria, transport properties, and possibly microturbulence in contemporary experiments, the low probability of penetration of neutral beams into dense hot plasmas suggests that extrapolation of these techniques to future fusion ignition experiments may be difficult, and alternative diagnostic techniques based on more classical emission line spectroscopy must also be explored.

INTRODUCTION

The use of atomic processes and spectroscopic observations continues to play a wide and expanding role in the study of magnetically confined fusion plasmas. A wide variety of atomic processes and spectroscopic techniques are presently in use or under steady development for the purpose of diagnosing thermonuclear fusion-grade plasmas. In this paper, we discuss an overview of the present spectroscopic techniques usually employed in the hot core of contemporary magnetically confined fusion plasmas. Ourdiscussion is weighted towards the exploitation of the interaction of high energy neutral particle beams and the background plasma since this area has seen the most development in the past few years. Inaddition, we discuss only the diagnosis of plasmas found in the tokamak confinement device, which is the main toroidal confinement device in use in contemporary fusion research. For such experiments, one usually deals with plasmas with particle densities ranging from 10^{13} to 10^{15} m^{-3}, temperatures of 1 to 30 keV, and magnetic fields of 1 to 10 T.

The purpose of most diagnostic techniques applied to high temperature plasmas is to address the central physics issue of contemporary magnetic confinement research: to measure and derive an understanding of the physics of cross-magnetic-field transport of energy and particles in magnetically confined plasmas. This task involves being able to determine the steady-state plasma equilibrium state (as characterized by electron and ion temperatures, electron and ion densities, plasma current densities, etc.), measure the particle and energy cross-field transport rates, and determine the nature of instabilities and turbulence in the plasma which gives rise to non-classical or anomalous transport. In all these tasks, spectroscopic diagnostic techniques are seen to play a central and expanding role. In fact, the rate of progress towards achieving these goals of fusion research has been directly dependent on our ability to develop reliable and accurate diagnostic techniques which are applicable to the hostile experimental environment.

The next section of this paper presents a short discussion of the continuing role of classical emission spectroscopy in high temperature plasma studies, while the following section discusses the diagnostic applications of beam-plasma interactions. Finally, we discuss possible future trends and problems confronting the community in the ignition-level experiments planned for the near future.

CLASSICAL EMISSION SPECTROSCOPY

The measurement of emissions from partially stripped ions in the hot plasma core provides a wealth of information on local plasma parameters and transport properties. By making quantitative measurements of impurity line radiation, continuum radiation and possibly working gas emissions from the cool edge, we can deduce particle confinement times, ion densities and plasma composition, radiated power losses, and ion particle transport coefficients. These emissions are usually excited by electron-ion collisions, and observations can range from the visible through the vacuum ultraviolet to the soft x-ray spectral ranges, including electric dipole, magnetic dipole, and intercombination lines as transitions of interest.

This tends to be a reasonably well-developed area of spectroscopic diagnostic techniques, and, coupled with appropriate impurity transport modelling codes still is the best means of studying metallic impurities in the plasma core. With the introduction of multichannel spectroscopic techniques and routine application of modelling codes in the 1980's, further progress in this area has been more incremental than revolutionary. However, significant questions do persist concerning the available rate coefficients for ionization, excitation, and recombination processes in the multiply charged ions.[1,2]

Even though metallic ions are usually of negligible densities in most tokamak experiments today, emission spectroscopy of intrinsic or artificially injected metallic ions allows us to compare the differences in particle transport rates between light and heavy ions in the plasma core. In particular, recent studies have shown that, under some circumstances at least, metallic and light ion impurities can exhibit considerably different confinement and transport properties in tokamak plasmas, as predicted by neoclassical transport theory.[3,4] In addition, diagnostic improvements, such as polarimetry measurements of intrinsic line emissions to measure local magnetic fields, continue to be actively pursued.[5]

ACTIVE SPECTROSCOPIC DIAGNOSTICS

By far the most active area of development in spectroscopic diagnosis of tokamak plasmas has revolved around the exploitation of emissions resulting from the interactions between plasma ions and high energy neutral beam atoms injected into the plasma core region. Depending on the particular installation, the beams employed range from high power hydrogenic beams (which are used to heat the plasma) to low power diagnostic beams whose atomic species (e.g., H, D, He, Li, etc.) is chosen to optimize a particular measurement technique. These measurements can yield a wealth of information on the plasma parameters and transport properties. For example, line profile measurements allow the determination of plasma ion temperatures and rotation speeds, line intensity measurements provide measures of fully stripped low-Z ion densities and hence plasma composition, polarization measurements provide information on internal magnetic field structures, and higher precision line intensity and/or profile measurements may provide crucial information on plasma fluctuation and turbulence levels.

A schematic diagram of a typical neutral beam spectroscopy system on a toroidal plasma experiment is shown in Fig. 1. The measured intensity is emitted only from the volume defined by the overlap of the neutral beam and the optical line of sight, which allows localized measurements and eliminates the need for uncertain inversion techniques to derive local emissivities from a series of chordally integrated intensity measurements. Theadvantages of good spatial resolution in the hot plasma core, easily accessible spectral ranges (usually the visible to near-UV regions have lines of sufficient intensities), and the ready availability of neutral beams on most tokamaks have made this approach the spectroscopic diagnostic technique of choice on the present generation of large fusion plasma experiments.

There are two main sources of line emission of interest when plasma particles and a neutral beam collide. The first process of interest is charge exchange recombination, wherein a (usually) fully stripped ion collides with the fast neutral and a charge transfer reaction takes place:

$$H^0 + A^q \rightarrow H^+ + (A^{q-1})^* \rightarrow H^+ + A^{q-1} + h\nu \qquad (1)$$

leaving the resultant recombined ion in an excited state. The radiative decays of the excited states give rise to line emission from the plasma ion. Since this is a resonance process which tends to populate high n and high llevels, long wavelength yrast transitions have sufficient intensities to allow precision measurements of line profiles, intensities, polarizations, etc. This technique of CHarge Exchange Recombination Spectroscopy (CHERS)[6,7] is routinely used on most major tokamaks for ion temperature and plasma rotation speed measurements, and has allowed unambiguous measurements of local thermal diffusion coefficients in near-thermonuclear plasmas. In addition, it has been employed in the radially and temporally resolved measurements of local ion densities in attempts at deriving local transport coefficients.

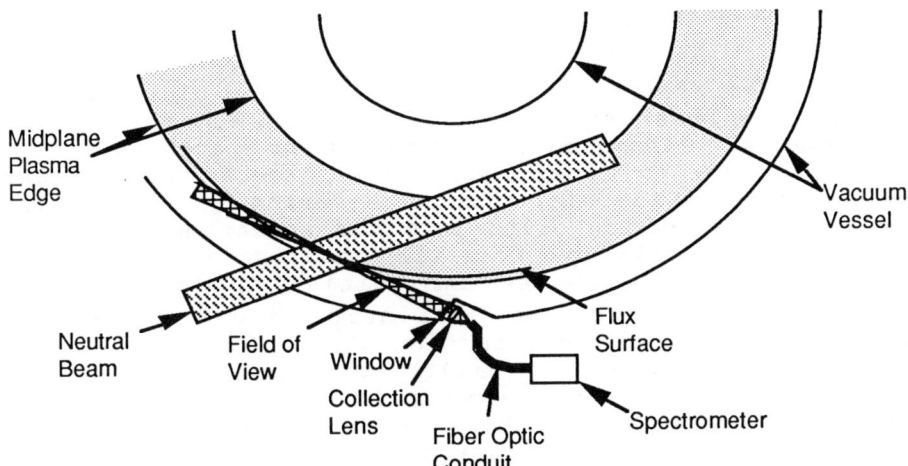

Fig. 1. Sample layout for active spectroscopy diagnostic using a high energy neutral beam for tokamak plasma studies.

As an example, Fig. 2 shows a radial profile of plasma ion temperature obtained via Doppler width measurements of the CVI n = 8-7 transition at 529 nm using the heating beam on TFTR for CHERS measurements. These results show a clear disagreement with the values expected from classical transport theory. The development of the CHERS technique and its application in the past few years has caused a significant change in the understanding of ion thermal transport in tokamaks.[8,9]

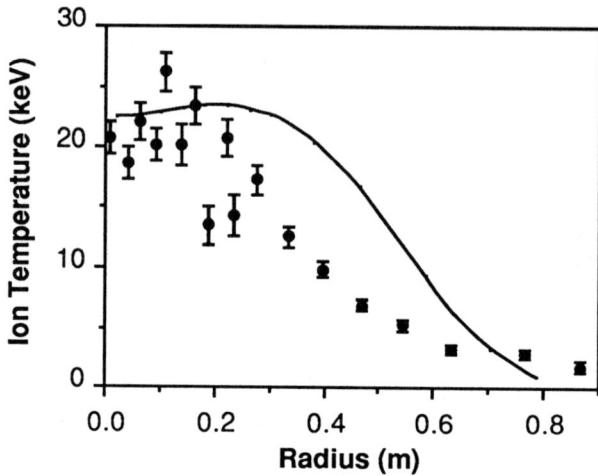

Fig. 2. Plasma ion temperature profile measured in the TFTR tokamak via CHERS. The solid line shows the profile shape predicted by neoclassical transport theory.

There are several unresolved atomic and plasma physics issues complicating the use of CHERS for plasma measurements. Plasma environmental effects such as ion-ion collisional mixing of excited levels, Zeeman and motional Stark mixing of levels, and interfering radiation from recombined ions which drift into the spectrometer line of sight before ionizing, can all complicate the cascade process and give rise to ambiguities in the net line excitation rate and the intrinsic line profile.[10,11] Thecharge exchange recombination rates for populating a given n,l level are not readily available for the full energy range of interest, especially for the highly excited states of interest well above the resonant level. These are the states which give rise to the long wavelength lines of diagnostic interest (since visible wavelengths allow high throughput complex optics with fiber optic access to high radiation environments). In addition, the presence of excited states in the beam neutral population can significantly change the net charge exchange recombination rate into a given level,[12] and hence must be modelled to some degree.

Two new issues arise in very high plasma temperature environments. First, the effect of the energy dependence of the excitation rate coefficient can cause distortions in the net Doppler-broadened and shifted line profiles.[13] Second, the presence of second generation (i.e., halo) neutrals in the vicinity of the neutral beam can provide an additional less-localized source of fast neutrals in the optical line of sight.[14] To assess and account for all of these effects, a thorough understanding and modelling of the excitation and cascade processes must be obtained for a given plasma and experimental condition. Obviously, this requires accurate charge exchange recombination cross sections for a wide range of excited states of both the neutrals and the target ions. When these various effects are included in the analyses, reasonable agreements (~ 50 % or so) between CHERS measurements and independent measures of ion species densities have been obtained[11,14], but more accurate measurements are clearly desirable.

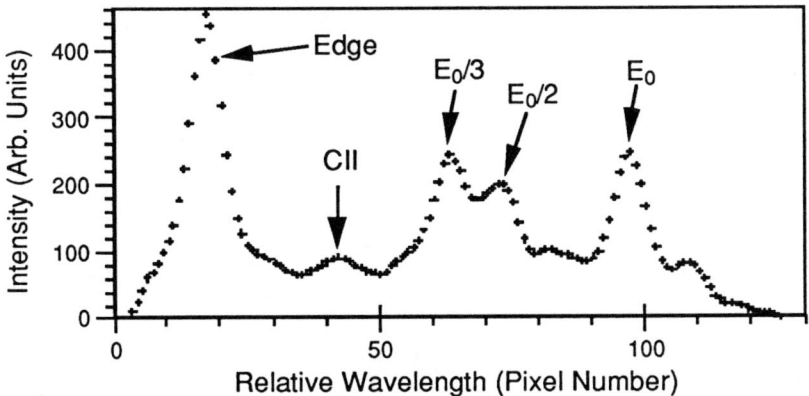

Fig. 3. Spectrum near H-alpha from TFTR when viewing across a neutral beam. The emissions from the three neutral beam species are Doppler shifted from the stationary edge emissions. The wings evident on the full energy component are due to the motional Stark spliting of H-alpha.

The second source of light from the beam-plasma interaction which has significant diagnostic potential is the collisionally induced neutral beam fluorescence, the measurement of which is sometimes referred to as Beam Emission Spectroscopy (BES). The collision

$$H^0 + A^q \rightarrow (H^0)^* + A^q \rightarrow H^0 + A^q + h\nu \qquad (2)$$

between a fast neutral atom and the background plasma species, A^q, leaves the neutral in an excited state which can then radiatively decay. Here, the resultant emission line profile is representative of the beam neutral velocity distribution rather than that of the plasma ions. With beam energies ≥ 10 keV/amu, this light can be Doppler shifted well away from any interfering background radiation, and is easily detected as seen in the example in Fig. 3. The first real diagnostic use of this emission has been the determination of plasma current density profiles in hot tokamak core regions, a measurement which has been notoriously difficult to make in the past. As the beam traverses the plasma, the light emitted from it can exhibit Zeeman splitting and polarization due to the strong magnetic fields, and this has been exploited for measurements of local magnetic field directions.[15] More recently, the strong polarization of the H-alpha line due to the motional Stark effect experienced by the neutrals traversing the magnetic field has been effectively exploited to provide precision measurements of the internal magnetic field in a dense hot plasma.[16,17] The intensity of light emitted by a beam in a plasma depends on collisions with all the plasma species, including impurity ions, and it has recently been pointed out that a combined measurement of the beam emission and the charge-exchange induced H-alpha emission gives a local measurement of the net plasma ionic charge.[17,18]

The use of BES as a quantitative diagnostic presupposes a detailed understanding of the beam excitation as it interacts with the plasma. While a variety of calculations have been done for H, He, and Li beams, the most detailed modelling has been done for hydrogenic beams since they are also of interest for plasma heating. Inparticular, one finds that multistep collisional excitation and ionization processes are nonnegligible, and must be accurately taken into account before such measurements can provide reliable plasma information.[19]

An alternative to the use of technically complex neutral beams for active low-Z impurity spectroscopy is the use of high speed solid pellets of low-Z materials such as Li or C injected into the hot plasma core. The most developed application to date has consisted of using the light emitted from the collisionally excited ions ablated from the solid pellet as it traverses the plasma to determine the local magnetic field direction via polarization measurements.[20] One significant advantage of this approach is that the pellets may be designed to penetrate the hot high density plasmas expected in the next generation of near-ignition tokamak devices. As discussed below, neutral beams will have a very difficult time penetrating into the hot plasma core of such plasmas.

RECENT DEVELOPMENTS AND FUTURE TRENDS

Before discussing future directions in the applications of atomic processes for fusion plasma diagnostics, we discuss two recent areas of activity which appear to hold enough promise that further work in the near future is anticipated in these areas. The first is the extension of the CHERS technique to hydrogen-like and helium-like medium-Z metallic ions such as Fe^{+25} and Ni^{+27}, while the second is the use of BES to measure internal plasma density fluctuations.

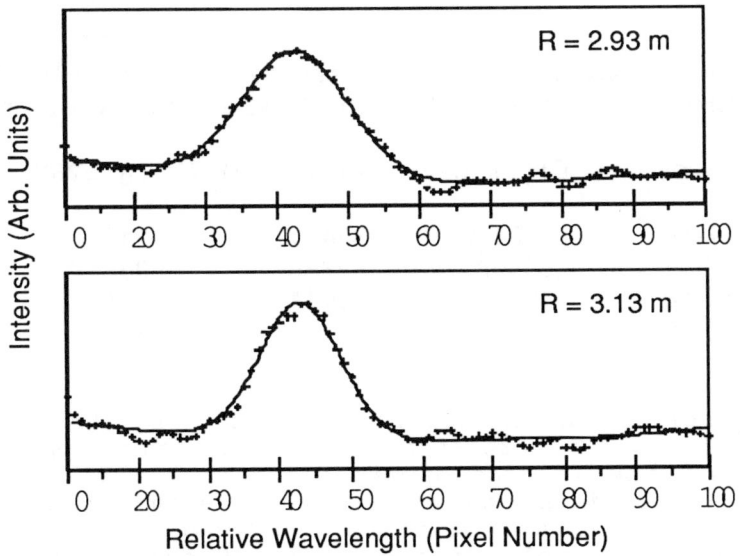

Fig. 4. Line profiles of the Fe XXIV n=18-17 Rydberg transition at 423.5 nm excited by charge exchange recombination in the TFTR tokamak. The line width is larger nearer the hot plasma core at R=2.93 m.

Similar to the low-Z cases exploited routinely, the basic approach for CHERS measurements of high-Z ions is to observe $\Delta n = 1$ or 2 Rydberg transitions in He-like, Li-like, or Be-like ions arising from charge-exchange recombination reactions with fast neutrals such as H° or He. The interest in this approach arises because for states with n = 15 to 20, the emitted radiation lies in the visible spectral range and hence complex radiation-hardened optical systems can be employed. It provides measurements of terminal charge states out to the plasma edge for studies of high-Z ion transport and confinement. Because the characteristic line radiation is not collisionally excited in the cooler plasma periphery, these ions have not been observable via conventional emission spectroscopy in that region. The virtue of this approach has already been demonstrated in low-Z ion transport studies.[21] Visible line radiation from highly-ionized species excited via charge exchange recombination have already been observed in present-day experiments[22,23] and measurable line intensities have been obtained. For example, Fig. 4 shows the line profile of the n = 18-17 transition in the Fe XXIV spectrum at 423.5 nm. The increased thermal broadening of the spectral line in the hot plasma core region is evident. A systematic search of the visible spectrum in the TFTR tokamak with neutral beam heating has yielded observations of a range of Rydberg transitions arising from charge exchange with He-like metallic ions (see Table I).

TABLE I. Rydberg transitions of metallic impurities observed on TFTR via CHERS: Li-like ions. (Bohr formula wavelengths with charge Z_e)

n_2-n_1	Cr(Z_e =22)	Fe(Z_e =24)	Ni(Z_e =26)
17-16	4224 Å	3549 Å	3024Å
18-17	5040Å	4235Å	3609Å
19-18	--	5004Å	4264Å
20-19	--	--	4994Å

Further use of this approach to study the transport of metallic ions is in progress, but quantitative analysis is hampered by the lack of atomic data. The net excitation rates of these lines from charge exchange and the resultant cascade processes are relatively unknown. In addition, fine structure splitting of the Rydberg levels is nonnegligible, with the result that the population distribution among the excited levels has to be carefully considered.

In addition to the use of BES for internal magnetic field measurements, as discussed earlier, considerable attention is being given to high speed (i.e., sampling rates up to 1 MHz) measurements of beam fluorescence to deduce localized levels of internal plasma-density fluctuations. To date, it has been very difficult to measure local plasma density fluctuations of moderate to long instability wavelengths in the hot plasma core. Since the collisionally induced beam fluorescence is monotonically related to the local plasma density, fluctuations in the intensity of light emitted from the beam may be used to measure local density fluctuations. Since the expected density fluctuation amplitudes are on the order of ≤ 1% of the steady-state level in the frequency range up to ~ 500 kHz, such measurements impose strict requirements on the attainable photon count rates and on the stability of the neutral beam intensity itself. Nonetheless, initial measurements have indicated that detectable plasma fluctuation levels can be obtained (see Fig. 5), and intensive developments in this area are in progress. Given a particular experimental facility and goals of the measurement, a variety of neutral beam species (e.g., H, He, Li) and atomic transitions can be employed for these measurements. In each case, a multistep excitation model must be employed to determine the exact relationship of the line intensity to the plasma density.

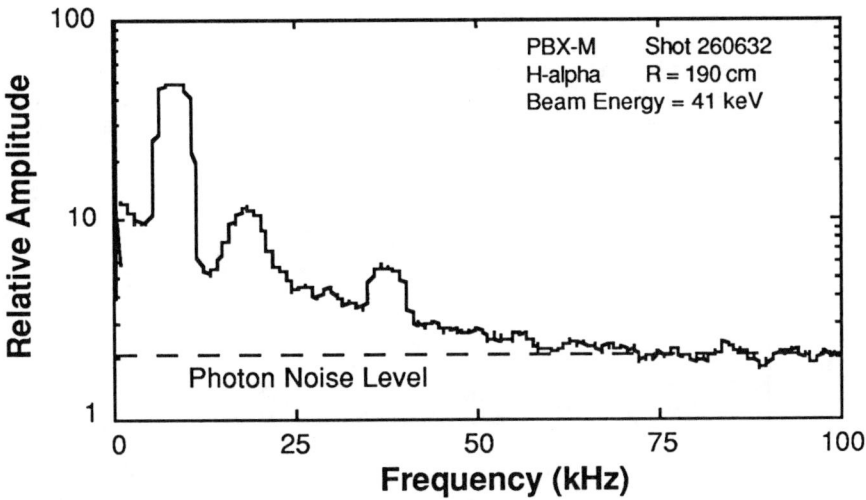

Fig. 5. Frequency spectrum of neutral beam fluorescence showing broadband low frequency plasma density turbulence. The two lower peaks in the spectrum are due to coherent MHD instablities in the plasma, while the peak at 35 kHz is due to a beam source instability.

With the wide range of measurements possible on present experiments using beam-based diagnostics and the continuing intensive development efforts in this area, it is somewhat ironic to note that there is good reason to doubt that many of these techniques will be applicable to the next generation of fusion plasma experiments. This situation arises because, as we progress to ignition-level plasma fusion experiments, the plasma density and the magnetic fields will be substantially higher than they are at present. For example, the CIT device will have B = 10 T and N = $10^{20} - 10^{21}$ m^{-3} with plasma cross-sectional widths on the order of 1 m. Under these conditions, conventional neutral beams will suffer unacceptably high attenuation before reaching the central plasma core, with the result that intensities for CHERS-like and possibly BES-like measurements will be substantially below the background plasma continuum radiation level.

For example, Fig. 6 shows a calculation of the intensity of a commonly used CHERS transition arising from fully stripped C for the planned CIT device. The cases shown are for flat and peaked density profiles and for two values of neutral beam energies, assuming a constant beam power of 1 MW. At low beam energy, where the excitation and charge exchange cross sections peak, the beam attenuation is so great in such a plasma that only edge plasmas may be accessible, depending on the details of the plasma profiles. At higher energies, beam attenuation is reduced, but so are all the relevant excitation/recombination rates, resulting in quite low signal levels. Further detailed studies are obviously needed here to see if some combination of beam species, energy, and observed transitions can result in widely applicable diagnostic measurements.

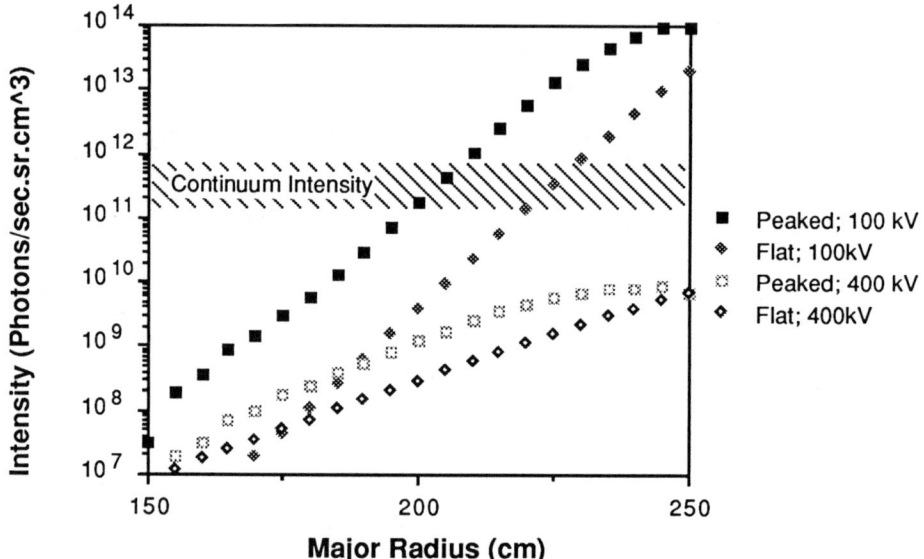

Fig. 6. Sample calculation of CVI n=8-7 intensities for a CHERS measurement in the CIT device. Results are shown for cases of peaked and flat densities profiles and high and low beam energies. The plasma center is assumed to be 200 cm.

Given the potential problems with beam attenuation in these ignition plasmas, it behooves us to take a fresh look at plasma spectroscopic diagnostics and consider new approaches or improved variations of our present techniques. For example, beam-based spectroscopic techniques would benefit from the development of very intense short-pulsed neutral beams which would allow at least single-time point measurements by instantaneously raising the signal above background bremsstrahlung. In addition, the development of high efficiency multi-layer soft x-ray optics may allow us to exploit more intense transitions at lower wavelengths.

Another approach is to give up on the use of beams altogether and return to the use of classical emission spectroscopy of minute amounts of high-Z trace elements for diagnostic applications. For the 10-20 keV plasma temperatures of interest, the use of magnetic dipole and/or intercombination lines in the deep vacuum ultraviolet region would probably be necessary. To effectively exploit this approach, a new generation of high throughput soft x-ray and VuV spectrometers at high resolution will be necessary. In addition, considerable work on the atomic spectra of very high-Z ions will be required to identify the best transitions for specific diagnostic applications.

SUMMARY

Conventional passive spectroscopy of highly ionized metallic ions is a common diagnostic on all large experiments, and has taken on added interest with the observations that, in some cases, ions of different charge exhibit different cross-field transport properties.

Over the past several years, most development activities have revolved around the interaction of high energy neutral atomic beams with the background plasma ions. In particular, production and excitation of hydrogenic light ions in hot plasmas via charge exchange recombination interactions has become the method of choice for the measurements of plasma ion temperatures and rotation velocities. The application of these measurements has recently led to a significantly deeper appreciation for the role of ion thermal transport in tokamak plasmas. Inaddition, charge exchange spectroscopy is being applied to light ion particle transport studied in large tokamaks to provide radially resolved particle transport rates. Most recently, efforts have begun to measure local concentrations of highly ionized metallic ions via observation of Rydberg transitions in the visible excited by charge exchange recombination. In all cases, a lack of reliable partial cross sections and interactions with the plasma environment during the cascade process limits the accuracy of these measurements.

Two areas of considerable present interest in the plasma diagnostic community involve searches for reliable means of measuring internal magnetic fields in the plasma and/or small-amplitude high-frequency fluctuations in various plasmas properties such as density and temperature. The solutions to both of these problems are obtained by the exploitation of neutral beam fluorescence induced by plasma-beam collisions.

There are indications, however, that the use of neutral beam and other low-Z spectroscopy may not be applicable to the next generation of dense hot fusion plasmas without further neutral beam technological developments. Depending on the particular details of the ignition plasma, renewed attention may be given to the passive observation of minute amounts of highly ionized metallic ions in the plasma core and/or new concepts for introducing partially ionized light ions in the hot plasma. Continued development of high-Z spectra and advances in spectroscopic instrumentation will be required to provide reliable diagnosis of these plasmas.

ACKNOWLEDGEMENTS

The author wishes to thank A. Boileau, R. Hulse, R. Isler, F. Levinton, G. Schilling, B. Stratton, E. Synakowski, and M. von Hellermann for useful discussions and suggestions. This work was supported by the U.S. Department of Energy Grant No. DE-FG02-89ER53296.

REFERENCES

1. B.C. Stratton, these proceedings.
2. M. Finkenthal, these proceedings.
3. K. Ida, et al., Nucl. Fusion $\underline{29}$, 231 (1989).
4. R.D. Petrasso, et al., Phys. Rev. Lett. $\underline{57}$, 707 (1986).
5. D. Wroblewski, L.K. Huang, H.W. Moos, and P.E. Phillips, Phys. Rev. Lett. $\underline{61}$, 1724 (1988).
6. R.J. Fonck, Rev. Sci. Instrum. $\underline{56}$, 885 (1985).
7. R.C. Isler, Physics Scripta $\underline{35}$, 650 (1987).
8. R.J. Fonck, et al., Phys. Rev. Lett. $\underline{63}$, 520 (1989).
9. R.J. Groebner, et al., Nucl. Fusion $\underline{26}$, 543 (1986).
10. R.J. Fonck, D.S. Darrow, and K.P. Jaehnig, Phys. Rev. A $\underline{29}$, 3288 (1984).
11. A. Boileau, et al., Plasma Phys. and Controlled Fusion, $\underline{31}$, 779 (1989).
12. R.C. Isler and R.E. Olson, Phys. Rev. A $\underline{37}$, 3399 (1988).
13. R.B. Howell, R. Fonck, K. Jaehnig, and R. Knize, Rev. Sci. Instrum. $\underline{59}$, 1521 (1988).
14. B.C. Stratton, et al., Princeton Plasma Physics Laboratory Report PPPL-2648, Sept. 1989, 31 pp. To be published in Nuclear Fusion.
15. W.P. West, D.M. Thomas, J.S. DeGrassie, and S.B. Zheng, Phys. Rev. Lett. $\underline{58}$, 2758 (1979).
16. F.M. Levinton, et al., Phys. Rev. Lett. $\underline{63}$, 2060 (1989).
17. A. Boileau, et al., J. Phys. B $\underline{22}$, L145 (1989).
18. M. von Hellermann, Private Communication and Bull. Am. Phys. Soc. $\underline{34}$, 2056 (1989). To be submitted for publication.
19. R.K. Janev, C.D. Boley, and D.E. Post, Nucl. Fusion $\underline{29}$, 2125 (1989).
20. E. Marmar, Bull. Am. Phys. Soc., $\underline{32}$, 1847 (1987).
21. R.J. Fonck, et al., Phys. Rev. Lett. $\underline{49}$, 737 (1982).
22. R.J. Knize, et al., Rev. Sci. Instrum. $\underline{59}$, 1518 (1988).
23. P.G. Carolan, et al., Phys. Rev. A $\underline{35}$, 3454 (1987).

MODELING OF IMPURITY EMISSIONS FROM TOKAMAK PLASMAS

B. C. Stratton
Princeton University, Plasma Physics Laboratory, Princeton, NJ 08543

ABSTRACT

Impurity concentrations, radiative losses, and particle transport coefficients in tokamak plasmas can be deduced from impurity emissions in the visible, ultraviolet, and x-ray regions of the spectrum. These measurements rely on impurity transport code modeling of the data; such codes generally require extensive ionization, recombination, and excitation atomic data to link the measured brightnesses to the quantities of interest. Thus, the availability and accuracy of the atomic data play an important role in determining the reliability of the modeling.

This paper reviews the use of impurity transport codes to model emissions from the core region of tokamak plasmas, with the emphasis on measurements of impurity transport and on the role of atomic data in these measurements. Following a brief discussion of the general features of impurity transport codes, examples of impurity transport measurements based on measurements of line and continuum emissions in the visible, vacuum ultraviolet, and x-ray regions are given. These examples represent the major techniques used. The atomic data required in each case are discussed, and areas in which improved data are needed are pointed out.

INTRODUCTION

It is well known that impurities are present in tokamak plasmas and can be detrimental to plasma performance: impurities cause power losses through radiation and cause dilution of the working gas at fixed electron density, and impurity issues such as helium ash accumulation in reactors will be crucial in the future. Understanding impurity behavior in tokamaks is therefore clearly important. In addition, measurements of the time and spatial evolutions of the densities of impurity ions provides a means of studying ion cross-field transport in tokamak discharges. This diagnostic use of impurities is valuable because transport in tokamak plasmas is poorly understood.

Impurity concentrations, radiative losses, and particle transport coefficients can be deduced from spectroscopic measurements of impurity emissions in the visible, ultraviolet, and x-ray regions. Impurity transport code modeling of the data is usually required to link the measured brightnesses to the quantities of interest. These codes require extensive ionization, recombination, and excitation atomic data. Thus, the availability and accuracy of the atomic data plays an important role in determining the reliability of the modeling.

This paper reviews the use of impurity transport codes to model emissions from the core region of tokamak plasmas (the region inside the last closed flux surface of the magnetic field). The focus is on measurements of impurity transport and on the role of atomic data in these measurements. Following a brief discussion of the general features of impurity transport codes and of the atomic data required by the codes, examples of transport measurements based on measurements of line and continuum emissions in the visible, vacuum ultraviolet (VUV), and x-ray regions of the spectrum are given. These examples are drawn from recent work on various tokamaks and illustrate different approaches to the study of impurity transport. (The literature on this subject is extensive and will not be cited exhaustively; work through 1984 is reviewed by Isler.[1]) The atomic data required in each case are discussed and areas in which improved data are needed are pointed out. Although the emphasis in this paper is on measurements of impurity transport, much of the discussion is directly applicable to spectroscopic measurements of impurity concentrations and radiative losses.

IMPURITY TRANSPORT CODE MODELING

Although different in detail, the impurity transport codes currently used to model tokamak data are conceptually similar. As an example, consider the MIST code.[2] MIST solves the impurity continuity equations assuming cylindrical symmetry:

$$\frac{\partial}{\partial t} n_q = -\frac{1}{r}\frac{\partial}{\partial r}(r\Gamma_q) + I_{q-1} n_{q-1} - (I_q + R_q) n_q + R_{q+1} n_{q+1} - n_q/\tau_q + S_q \tag{1}$$

where n_q is the impurity density in ionization state q, Γ_q is an impurity flux density which describes radial (cross-field) motion of the particles, and I_q and R_q are ionization and recombination rates, respectively. S_q describes the impurity source and particle losses are described by τ_q, the confinement time in the scrape-off region (the region outside the last closed flux surface of the magnetic field). Because particle confinement in the scrape-off region is poor compared to that in the plasma core, modeling of core transport in large tokamaks is not very sensitive to the specific value of τ_q used.

Γ_q is written as the sum of diffusive and convective terms:

$$\Gamma_q = -D(r)\frac{\partial}{\partial r} n_q(r) + v_q(r) n_q(r) \tag{2}$$

where D_q is a diffusion coefficient and v_q is a convective velocity. This form is used because theoretical expressions for particle flux densities in tokamak plasmas, such as those of neoclassical theory, are often of this form. Because theoretical predictions of the impurity flux densities for a given tokamak discharge are not often available, the approach usually taken is to measure the diffusion coefficient and convective velocity by modeling the data, rather than simulating the data using a theory of impurity transport. To do this, assumptions on the functional forms of the diffusion coefficient and convective velocity are required. The most common assumptions are that D and v are independent of impurity charge state, q, and that D is constant as a function of radius.

Two parameterizations of the convective velocity are used:

$$v(r) = -c_1 2 D(r)/a\, (r/a) \tag{3}$$

$$v(r) = c_2 D(r)\frac{\partial}{\partial r} \ln(n_e(r)) \tag{4}$$

where c_1 and c_2 are are constant parameters which are related to the degree of central peaking of the impurity density profile and $n_e(r)$ is the measured electron density. The motivation behind these forms can be seen by examining an equilibrium solution ($\partial n_q/\partial t=0$ for all q) to Eq. 1. In equilibrium, the shape of the total impurity density profile, $n_z(r) = \Sigma n_q(r)$, is simply related to v(r)/D(r), and since D is assumed to be constant, the shape of $n_z(r)$ is determined by v(r). If v(r) is parameterized according to Eq. 3, $n_z(r)$ in equilibrium has the shape of a Gaussian to the c_1 power. If Eq. 4 is used, $n_z(r)$ in equilibrium has the shape of $n_e(r)$ to the c_2 power. Equation 3 is the simplest form for v(r) which satisfies the boundary condition that v(0)=0 in the absence of sources and sinks at the plasma axis. Equation 4 is useful because it allows the shape of the impurity density profile to be compared with the electron density profile in a simple way.

ATOMIC DATA REQUIREMENTS

To evaluate Eq. 1, ionization and recombination rates between adjacent ionization states are required. Ionization of impurities in tokamak plasmas is primarily due to electron impact, with both direct ionization and excitation followed by autoionization being important. (Excitation-autoionization rates can be comparable to the direct rates for ionization states up to Na-like.) The total recombination rate is the sum of the rates for radiative recombination, dielectronic recombination, and charge exchange with neutral hydrogen. Of the electron-impact processes, the dielectronic rates are considerably larger than the radiative rates for all but the highest ionization states of an element. Charge exchange recombination does not have a large effect on the ionization balance in discharges with ohmic heating only, but it can be important in discharges heated by injection of neutral hydrogen beams.[3] An extensive discussion of the status of atomic data for fusion applications is given by Janev and Katsonis.[4]

As a result of the large amount of atomic data required, experimental values are not usually used. Ab initio calculations of the rate coefficients are available in some cases but in general it is necessary to use simple formulae such as the Lotz formula[5] for ionization and the Burgess-Merts formula[6] for dielectronic recombination. The rates for electron impact processes are evaluated using measured electron density and temperature (T_e) profiles and, in the case of charge exchange recombination, calculated beam and thermal neutral hydrogen densities and energies.

Excitation rates are required in order to calculate line and continuum intensities from the impurity ion radial distributions calculated by the code. The important processes for line excitation in tokamak discharges are electron impact excitation and charge-exchange recombination in the visible and VUV regions of the spectrum and electron impact excitation and dielectronic recombination in the x-ray region. For continuum emission, the important processes are free-free transitions (bremsstrahlung) in the visible and free-free and free-bound transitions (radiative recombination) in the x-ray region. The situation for excitation rates is more tractable than for ionization and recombination because the amount of atomic data required for a specific diagnostic application is smaller. Ab initio calculations of excitation rates are available for many cases of interest; formulae for electron impact line excitation, such as the Van Regemorter formula with the Gaunt factors of Mewe,[7] are used where such calculations are not available.

The following sections present examples of impurity transport measurements which illustrate the major techniques used.

IMPURITY TRANSPORT IN OHMICALLY HEATED TFTR PLASMAS

Modeling of emissions from impurities injected into a tokamak discharge via the laser blow-off method[8] is a widely used technique for the study of impurity transport. An example is a measurement[9] of impurity transport in an ohmically

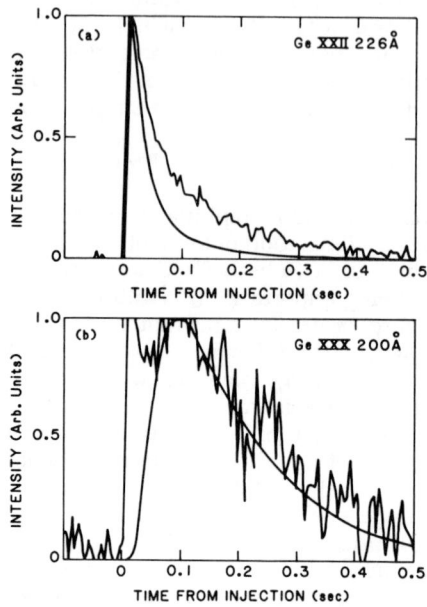

FIG. 1. Measured and calculated time evolutions of lines from germanium injected into TFTR. $D=1.3$ m^2/s and $c_2=0.3$ in Eq. 4 were used in the calculations. (From ref. 9.)

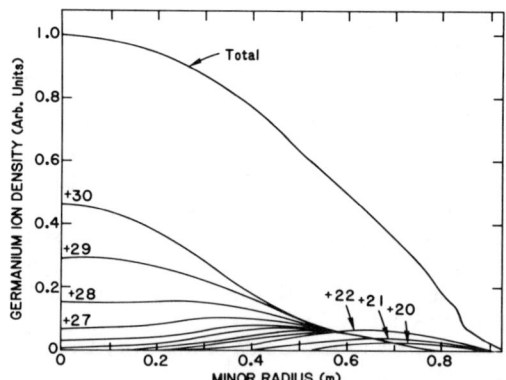

FIG. 2. Distribution of germanium ionization states in TFTR calculated assuming $D=1.3$ m^2/s and $c_2=0.3$ in Eq. 4. (From ref. 9.)

heated TFTR discharge with $T_e(0)$=4.2 keV, a line average electron density of 1.2×10^{19} m^{-3}, and major and minor radii of 2.45 m and 0.79 m, respectively. Because low-Z elements are fully ionized over most of the hot plasma core of modern tokamaks and therefore do not emit line radiation, elements with Z in the range 20-40, which are ionized through the L shell, are usually used in these studies. A thin metallic film on a glass slide, germanium in this case, is illuminated from behind by a powerful laser pulse and material is ablated. Both individual atoms and clusters of atoms are produced and enter the plasma edge with an energy of several eV. As a result of the centrally peaked T_e profile, impurity atoms are ionized to successively higher states as cross-field transport moves them toward the plasma center.

The time evolutions of lines from a number of ionization states of germanium were observed in this experiment. Because metallic impurities do not recycle significantly (re-enter the plasma after being deposited on the limiter or divertor plates), the total number of impurity ions in the discharge decreases with time following the injection. Thus, emission from a given ionization state peaks as the number of particles in that state reaches a maximum and then decays. Because higher ionization states exist closer to the plasma axis, the time to reach peak intensity and the decay time are longer for lines emitted by these states. This is seen in Fig. 1, which shows the measured time evolutions of two VUV lines, one from an ionization state near the plasma edge, Ge XXII, and one from a state which peaks on axis, Ge XXX. Figure 2 shows the calculated radial distribution of ionization states at 0.2 s after injection, by which time the final radial distribution of ionization states has been reached. The calculations were performed using the MIST code and a transport model of D=1.3 m^2/s and c_2 in Eq. 4 equal to 0.3. This value of the diffusion coefficient is over an order of magnitude larger than that predicted by neoclassical theory. This level of disagreement is observed in many tokamaks and is thought to be due to particle transport driven by plasma turbulence, which is not included in neoclassical theory. Also shown in Fig. 1 is the time evolution of the lines calculated assuming this transport model. Although the observed time evolution of the Ge XXX line is accurately reproduced, that of the Ge XXII line is not. A wide variety of transport models were tried, including spatially varying diffusion coefficients, and it was not possible to simultaneously reproduce the time evolutions of both lines with one transport model. Transport models which reproduce the central line data well always predicted a more rapid decay of the edge line than observed.

This discrepancy was observed for a number of elements injected into a wide variety of TFTR discharges and can be explained by inaccurate ionization and recombination rates in the code. Few ab initio calculations of total electron impact ionization rates and dielectronic recombination rates are available for elements such as germanium (Z=32); the Lotz[5] and Burgess-Merts[6] formulae were therefore used. It is well known[4] that these expressions are inaccurate for elements with Z $\geq$20, and it is therefore reasonable to vary the values by a factor of two in order to test the hypothesis that inaccurate rates are the cause of the discrepancy seen in Fig. 1. The results of multiplying the rates by 2 and 0.5 are shown in Figs. 3 and 4. Because the evolution of the higher ionization states is dominated by transport, the effect of these variations on the Ge XXX line is small. However, the effect on the lower states such as Ge XXII is dramatic: halving the ionization rates or doubling the recombination rates produces good agreement between the measured and calculated time evolutions of the Ge XXII line. While the Lotz formula is not believed to systematically overestimate the total ionization rates in medium-Z elements by a factor of two, the Burgess-Merts

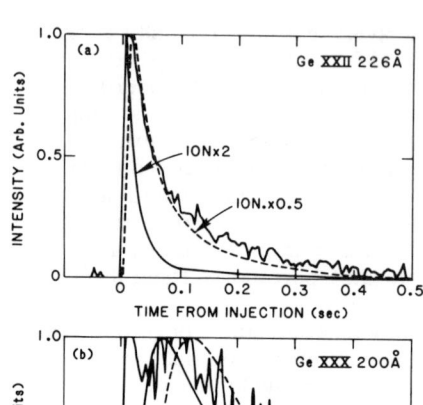

FIG. 3. As in Fig. 1 with ionization rates multiplied by 2.0 and 0.5. (From ref. 9.)

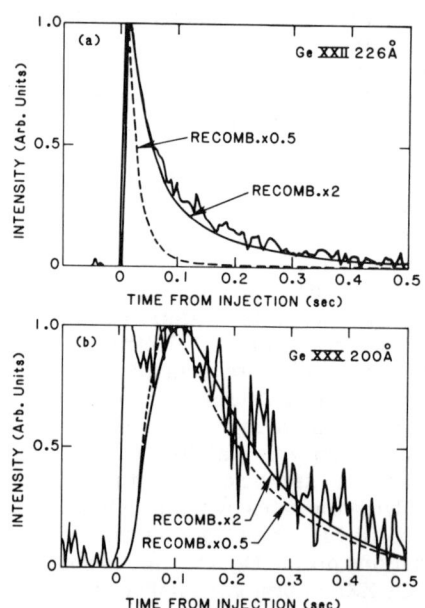

FIG. 4. As in Fig. 1 with recombination rates multiplied by 2.0 and 0.5. (From ref. 9.)

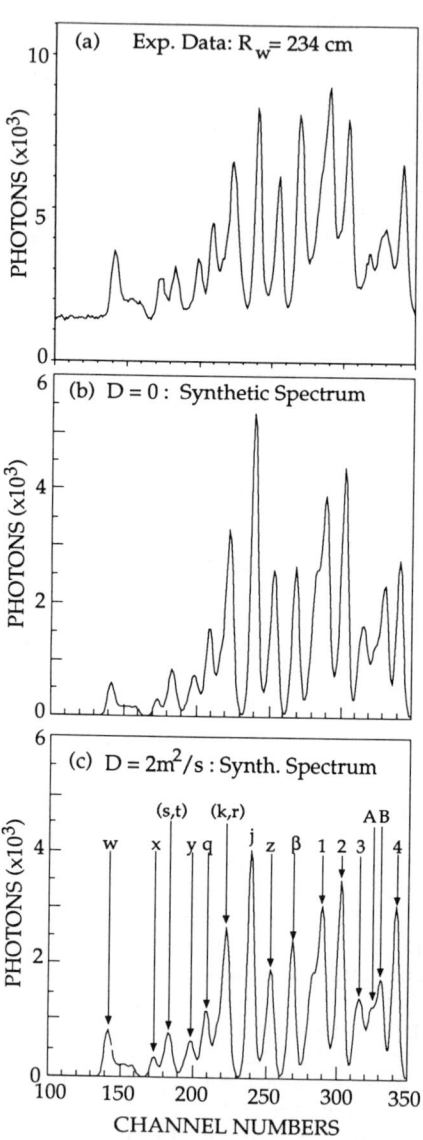

FIG. 5. Observed (a) and synthetic (b and c) soft x-ray (1.85-1.88 Å) iron spectra from TFTR. (From ref. 10.)

formula does underestimate the dielectronic recombination rates, often by a factor of two or more.[4] Thus, it is clear that modeling time evolutions of lines emitted by impurities injected using the laser blow-off technique could be significantly improved by better rates. In particular, it may be possible to discern spatial variations in the diffusion coefficient by modeling lines emitted by ions which exist at different spatial locations in the plasma, and the relative intensities of the lines, which are also sensitive to the rates, could be used as an additional constraint on the modeling.

Despite the sensitivity of the edge states to the atomic rates, this is a useful technique for the study of impurity transport because the time evolutions of lines emitted by the core ionization states are not very sensitive to the rates used in the modeling. Because it is simple to apply, a spatially-averaged diffusion coefficient and parameterized convective velocity can be measured in a wide variety of discharges and correlations with changes in other plasma parameters can be observed.[9]

HIGH RESOLUTION SOFT X-RAY SPECTROSCOPY ON TFTR

The radial distributions of the ionization states of an impurity can be sensitive to changes in impurity transport. Modeling measured profiles of impurity ions or relative intensities of lines that are sensitive to variations in the ionization balance therefore provides another means of studying impurity transport. In the core region of large tokamaks, the ionization balance tends to be close to coronal equilibrium; changes in the ion spatial distributions are therefore small for reasonable variations in the transport. As a result, this approach requires high quality data, but it does have the advantage that emissions from naturally occurring impurities in steady state may used, eliminating the perturbation of the plasma caused by impurity injection.

An example of this technique is the modeling by Bitter, et al.[10] of soft x-ray spectra of Fe XXII-Fe XXV emitted by ohmically heated TFTR plasmas. Figure 5 shows a spectrum of the 1.85-1.88 Å region measured by a high resolution crystal spectrometer that views five vertical chords through the plasma. Data from a detector with a line of sight intersecting the plasma midplane at a major radius of 2.34 m are shown. The data were averaged over the steady state period of 10 identical discharges.

The lines in Fig. 5 are labeled according to the notation of Gabriel.[11] These lines are excited by the electron impact processes of direct excitation, inner-shell excitation, or dielectronic recombination. Extensive theoretical studies of the rates for these processes have been made in attempts to model solar flare and tokamak spectra. The spectra of Fig. 5 were modeled using the MIST code, with the excitation rates of Bely-Dubau, et al.,[12] the ionization and recombination rates as described previously, and the measured T_e and n_e profiles. The results of the modeling for zero diffusion and for a transport model of $D=2$ m^2/s and $c_2=1.0$ in Eq. 4 are shown in Fig. 5. The zero-diffusion model yields the same radial distributions of the ions as in coronal equilibrium, but, in order to permit direct comparison of the two models with the data, the radial profile of the total iron density was forced to be the same as the n_e profile, as is the case for the non-zero diffusion model. Three lines of particular interest are: w (Fe XXV $1s^2$ $^1S_0$ - $1s2p$ $^1P_1$), q (Fe XXIV $1s^22s$ $^2S_{1/2}$ - $1s2p2s$ $^2P_{3/2}$), and β ($1s^22s^2$ $^1S_0$ - $1s2s^22p$ $^1P_1$). The measured ratio of the intensity of the q line to that of the w line is 1.36; the calculated ratios are 2.58 for zero diffusion and 1.41 for $D=2$ m^2/s. The measured intensity ratio for the β and w lines is 2.88, while the calculated ratios

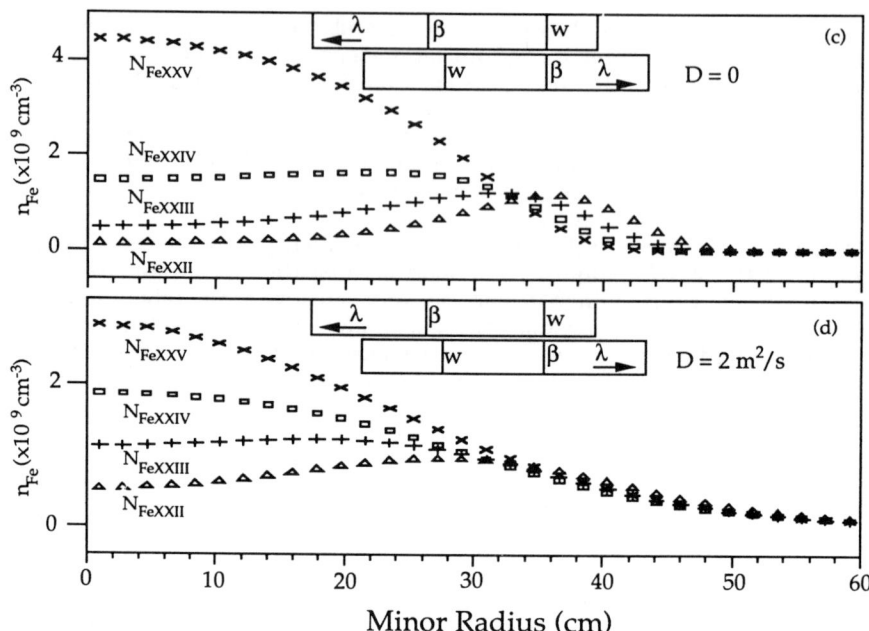

FIG. 6. Calculated ionization balance for iron in TFTR with zero diffusion (top) and with D=2 m^2/s (bottom). (From ref. 10.)

are 4.50 for zero diffusion and 2.94 for D=2 m^2/s. Thus, a diffusion coefficient of 2 m^2/s gives considerably better agreement with the measured relative intensities of the lines than does zero diffusion; this is true for the other lines shown in Fig. 5 and for spectra from the other spatial channels of the spectrometer. The calculated radial distributions of the lines are shown in Fig. 6, along with the regions observed by two spatial channels of the spectrometer for the w and β lines. Note that large changes in the relative intensities of the lines result from the small changes in the ionization balance caused by varying the diffusion coefficient, making this a sensitive technique.

As in the impurity injection experiment, the primary source of uncertainty in the modeling arises from uncertainties in the ionization and recombination rates, resulting in an estimated uncertainty in the measured diffusion coefficient of a factor of two. As a result of its astrophysical importance, the availability of atomic data for iron is better than for many other medium-Z elements. In particular, Pindzola, et al.[13] have compiled total ionization rates (direct plus excitation-autoionization) for iron. However, there is no correspondingly complete compilation of dielectronic recombination rates in iron. In order to permit reliable modeling of these spectra in neutral beam-heated discharges, future work will include charge exchange as an additional recombination mechanism.

BROADBAND SOFT X-RAY MEASUREMENTS ON ALCATOR-C

The modeling in the preceding examples is sensitive to the ionization and recombination rates because line intensity measurements are charge-state specific: the rates link the densities of the observed states to the temporal and spatial evolution of the total impurity density, which is determined by transport. Thus, non-state-specific measurements should be less dependent on the rates.

The study by Petrasso, et al.[14] of impurity transport in Alcator-C discharges fueled by injection of a frozen hydrogen pellet is an example of such a measurement. In this experiment, two arrays of filtered soft x-ray (1-10 keV) detectors were used to observe the time evolutions of the profiles of the two dominant impurity species in the discharge, carbon and molybdenum. Because carbon is fully-ionized over the core region of tokamak discharges, it emits only continuum radiation consisting in the soft x-ray region of bremsstrahlung and recombination radiation in approximately equal amounts. The molybdenum radiation in this region is dominated by 2p-3d transitions emitted around 2.5 keV by Mo XXIX - Mo XXXIII.[15] Array A had significant sensitivity down to approximately 1.5 keV; its signals were therefore due to both carbon and molybdenum radiation. Array B had little sensitivity below 4 keV and thus was sensitive only to carbon radiation. Both arrays were sensitive to continuum radiation from the working gas protons.

The chordal intensities were Abel inverted to obtain emissivities as a function of radius. The impurity densities were then derived from the relations

$$\varepsilon^A / n_e - P^A_H n_H \approx P^A_C n_C + P^A_{Mo} n_{Mo} \tag{5}$$

$$\varepsilon^B / n_e - P^B_H n_H \approx P^B_C n_C \tag{6}$$

$$n_H = n_e - Z_C n_C - Z_{Mo} n_{Mo} \tag{7}$$

where ε^A and ε^B are emissivities measured by arrays A and B and P^j_k is the calculated spectral power function for array j and element k. Equation 7 represents local charge neutrality and was evaluated with $Z_C=6$ and $Z_{Mo}=30$.

Figure 7 (a) shows the signals from five of the array-A detectors as a function of time. The pellet was injected 223 ms into the discharge and an event known as a giant impurity disruption (GID), which expels impurities from the core region, occurred at 259 ms. The signals fall at the time of pellet injection as a result of the drop in $T_e(0)$ from 1.6 keV to 0.6 keV caused by the pellet and then begin to rise due to reheating of the plasma and peaking of the impurity profiles on axis. The modulations of the signals during this time are due to sawteeth, a plasma instability which causes periodic flattening and repeaking of the particle density profiles. Figure 7 (b) is an expansion in time of the region around the GID. Figure 7 (c) shows the emissivity obtained from the array-A signals at three times: before pellet injection, immediately before the GID and after the GID. The latter two times are indicated by arrows in Fig. 7 (b). The emissivity following pellet injection is more peaked than before injection or following the GID.

Figure 8 shows the derived profiles of the carbon and molybdenum densities in the core region of the discharge before and after the GID. (The major and minor radii of the discharge were 0.64 m and 0.0165 m.) The profiles are considerably more peaked before the GID than after it. The time evolution of the carbon profile in the 12 ms period before the GID was modeled using an impurity transport code to obtain a diffusion coefficient of 0.03 m^2/s and a convective velocity of 10 (r/a) m^2/s, which are within a factor of two of the values predicted by neoclassical theory. The neoclassical predictions are shown as dashed lines in Fig. 8. As shown in Fig. 8, similar modeling of the molybdenum profiles did not produce as good agreement with the standard neoclassical treatment; this was attributed to

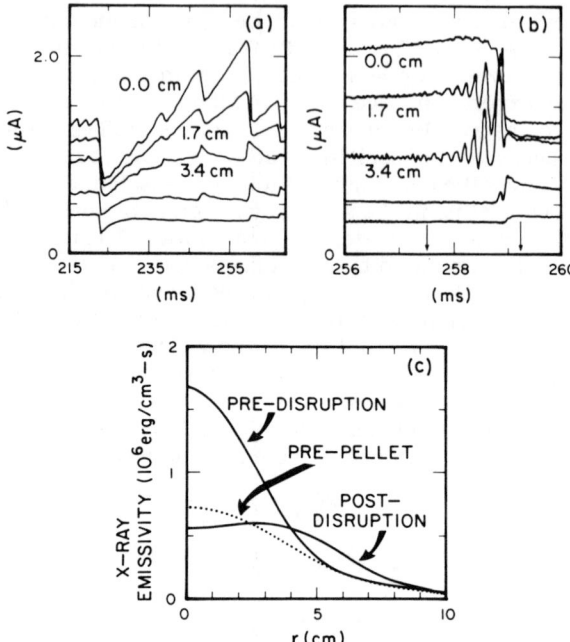

FIG. 7. (a) Array-A x-ray signals following pellet injection into Alcator-C. (b) Expansion of (a) showing region around giant impurity disruption (GID). (c) Array-A emissivities before pellet injection and just before and after the GID. (From ref. 14.)

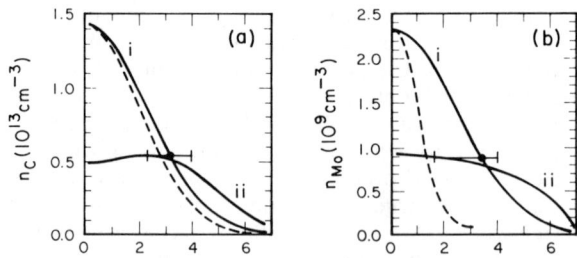

FIG. 8. Carbon (a) and molybdenum (b) profiles before (curves i) and after (curves ii) the giant impurity disruption. The dashed curves are the predictions of neoclassical theory. (From ref. 14.)

the neglect of impurity-electron friction in the theory, an important effect for molybdenum but not for carbon.

This technique for studying impurity transport has the advantage that the modeling is insensitive to the ionization balance in the core region[16] and therefore does not depend on detailed knowledge of ionization or recombination rates. Disadvantages are that it is not always possible to clearly separate the contributions to the spectrum of the different impurity species present in the plasma using filters and that the data must be Abel inverted, which can lead to uncertainties in flat or hollow impurity profiles.

CHARGE EXCHANGE RECOMBINATION SPECTROSCOPY ON TEXT

Another approach to the study of impurity transport makes use of the line radiation emitted by impurity ions following charge exchange reactions with particles in an energetic neutral hydrogen beam. Electrons captured into excited levels of the impurity ion emit visible and VUV lines as they cascade to the ground state. The emissivity at a given location in the plasma is the product of the impurity density, the beam density, and the line excitation rate. The latter two quantities can be calculated, making possible a measurement of the impurity density profile. As a result of the presence of molecular hydrogen as in the ion source, neutral hydrogen beams have three energy components. Thus, the beam densities are calculated for each component and the products with the excitation rates summed to obtain a total predicted line brightness. A detailed discussion of charge exchange recombination spectroscopy is given by Fonck, et al.[17,18]

The measurement by Synakowski, et al.[19] of fully-ionized carbon and oxygen profiles in TEXT tokamak discharges with pellet and gas puff fueling is an example of this technique. VUV lines were observed by a spectrometer viewing 20 chords in a vertical fan, and a diagnostic beam of 30-45 keV neutral hydrogen was injected vertically. The radial profile of the line emissivity was measured, with the position of each spatial point determined by the intersection of the beam and one of the spectrometer chords.

The primary beam attenuation processes are charge exchange with plasma protons and impurity ions and ionization due to collisions with protons and electrons. The local beam densities were calculated using a beam attenuation code with measured T_e and n_e profiles as inputs. Excitation of beam atoms was also included because electron-loss cross sections for beam atoms in excited states are large.[20]

The line excitation rates were calculated by correcting the cross sections for electron capture into each level n, ℓ for the effects of cascades and assuming the appropriate degree of ℓ-mixing between the levels for a given n. The charge exchange cross sections were calculated using the perturbed stationary state method[21,22] at low energies and the classical trajectory Monte Carlo approximation[23] at high energies.

O^{+8} profiles in two discharges, one with pellet fueling and the other with gas puff fueling, are shown in Fig. 9. These discharges had similar plasma parameters, with the line average density being 6×10^{19} m^{-3} in the pellet-fueled discharge and 5×10^{19} m^{-3} in the gas-puff fueled plasma. The major and minor radii were 1.0 m and 0.26 m, respectively. MIST code calculations show that ionization states lower than O^{+8} comprise less than 10% of the total oxygen profile for radii less than 0.12 m; the O^{+8} profile is therefore an accurate representation of the total oxygen profile inside this radius.

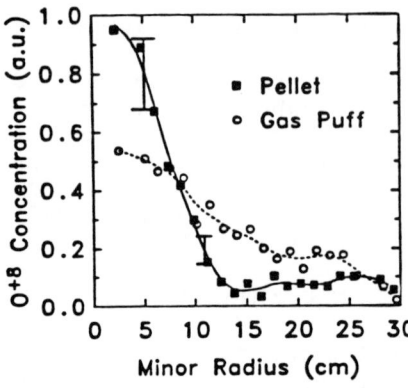

FIG. 9. O^{+8} profiles in TEXT discharges with gas puff fueling and pellet fueling normalized to the central O^{+8} density in the pellet case. (From ref. 19.)

It is clear from Fig. 9 that the oxygen profile is more peaked in the pellet-fueled discharge than in the gas-puff fueled one. This was also true of the electron density profiles, but the ratio of the electron density and oxygen profile scale lengths was less than 1.5 in the gas-puff fueled case, while it was 3-4 in the pellet-fueled case, indicating impurity accumulation in the pellet-fueled discharge but not in the gas-puff fueled one. Similar results were obtained for carbon.

The time evolutions of the low-Z impurity profiles after a non-terminal disruption which sometimes followed pellet injection were modeled to determine the diffusion coefficient and convective velocity. The results are $D=0.15\pm0.02$ m^2/s and $v(r)=(0.25\pm0.04)r$ m/s. These values are larger by nearly an order of magnitude than the neoclassical transport parameters for these discharges. The diffusion coefficient is smaller than the value of approximately 1 m^2/s obtained from impurity injection into discharges similar to the gas-puff fueled discharge, but the convective velocity is similar. Thus, the primary cause of impurity peaking in the pellet-fueled discharges was a reduction of the diffusion coefficient compared to the gas-puff fueled case, not an increase in convection.

Charge exchange recombination spectroscopy has several advantages over the other techniques for studying impurity transport discussed thus far. Because it is a crossed-beam technique, no Abel inversion of the data is required and small changes in the profiles, as indicated by the representative error bars in Fig. 9, can be measured. To date, the technique has been applied only to measurements of impurity transport in low-Z impurities which are fully-ionized in the core region of tokamak discharges; thus, ionization and recombination rates are not a source of uncertainty. This technique also has the potential to be applied to metallic impurities such as helium-like iron.[24] Because medium-Z impurities are not fully-ionized in the core region, such measurements would, however, be sensitive to the values of the ionization and recombination rates. The primary drawbacks are that a beam is required and that problems[17,18] with the line excitation rates such as the reliability of the cross sections at low energy for visible transitions, difficulties of an accurate ℓ-mixing treatment, and issues such as the contribution of thermal neutrals to the line brightness,[25] background light due to electron impact excitation,[17] and the effects of excited states[26] in the beam have yet to be fully addressed. It is also worth mentioning that beam-based techniques will be very difficult on future ignition tokamaks due to the beam attenuation caused by the high densities at which these devices will operate.[18]

IMPURITY TRANSPORT IN PBX H-MODE DISCHARGES

The impurity transport measurements discussed thus far are based on a single impurity element and one type of spectroscopic measurement. Considerable information can be obtained by combining measurements of several impurity species. This was done by Ida, et al.[27,28] in an impurity transport study on the PBX tokamak based on profiles of Z_{eff}, the average ion charge in the plasma, derived from measurements of the visible bremsstrahlung emission.

Z_{eff} is defined as $\Sigma Z_i^2 n_i/n_e$ where the sum is over all charge states of all elements present in the plasma. The visible bremsstrahlung emissivity is related to Z_{eff} by the expression

$$\partial \xi / \partial \lambda = Z_{eff} (n_e^2 / T_e^{1/2}) \overline{g}_{ff} \tag{8}$$

where $\overline{g}_{ff}$ is the energy-averaged Gaunt factor[29] for free-free transitions.

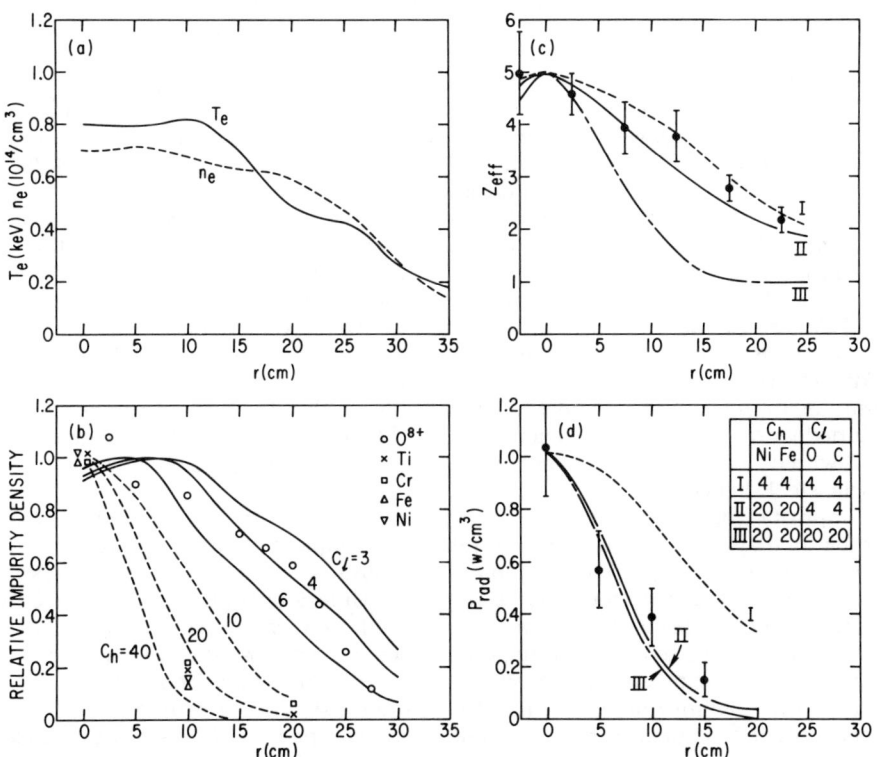

FIG. 10. Impurity and plasma parameter profiles in a PBX H-mode discharge: (a) T_e and n_e, (b) Z_{eff}, (c) impurity densities, and (d) radiated power. The dashed lines are profiles calculated assuming the indicated values of the peaking parameters c_ℓ and c_h in Eq. 3. (From ref. 27.)

The PBX tokamak produces a plasma with a kidney-bean-shaped cross section with a midplane half-width of 0.36 m. A tangentially-viewing array of detectors was used to measure the visible bremsstrahlung emission in a 10 Å wide region centered on 5235 Å. The brightnesses were Abel inverted to obtain emissivity profiles which were converted to Z_{eff} profiles using measured T_e and n_e profiles. An example is shown in Fig. 10 for an H-mode discharge with long period sawteeth; due to the long sawtooth period, the profiles can be considered steady state. (The H-mode is a regime of enhanced particle and energy confinement observed on PBX and other tokamaks.) The primary impurity species in PBX were carbon, oxygen, titanium, chromium, iron, and nickel. Also shown in Fig. 10 are profiles of the metallic impurity densities derived from measurements of K_α radiation using soft x-ray pulse-height-analysis, O^{+8} profiles from charge exchange recombination spectroscopy, and profiles of the power radiated by the discharge measured using a bolometer array. The Z_{eff}, metallic impurity density, and radiated power profiles are peaked on axis, indicating impurity accumulation relative to the n_e profile; the degree of peaking is larger for metallic impurities than for low-Z impurities.

These profiles were modeled using the MIST code with the convection parameterized as in Eq. 3 to obtain peaking parameters c_i for each impurity element. The results are the dashed curves in Fig. 10. The curves assume different degrees of peaking for the low-Z and metallic impurities, with the values of the peaking parameters c_ℓ for low-Z impurities and c_h for metallic impurities shown in Fig. 10 (b) and the legend of Fig. 10 (d). Although each of the transport models fit some of the profiles well, only a model with $c_h=20$ and $c_\ell=4$ fits all of the data. Thus, it is clear that the ratio of convection to diffusion in steady state, v/D, is larger for metallic impurities than for low-Z impurities.

The transport coefficients for the core region were obtained by modeling the time evolution of the Z_{eff} profile during the sawteeth using these values of the peaking parameters. The time evolution of Z_{eff} on axis is shown in Fig. 11 (a) and the profiles before and after the sawteeth are shown in Fig. 11 (b). Also shown in Fig. 11(a) are the time evolutions of the central iron and nickel concentrations. The sawteeth were modeled by flattening the impurity profiles inside the sawtooth mixing radius at each sawtooth crash and then allowing the profiles to evolve according to the transport model. Figure 11 shows that the best agreement is obtained with a diffusion coefficient of 0.1 m^2/s and convective velocity given by Eq. 3 with $c_\ell=4$ and $c_h=20$. These values are close to those predicted by neoclassical theory.

Impurity transport studies of this type, which integrate measurements from a number of diagnostics, can provide a relatively complete picture of impurity transport in a particular regime of tokamak operation. The obvious drawback is that considerable instrumentation is required. Unless only one impurity species is present in the plasma, the visible bremsstrahlung measurement of Z_{eff} alone can not be modeled to obtain transport coefficients; additional spectroscopic information on the plasma composition is needed. The visible bremsstrahlung data require Abel inversion, which, although not a significant source of uncertainty for the peaked profiles shown here, can be uncertain for flat or hollow profiles. Also, accurate electron density profiles are required as a result of the n_e^2 factor in Eq. 8. The free-free Gaunt factors have been accurately calculated[29] and therefore are not a source of uncertainty.

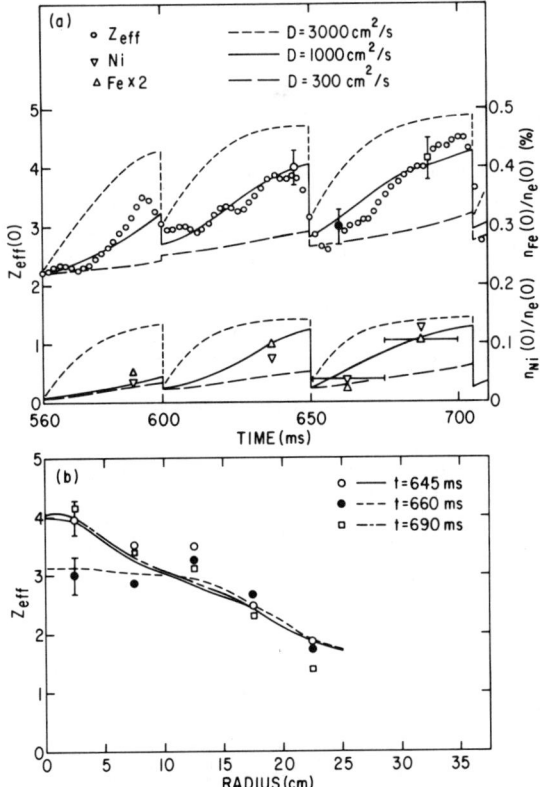

FIG. 11. (a) Measured and calculated time evolutions of central Z_{eff} and metallic impurity concentrations in a PBX discharge with strong sawteeth. (b) Measured and calculated Z_{eff} profiles before and after a sawtooth crash. (From ref. 27.)

SUMMARY

A variety of approaches to the study of impurity transport in tokamak plasmas have been described, all of which rely to some extent on impurity transport code modeling. All of the measurements require excitation rates for the radiation being observed; uncertainties in these rates therefore affect the accuracy of the deduced transport coefficients. Although more work on cascade-corrected charge exchange recombination rates for line excitation in the visible is needed, adequate excitation data does exist for many applications.[4]

Many of the impurity transport measurements described here are also sensitive to the values of the ionization and recombination rates used, making it important to have accurate values of these rates. Systematically compiled rate coefficients in a standard format[30] are the most useful form of atomic data for impurity transport codes. This is a difficult problem due to the complexity of the calculations and the quantity of data required. However, some progress has been made in recent years: direct electron impact ionization rates for all states of the elements from hydrogen to nickel have been compiled by the group at The Queen's University of Belfast,[31,32] Pindzola et al.[13] have compiled total (direct plus excitation-autoionization) electron impact ionization rates for iron, and reliable total charge exchange recombination rates are available.[4] Thus, although there is still a need for more work on ionization rates, particularly on multi-step processes such as excitation-autoionization, the most serious need at this point is for a compilation of dielectronic recombination rates. Of course, the difficulty of calculating these rates is the reason such a compilation does not already exist. The close-coupling calculations required to obtain accurate results are time-consuming and are usually performed on a case-by-case basis. An additional complication is the effect of external fields on the rates, which is important in some cases. Nevertheless, any progress in the direction of a compilation of dielectronic recombination rates would be of great value to current work in modeling of impurity emissions from tokamak plasmas.

ACKNOWLEDGMENTS

It is a pleasure to thank M. Bitter, R. Fonck, R. Hulse, R. Petrasso, and E. Synakowski for valuable discussions of their work and to acknowledge the continuing support of H. Furth, D. Meade, and K. Young of the Princeton Plasma Physics Laboratory. This work is supported by U. S. Department of Energy Contract No. DE-AC02-76-CHO-3073.

REFERENCES

1. R. C. Isler, Nucl. Fusion 24, 1599 (1984).
2. R. A. Hulse, Nucl. Technol./Fusion 3, 259 (1983).
3. R. A. Hulse, D. E. Post, and D. R. Mikkelsen, J. Phys. B 13, 3895 (1980).
4. R. K. Janev and K. Katsonis, Nucl. Fusion 27, 1493 (1987).
5. W. Lotz, Astrophys. J. Suppl. 14, 207 (1967).
6. A. L. Merts, R. D. Cowan, N. H. Magee, Los Alamos Scientific Laboratory Report No. LA-6220-MS (1976) (unpublished).
7. R. Mewe, Astron. Astrophys. 20, 215 (1972).
8. E. S. Marmar, J. L. Cecchi, and S. A. Cohen, Rev. Sci. Instrum. 46, 1149 (1975).
9. B. C. Stratton, et al., Nucl. Fusion 29, 437 (1989).
10. M. Bitter, et al., Princeton Plasma Physics Laboratory report No. PPPL-2610 (1989) (unpublished).
11. A. H. Gabriel, Mon. Not. R. Astron. Soc. 160, 99 (1972).
12. F. Bely-Dubau, J. Dubau, P. Faucher, and A. H. Gabriel, Mon. Not. R. Astron. Soc. 198, 239 (1982).
13. M. S. Pindzola, D. C. Griffin, C. Bottcher, S. M. Younger, and H. T. Hunter, in Nucl. Fusion Suppl. "Recommended Data on Atomic Collision Processes Involving Iron and its Ions", p. 21 (1987).
14. R. D. Petrasso, et al., Phys. Rev. Lett. 57, 707 (1986).
15. E. Källne, J. Källne, and R. D. Cowan, Phys. Rev. A 27, 2682 (1983).
16. R. D. Petrasso, private communication, 1989.
17. R. J. Fonck, D. S. Darrow, and K. P. Jaehnig, Phys. Rev A 29, 3288 (1984).
18. R. J. Fonck, these proceedings.
19. E. J. Synakowski, R. D. Bengston, A. Ouroua, A. J. Wootton, and S. K. Kim, Nucl. Fusion 29, 311 (1989).
20. C. D. Boley, R. K. Janev, and D. E. Post, Phys. Rev. Lett. 52, 534 (1984).
21. E. J. Shipsey, T. A. Green, and J. C. Brown, Phys. Rev. A 27, 821 (1983).

22. T. A. Green, E. J. Shipsey, and J. C. Brown, Phys. Rev. A $\underline{25}$, 1364 (1982).
23. R. E. Olson, Phys. Rev. A $\underline{24}$, 1726 (1981).
24. R. J. Knize, R. J. Fonck, R. B. Howell, R. A. Hulse, and K. P. Jaehnig, Rev. Sci. Instrum. $\underline{59}$, 1518 (1988).
25. B. C. Stratton, et al., Nucl. Fusion (in press).
26. R. C. Isler and R. E. Olson, Phys. Rev. A $\underline{37}$, 3399 (1988).
27. K. Ida, R. J. Fonck, S. Sesnic, R. A. Hulse, and B. LeBlanc, Phys. Rev. Lett. $\underline{58}$, 116 (1987).
28. K. Ida, R. J. Fonck, S. Sesnic, R. A. Hulse, B. LeBlanc, and S. F. Paul, Nucl. Fusion $\underline{29}$, 231 (1989).
29. W. J. Karzas and R. Latter, Astrophys. J. (Suppl.) $\underline{6}$, 167 (1961).
30. R. A. Hulse, these proceedings.
31. K. L. Bell, H. B. Gilbody, J. G. Hughes, A. E. Kingston, and F. J. Smith, J. Phys. Chem. Ref. Data $\underline{12}$, 891 (1983).
32. M. A. Lennon, K. L. Bell, H. B. Gilbody, J. G. Hughes, A. E. Kingston, M. J. Murray, and F. J. Smith, Culham Laboratory Report CLM-R270 (1986) (unpublished).

LIQUID AND SOLID ATOMIC ION PLASMAS*

J.J. Bollinger, S.L. Gilbert, D.J. Heinzen,
W.M. Itano, and D.J. Wineland†
National Institute of Standards and Technology,
325 Broadway, Boulder, CO 80303

ABSTRACT

Atomic ions which are stored in electromagnetic traps can be viewed as one component plasmas. When the ions are laser cooled they become strongly coupled ($\Gamma > 100$) and exhibit liquid-like and solid-like behavior. Experimental evidence for this behavior is discussed for nonneutral $^9Be^+$ ion plasmas stored in a Penning ion trap.

INTRODUCTION

We have observed[1] spatial correlations with up to 15 000 Be^+ ions in a Penning trap with a coupling (defined below) of $\Gamma > 100$. These correlations are strongly affected by the boundary conditions and take the form of concentric shells as predicted by computer simulations.[2-4] In this paper we briefly describe the experimental confinement geometry and the method of producing low temperature ions. The relatively large spacings between the ions (~ 20 μm) permit the shells to be directly viewed by imaging the Be^+ laser-induced fluorescence onto a photon-counting camera. Diagnostic techniques capable of measuring the ion diffusion are then discussed. Qualitative observations of the ion diffusion are compared with theoretical predictions.

CONFINEMENT GEOMETRY

The Penning trap uses a static, uniform magnetic field and a static, axially symmetric electric field for the confinement of charged particles. The magnetic field, which is directed along the z axis of the trap, provides confinement in the radial direction. The ions are prevented from leaving the trap along the z axis by the electric field. In the work described here, the electric field was provided by three cylindrical electrodes as shown in Fig. 1. The dimensions of the trap electrodes were chosen so that the first anharmonic term (i.e. fourth order term) in the expansion of the trapping potential was zero. Over the region near the trap center, the potential can be expressed (in cylindrical coordinates) as $\Phi \simeq AV_o(2z^2-r^2)$ where $A = 0.146$ cm^{-2}. A background pressure of 10^{-8} Pa ($\approx 10^{-10}$ Torr) was maintained by

---

*Contribution of the U.S. Government; not subject to copyright.
†Presenting author

a triode sputter-ion pump. The confinement geometry is similar to that used by the electron plasma group of the University of California at San Diego (UCSD)[5] with the exception that our trap is smaller than the UCSD traps.

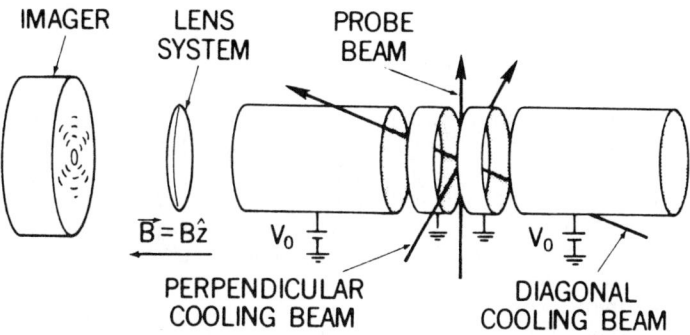

Fig. 1 Schematic drawing of the trap electrodes, laser beams, and imaging system (not to scale). The overall length of the trap is 10.2 cm. The trap consists of two end cylinders and two electrically connected central cylinders with 2.5 cm inner diameters. Ion clouds are typically less than 1 mm in both diameter and axial length. The diagonal cooling beam crosses the cloud at an angle of 51° with respect to the z axis. In the experiments, B = 1.92 T or 0.82 T and V_o ranged between 20 V and 200 V.

The stored ions can be characterized by a thermal distribution where the "parallel" (to the z axis) temperature $T_\parallel$ is approximately equal to the "perpendicular" temperature $T_\perp$. This thermal distribution is superimposed on a uniform rotation of the cloud[6-9] at frequency ω which, at the low temperatures of this experiment, is due to the $\vec{E} \times \vec{B}$ drift, where $\vec{E}$ is the electric field due to the trap voltage and the space charge of the ions. In a frame of reference rotating with the ions, the static thermodynamic properties of an ion cloud confined in a Penning trap are identical to those of a one-component plasma (OCP).[7] An OCP consists of a single species of charge embedded in a uniform-density background of opposite charge. For the system of ions in a Penning trap, the trapping fields play the role of the neutralizing background charge. An OCP is characterized by the Coulomb coupling constant,[7,10]

$$\Gamma \equiv q^2/(a_s k_B T),$$

which is a measure of the nearest-neighbor Coulomb energy divided by the thermal energy of a particle. The quantities q and T are the ion charge and temperature. The Wigner-Seitz radius a_s is defined by $4\pi a_s^3 n_0/3 = 1$, where $-qn_0$ is the charge density of the neutralizing background. In the Penning trap the background density n_0 depends on the rotation frequency ω and the cyclotron frequency Ω and is given by[6-9]

$$n_0 = m\omega(\Omega-\omega)/(2\pi q^2). \qquad (1)$$

LASER COOLING AND COMPRESSION

The ion density that can be achieved in a Penning trap is limited by the magnetic field strength that is available in the laboratory. Consequently to obtain large values of Γ and therefore strong couplings, a technique to obtain low ion temperatures is necessary. In our work, radiation pressure from lasers is used to reduce the temperature of the stored ions to less than 10 mK. This technique, known as laser cooling,[11-13] uses the resonant scattering of laser light by atomic particles. The laser is tuned to the red, or low-frequency side of the atomic "cooling transition" (typically an electric dipole transition like the D lines in sodium). Ions with a velocity component opposite to the laser beam propagation ($\vec{k} \cdot \vec{v} < 0$) will be Doppler shifted into resonance and absorb photons at a relatively high rate. Here, $\vec{k}$ is the photon wave vector ($|\vec{k}|=2\pi/\lambda$, where λ is the wavelength of the cooling radiation). For the opposite case ($\vec{k} \cdot \vec{v} > 0$), the ions will be Doppler shifted away from the resonance and the absorption rate will decrease. When an ion absorbs a photon, its velocity is changed by an amount $\Delta\vec{v} = \hbar\vec{k}/m$ due to momentum conservation. Here $\Delta\vec{v}$ is the change in the ion's velocity, m is the mass of the ion, and $2\pi\hbar$ is Planck's constant. The ion spontaneously reemits the photon symmetrically. In particular, when averaged over many scattering events, the reemission does not change the momentum of the ion. The net effect is that for each photon scattering event, the ion's average velocity is reduced by $\hbar\vec{k}/m$. To cool an atom from 300 K to millikelvin temperatures takes typically 10^4 scattering events but, since scatter rates can be 10^8/s, the cooling can be rapid.

In our work with Be$^+$, the $2s\ ^2S_{1/2} \to 2p\ ^2P_{3/2}$ "D$_2$" transition was used as the cooling transition as indicated in Fig. 2. Cooling laser beams were directed both perpendicularly and at an angle with respect to the magnetic field as indicated in Fig. 1. This enabled us to control[14,15] the cloud size and obtain the lowest possible temperatures. The 313 nm radiation required to drive this transition was obtained by frequency doubling the output of a continuous wave, narrow band (3 MHz) dye laser. The 313 nm power was typically 50 μW. The theoretical cooling limit,[11-13] due to photon recoil effects, is given by a temperature equal to $\hbar\gamma/(2k_B)$ where γ is the radiative linewidth of the atomic transition in angular frequency units. For the Be$^+$

cooling transition ($\gamma = 2\pi \times 19.4$ MHz), the theoretical minimum temperature is 0.5 mK.

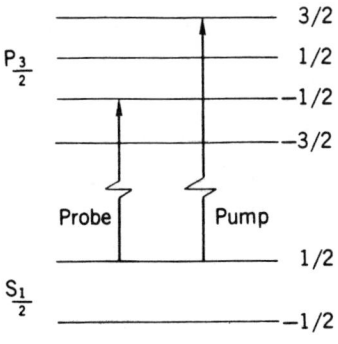

Fig. 2 Energy level structure of the $^9Be^+$ $^2S_{1/2}$ ground state and the first excited $^2P_{3/2}$ state. The magnetic field splits each state into its m_J sublevels. The laser cooling (pump) and depopulation (probe) transitions are shown.

Laser scattering can also be used to change the angular momentum and compress the stored ion plasma.[8,9] The z component of the canonical angular momentum for an individual ion in the plasma is

$$\ell_z = mv_\theta r + \frac{qBr^2}{2c}. \qquad (2)$$

The two terms in Eq. (2) are the ion's mechanical angular momentum and the field angular momentum. The total z component of the angular momentum of the plasma is

$$L_z = m(\Omega/2-\omega)N\langle r^2\rangle. \qquad (3)$$

Here N is the total number of ions and $\langle r^2 \rangle$ is the mean-squared radius of the plasma. For most of the work described in this paper $\omega \ll \Omega$ and

$$L_z \approx \frac{m\Omega N}{2} \langle r^2 \rangle > 0. \qquad (4)$$

Therefore the total angular momentum is dominated by the field angular momentum. Suppose the cooling laser beam is directed normal to the z axis but at the side of the plasma which is receding from the laser beam due to the plasma rotation. Because the rotation of the positive ions is in the $-\hat{\theta}$ direction, the torque of the laser on the ions will also be negative. Consequently, angular momentum is removed from the plasma and according to Eq. (4) the radius of the plasma must decrease. In general, the plasma is compressed until the torque due to the cooling laser is balanced by another external torque. As the radius decreases, the density of the plasma increases.

Even in the absence of external torques, there is a limit to how far the plasma can be compressed. From Eq. (1), the maximum density, known as the Brillouin density, occurs when the rotation frequency $\omega = \Omega/2$. The Brillouin density is given by

$$n_{max} = \frac{m\Omega^2}{8\pi q^2}.$$

We have recently been able to achieve densities at the Brillouin limit. We have also been able to achieve rotation frequencies $\omega > \Omega/2$ where according to Eq. (1) the ion density decreases. In these experiments, the temperature was not determined. At the magnetic field of 1.92 T used in some of the work discussed here, the Brillouin density is 1.1×10^9 cm^{-3}. This density with the theoretical minimum 0.5 mK temperature, results in a coupling $\Gamma \sim$ 5500. For the work reported here, we have been able to obtain ion temperatures in the 1-10 mK range with densities 5-10 times less than the Brillouin density. This results in couplings Γ of a few hundred.

We measured[8,14,15] the ion density and temperature by using a second laser, called the probe laser, to drive the "depopulation" transition as indicated in Fig. 2. The cooling laser optically pumps the ions into the $^2S_{1/2}$ $m_J=+1/2$ state. The resonance fluorescence (i.e. laser light scattered by the ions) from this transition is used as a measure of the ion population in the $m_J=+1/2$ state. The probe laser drives some of the ion population from the $^2S_{1/2}$ $m_J=+1/2$ state to the $^2P_{3/2}$ $m_J=-1/2$ state where the ions decay with 2/3 probability to the $^2S_{1/2}$ $m_J=-1/2$ state. This causes a decrease in the observed ion fluorescence because the $^2S_{1/2}$ $m_J=-1/2$ state is a "dark" state a (state which does not fluoresce in the cooling laser). The ion temperature is obtained from the Doppler broadening of the resonance lineshape when the probe laser is scanned through the depopulation transition. The ion rotation frequency is measured from the shift in the depopulation transition frequency as the probe laser is moved from the side of the plasma rotating against the laser beam to the side of the plasma rotating with the laser beam. From the measured rotation frequency, the density is calculated from Eq. (1). The measured density and temperature is then used to calculate the coupling Γ.

OBSERVED CORRELATIONS

With measured couplings $\Gamma > 100$, we anticipate that the ions will exhibit correlated behavior. If the number of stored ions is large enough for infinite volume behavior, the ions may be forming a bcc lattice.[10] Until now we have cooled and looked for spatial correlations with up to 15 000 Be$^+$ ions stored in the Penning trap of Fig. 1. A currently unanswered question is how many stored ions are required for infinite volume behavior, i.e. the appearance of a bcc lattice for $\Gamma > 178$. For a finite plasma consisting of a hundred to a few thousand ions, the boundary conditions are predicted to have a significant effect on the plasma state. Simulations involving these numbers of ions in a spherical trap potential predict that the ion cloud will separate into concentric spherical shells.[2-4] Instead of a sharp phase transition, the system is expected to evolve gradually from a liquid state characterized by short-range order and diffusion in all directions, to a state where there is diffusion within a shell but no diffusion between the shells (liquid within a shell, solid-like in the radial direction), and ultimately to an overall solid-like state.[4] These conclusions should apply to a nonspherical trap potential as well if the spherical shells are replaced with shells approximating spheroids. Independent theoretical investigations[16,17] of the nonspherical case support this conjecture.

We have observed shell structures with ^{9}Be$^+$ ions stored in a Penning trap by imaging the laser induced fluorescence from the cooling transition. This technique is sensitive enough to observe the structures formed with only a few ions in a trap.[18-21] About 0.04% of the 313-nm fluorescence from the decay of the $^2P_{3/2}$ state was focused by f/10 optics onto the photocathode of a resistive-anode photon-counting imaging tube (see Fig. 1). The imager was located along the z axis, about 1 m from the ions. The imaging optics was composed of a three-stage lens system with overall magnification of 27 and a resolution (FWHM) of about 5 µm (specifically, the image of a point source when referred to the position of the ions was approximately 5 µm in diameter). Counting rates ranged from 2 to 15 kHz. Positions of the photons arriving at the imager were displayed in real time on an oscilloscope while being integrated by a computer. The probe laser could be tuned to the same transition as the cooling laser and was directed through the cloud perpendicularly to the magnetic field. With the probe laser turned on continuously, the cooling laser could be chopped at 2 kHz (50% duty cycle) and the image signal integrated only when the cooling laser was off. Different portions of the cloud could then be imaged by the translation of the probe beam, in a calibrated fashion, either parallel or perpendicular to the z axis. Images were also obtained from the ion fluorescence of all three laser beams.

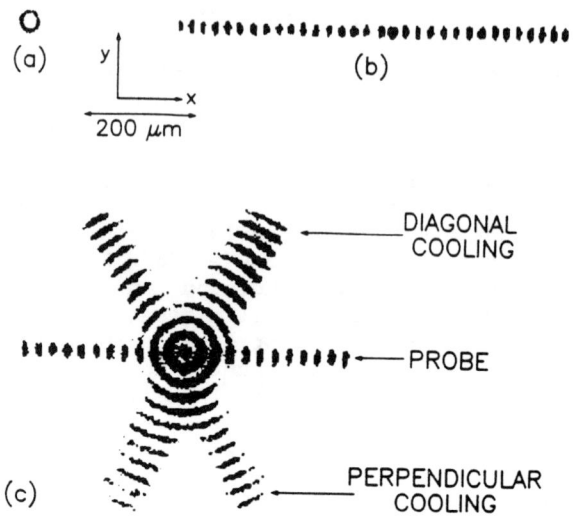

Fig. 3 Images of shell structures obtained with B = 1.92 T. (a) A single shell in a cloud containing approximately 20 ions. Trap voltage V_0 = 14 V and cloud aspect ratio a_r (axial length/diameter) $\simeq$ 6.5. This image was obtained from the ion fluorescence of the perpendicular and diagonal cooling beams. (b) Sixteen shells (probe-beam ion fluorescence only) in a cloud containing about 15 000 ions with V_0 = 100 V and $a_r \simeq$ 0.8. (c) Eleven shells plus a center column in the same cloud as (b), with V_0 = 28 V and $a_r \simeq$ 2.4. This image shows the ion fluorescence from all three laser beams. Integration times were about 100 s for all images.

We have observed shell structure in clouds containing as few as 20 ions (one shell) and as many as 15 000 ions (sixteen shells). Images covering this range are shown in Fig. 3. Even with 15 000 ions in the trap there is no evidence for infinite volume behavior. We measured the coupling constant Γ for several clouds containing about 1000 ions. Drift in the system parameters was checked by verifying that the same images were obtained before and after the cloud rotation frequency and ion temperatures were measured. Figure 4 shows examples of shell structures at two different values of Γ. The first image is an example of high coupling ($\Gamma \approx$ 180) and shows very good shell definition in an intensity plot across the cloud. The second image is an example of lower coupling ($\Gamma \approx$ 50) and was obtained with cooling only perpendicular to the magnetic field. Variations in peak intensities equidistant from the z axis are due to signal-to-noise limitations and imperfect alignment between the imager x axis and the probe beam.

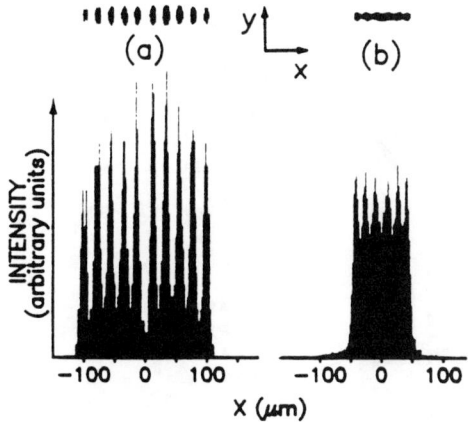

Fig. 4 Intensity plots along the imager x axis (approximately parallel to the probe beam) through the center of the ion cloud with corresponding images (above). (a) $\Gamma = 180^{+90}_{-70}$ ($T = 6^{+4}_{-2}$ mK, $n_0 \simeq 7 \times 10^7$ ions cm^{-3}). Cloud aspect ratio $a_r \simeq 3.5$. (b) $\Gamma = 50^{+30}_{-20}$ ($T = 33^{+17}_{-13}$ mK, $n_0 = 2 \times 10^8$ ions cm^{-3}), $a_r \simeq 5$. The clouds contained about 1000 ions and $B = 1.92$ T in both cases.

We obtained three-dimensional information on the shell structure by taking probe images at different z positions; two types of shell structure were present under different circumstances. The first type showed shell curvature near the ends of the cloud, indicating that the shells may have been closed spheroids. Shell closure was difficult to verify because of a lack of sharp images near the ends of the cloud where the curvature was greatest. This may have been due to the averaging of the shells over the axial width of the probe beam. In the other type of shell structure, it was clear that the shells were concentric cylinders with progressively longer cylinders near the center. An example of these data is shown in Fig. 5. Other evidence for cylindrical shells was obtained from the observation that shells in the diagonal-beam images occurred at the same cylindrical radii as those from the perpendicular beams. This can be seen in the three-beam images such as that shown in Fig. 3(c). Systematic causes of these two different shell configurations have not yet been identified.

One comparison which can be made between the theoretical calculations and our experimental results is the relationship between the number of shells and the number of ions, N, in a cloud. For a spherical cloud, two independent approaches[2,22] estimate $(N/4)^{1/3}$ and $(3N/4\pi)^{1/3}$ shells. For the nearly spherical cloud of Fig. 3(b) ($N \simeq 15\,000$), these formulae predict

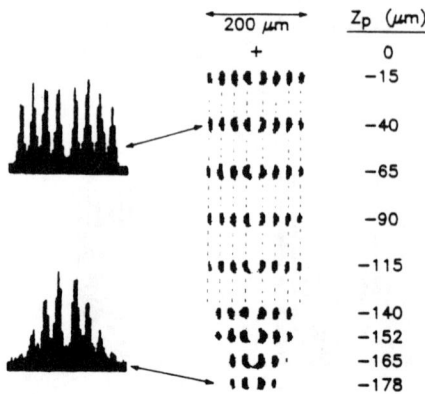

Fig. 5 Data showing evidence for concentric cylindrical shells. On the right is a series of images obtained with the probe beam for different z positions z_p of the probe beam (lower half of the cloud only). Intensity plots for z_p = -40 μm and z_p = -178 μm are shown on the left. The cloud aspect ratio a_r was about 1.9 and B ≃ 1.92 T.

15.5 and 15.3 shells and we measure 16. At present, it is difficult to make further quantitative comparisons between our data and the theoretical calculations. For example, there is substantial uncertainty in our measurement of Γ due to uncertainty in the temperature measurement. Our data do agree qualitatively with the simulations with the exception, in some cases, of the presence of an open-cylinder shell structure as opposed to the predicted closed spheroids. Shear (that is, different rotation frequencies) between the shells could possibly account for this discrepancy. In our experiment, shear could be caused[8] by differential laser torque or the presence of impurity ions. For the data here, we have determined that the rotation frequency does not vary by more than 30% across the cloud.

ION DIFFUSION

The probe laser can be used to optically tag ions and observe the ion diffusion.[14] With the probe laser tuned to the depopulation transition (see Fig. 2), ions in the path of the probe laser beam are put into the "dark" m_J=-1/2 ground state. These ions will not fluoresce when they pass through the cooling laser until they are optically pumped back into the m_J=+1/2 ground state. This repumping time is typically on the order of 1.0 s. By pulsing the probe laser on and measuring the length of time it takes the dark ions to diffuse from the probe beam to the cooling laser beam, it should be possible to measure the ion diffusion. By directing the probe laser beam to the radial edge of the plasma so that only the outer shell is intersected by the probe beam, it should be possible to observe the diffusion of ions in the radial direction or between shells. By directing the probe beam to the axial edge of the cloud it should be possible to observe the diffusion of the ions in the axial direction or within a shell. According to the simulations of Dubin and O'Neil,[4] for intermediate values of the coupling ($\Gamma \sim 100$) we expect to observe that the diffusion between shells is much slower than the ion diffusion within a shell (solid-like behavior between shells, liquid-like within a shell). As the temperature is lowered and the coupling Γ increases, the diffusion within a shell should smoothly slow down. At high enough couplings (i.e. $\Gamma > 400$) the diffusion should be very slow both between and within shells, indicative of solid-like behavior.

We have qualitatively observed the ion diffusion at intermediate values of Γ ($\Gamma \sim 100\text{-}200$). We observed that the diffusion of ions between shells is slow compared to the optical repumping time (~ 1 s) but that the diffusion within a shell (i.e. from the axial end of a shell to the $z = 0$ plane) is fast compared to this repumping time. We have also observed states with higher couplings (the couplings Γ were not measured) where the diffusion of ions both between and within a shell was slow compared to the optical repumping time. In the future we plan to make quantitative measurements of the ion diffusion.

ACKNOWLEDGEMENT

We gratefully acknowledge the support of the U.S. Office of Naval Research and the Air Force Office of Scientific Research. We thank M. Raizen and F. Moore for carefully reading the manuscript.

REFERENCES

1. S.L. Gilbert, J.J. Bollinger, and D.J. Wineland, Phys. Rev. Lett. **60**, 2022 (1988).

2. A. Rahman and J.P. Schiffer, Phys. Rev. Lett. **57**, 1133 (1986); J.P. Schiffer, Phys. Rev. Lett. **61**, 1843 (1988).

3. H. Totsuji, in Strongly Coupled Plasma Physics, eds. F.J. Rogers and H.E. DeWitt (Plenum, New York, 1987) pp. 19-33.

4. D. Dubin and T. O'Neil, Phys. Rev. Lett. **60**, 511 (1988).

5. C.F. Driscoll, J.H. Malmberg, and K.S. Fine, Phys. Rev. Lett. **60**, 1290 (1988); J.H. Malmberg and J.S. deGrassie, Phys. Rev. Lett. **35**, 577 (1975).

6. T.M. O'Neil, in Non-Neutral Plasma Physics, eds. C.W. Roberson and C.F. Driscoll (American Institute of Physics, New York, 1988) pp. 1-25.

7. J.H. Malmberg and T.M. O'Neil, Phys. Rev. Lett. $\underline{39}$, 1333 (1977).

8. L.R. Brewer, J.D. Prestage, J.J. Bollinger, W.M. Itano, D.J. Larson, and D.J. Wineland, Phys. Rev. A$\underline{38}$, 859 (1988).

9. D.J. Wineland, J.J. Bollinger, W.M. Itano, and J.D. Prestage J. Opt. Soc. Am. B$\underline{2}$, 1721 (1985).

10. S. Ichimaru, H. Iyetomi, and S. Tanaka, Phys. Rep. $\underline{149}$, 91 (1987) and references therein.

11. D.J. Wineland and W.M. Itano, Phys. Rev. A$\underline{20}$, 1521 (1979).

12. W.M. Itano and D.J. Wineland, Phys. Rev. A$\underline{25}$, 35 (1982).

13. D.J. Wineland and W.M. Itano, Phys. Today $\underline{40}$(6), 34 (1987); S. Stenholm, Rev. Mod. Phys. $\underline{58}$, 699 (1986).

14. L.R. Brewer, J.D. Prestage, J.J. Bollinger, and D.J. Wineland, in Ref. 3), pp. 53-64.

15. J.J. Bollinger and D.J. Wineland, Phys. Rev. Lett. $\underline{53}$, 348 (1984).

16. J.P. Schiffer, in *Proceedings of the Workshop on Crystalline Ion Beams, Wertheim, Germany, 1988*, ed. by R.W. Hasse, I. Hofmann, and D. Liesen (GSI Report 89-10, GSI Darmstadt) p. 2-32.

17. D.H.E. Dubin, Dept. of Physics, UCSD, La Jolla, CA, private communication.

18. F. Diedrich, E. Peik, J.M. Chen, W. Quint, and H. Walther, Phys. Rev. Lett. $\underline{59}$, 2931 (1987).

19. D.J. Wineland, J.C. Bergquist, W.M. Itano, J.J. Bollinger, and C.H. Manney, Phys. Rev. Lett. $\underline{59}$, 2935 (1987); D.J. Wineland, W.M. Itano, J.C. Bergquist, S.L. Gilbert, J.J. Bollinger, and F. Ascarrunz, in Ref. 6), pp. 93-108.

20. J. Hoffnagle, R.G. DeVoe, L. Reyna, R.G. Brewer, Phys. Rev. Lett. $\underline{61}$, 255 (1988).

21. Th. Sauter, H. Gilhaus, I. Siemers, R. Blatt, W. Neuhauser, P.E. Toschek, Z. Phys. D$\underline{10}$, 153 (1988).

22. D.H.E. Dubin, Phys. Rev. A$\underline{40}$, 1140 (1988).

CRITICAL QUANTITIES FOR SOLAR SYSTEM SCIENCE IN ATOMIC AND MOLECULAR REACTIONS

D.E. Shemansky
Lunar and Planetary Laboratory
University of Arizona

ABSTRACT

This paper reviews some of the more critical issues in planetary science that depend on our determination of basic reaction parameters. The discussion is not meant to be comprehensive, but three research domains have been selected because of their distinct dependence for advancement on basic physical and chemical properties of common atomic and molecular species. (1) The giant outer planets, with atmospheres dominated by hydrogen all show temperatures at the top of the thermospheres hundreds of degrees higher than expectation on the basis of solar energy deposition. The search for an explanation is a current major issue in atmospheric research. One of the most serious problems in this pursuit is in the determination of basic ionospheric physical chemistry; the identification of the major ion species. We are in need of rates for reactions critical to this issue. (2) The sulfur/oxygen plasma torus at the orbit of Jupiter's satellite Io dominates the magnetosphere, and has been a persistent relatively stable feature for at least a decade. The mechanisms by which energy is inserted to maintain the system, have remained unresolved since the discovery of the system. Heterogenous sources must be invoked in order to explain the plasma temperature. The central problem in determining radial energy input distributions is in the conflict that has developed recently between calculated charge exchange rate coefficients, observed ion partitioning, and calculated diffusion rates. It appears that we need an accurate determination of charge exchange and dielectronic recombination rate coefficients in order to advance our understanding of the system. (3) Recent observations of the Moon and Mercury have revealed the presence of the minor elements sodium and potassium. One of the consequences of this remarkable discovery is the realization that a major cosmic element, oxygen, has never been detected on the Moon, although the known loss processes for sodium and potassium are substantially faster than for oxygen. The discrepancy is at least a factor of 100 in predicted abundance, and we conclude that the solution must lie with preferential loss processes in the interaction of oxygen and its' parent species with the solid surface. An investigation of possible processes is needed.

INTRODUCTION

One of the more frequent obstacles to progress in planetary atmospheric and magnetospheric research is lack of absolute quantities for rate processes, or even a lack of knowledge of the existence of reaction paths that may be critical to the state of the system. I briefly discuss three current areas of research that appear to be limited by what we know of basic atomic and molecular physics and physical chemistry.

The first concern is with the hydrogen atmospheres of the giant outer planets. All of these planets, now including Neptune with the recent Voyager spacecraft flyby[1], have upper thermospheric temperatures far above expected values calculated from estimates of solar deposition. Predictions were spectacularly in error when the Voyager occultation experiments obtained direct measurements of the expanded thermospheres. This is an exceedingly important fact, since there is no doubt about the integrity of the observations, and the physical quantity points directly to the presence of major heating processes driven by mechanisms that we do not understand. The thermospheric temperature structure on Jupiter has received more attention mainly because the very high value at the top of the thermosphere (T $\sim$ 1100 K) determined at Voyager encounter in 1979, came as a complete surprise. Heating can be obtained from inertia-gravity wave propagation (see Ref 2) at levels in excess of solar input, but dissipation is concentrated just above the homopause, and the predicted temperature profile and magnitude is in contradiction to observation (see Ref. 3). Hunten and Dessler[4] have discussed the precipitation of soft protons and electrons as a source, but the heating rate is deficient by at least an order of magnitude. Auroral activity on Jupiter deposits substantial energy[5] but deposition is in or near the homopause and does not contribute significantly to the heating of the thermosphere.

In the exploration of possible sources of energy deposition, and the formation of the ionosphere we must have a good basic understanding of the physical chemistry of hydrogen. Some critical reactions that do not have satisfactory rate determinations will be discussed.

A second object of considerable interest that is not at all clearly understood is the plasma torus surrounding Jupiter at the orbit of the satellite Io. The plasma is mainly composed of sulfur and oxygen ions, with lesser amounts of sodium, potassium, hydrogen(protons), and other trace species. The system loses energy by radiation alone at a rate of $\sim 4 \times 10^{12}$ Watts. Energy and mass are also lost through charge exchange with the neutral species that mix with the plasma from the source. The torus is confined by Jupiter's magnetic field, but shows a complicated bimodal and at times trimodal radial structure (see Figure 1). The major feature containing most of the energy in the system, has a peak electron density [e] $\sim$ 2000 cm^{-3} - 10000 cm^{-3} located at 5.75 Jupiter radii (R_J) from the center of the system, with a torus minor radius of about 1.2 R_J ($\sim$ 85000 km). The radiated energy and the electron temperature are too high to be explained by the simple process of ion pickup in the rotating magnetic field[6]. The major source mechanism for the injection of energy into the system has not been satisfactorily identified although possible processes have been proposed[6,7]. In attempting to determine basic Io torus plasma properties we have found a several point failure in our ability to model the system. Diffusive loss processes are generally difficult to establish quantitatively, and the Io torus is no exception. The lack of definitive information on the distribution of neutral species in the plasma compounds the problem. As a result there are too many ill defined parameters to determine the source of discrepancies between model and observations. There is a current conflict between recent calculations of charge exchange reaction rates[8] and observed partitioning of the plasma species for reasonable ranges of the other adjustable parameters. However, it appears there is now another uncertainty in the rates for dielectronic recombination, particularly for sulfur species, leaving the necessity for some careful research on basic atomic properties in order to allow progress to be made on the most fundamental aspects of the torus.

The third area of growing interest is in atmospheric evolution of planetary bodies such as Mercury, the Moon and icy satellites of the outer planets, in which conditions are such that gas densities are too low to establish homogeneous collisional redistribution. The atmospheres are controlled entirely by collisions with the surface, and the characteristics of source processes. The atmospheres of the Moon and Mercury are of particular interest because of the recent discovery of sodium and potassium components (see for example Ref. 9). The physical chemistry of gas-surface reactions for these species and others such as atomic oxygen and OH with source rates expected to be much larger than the alkali metals, needs to be understood and determined quantitatively in order to obtain basic understanding of the controlling evolutionary processes.

THE HYDROGEN ATMOSPHERES OF THE OUTER PLANETS

The role of H_3^+ in the outer planet excited atmospheres remains a critical issue in ionospheric, dayglow and auroral research. The understanding of ionospheric physics, if anything has declined in the past few years because of uncertainties that have developed in the physical chemistry of H_3^+. Our interest has been piqued recently by the observations of Jupiter auroral emission[10] for the first time in the $2\nu_2$ bands of H_3^+ in the 4800 cm^{-1} region. It is well known that the primary direct reactions on electron excitation of hydrogen are in the following sequence. The electrons ionize H_2 to produce H_2^+, in a vibrationally excited state. $H_2^+(*)$ immediately reacts with H_2 to form H_3^+, also vibrationally excited. This sequence is shown as reactions R1 and R3a,b in Table 1.

TABLE 1
HYDROGEN REACTIONS

	Reactions	κ ^{a}(10^{-9} cm^3 s^{-1})	
R1	$e + H_2 \rightarrow H_2^+(*) + 2e$	38.	b
R2	$e + H_2 \rightarrow H + H^+ + 2e$	1.9	c
R3a	$H_2 + H_2^+ \rightarrow H_3^+(*) + H$	2.	d
R3b	$H(2s) + H_2 \rightarrow H_3^+ + e$	4.	e
R3c	$H(2s) + H_2 + H_2 \rightarrow H(2p) + H_2$	3.	e
R4	$e + H_3^+(*) \rightarrow H_2 + H(nl)$	46.	f
R5	$e + H_3^+(*) \rightarrow 3H(1s)$	109.	f
R6	$H_3^+(*) \rightarrow H_3^+(0) + h\nu$		g
R7	$e + H_3^+(0) \rightarrow H_2 + H$	?	h
	H$^+$ LOSS MECHANISMS		
R8	$e + H^+ \rightarrow H(1s) + h\nu$	2.2 x 10^{-3}	i
R9	$H^+ + H_2 + M \rightarrow H_3^+(0) + M$ for $[H_2] > 10^{11}$ cm^{-3}		j
R10	$H^+ + H_2(v \geq 4) \rightarrow H_2^+ + H$	?	
	H$_3^+$ CHEMISTRY		
R11	$H_3^+ + H_2 \rightarrow H_5^+ + h\nu$	?	
R12	$H_3^+ + H_2 + M \rightarrow H_5^+ + M$	3.2 x 10^{-30} cm^6 s^{-1}	k
R13	$H_5^+ + e \rightarrow nH_2 + mH$	?	

^{a} $T_e = 6. \times 10^5$ K (R1) - (R5)
$T_e = 10^3$ K (R7) - (R8)

^{b} M. Alexander, Chem. Phys., **20**, 83 (1977); L. J. Kieffer, and G. H. Dunn, Rev. Mod. Phys., **38**, 1 (1966).

^{c} A. Crowe, and J. W. McConkey, J. Phys. B: At. Mol. Phys., **6**, 2088 (1973); see C. Bottcher, J. Phys. B: At. Mol. Phys., **7**, L352 (1974); A. U. Hazi, J. Phys. B: At. Mol. Phys., **8**, L262 (1975).

^{d} H. Eyring, H. Gershinowitz, and C. Sun, J. Chem. Phys., **3**, 786 (1935); W. T. Huntress, Adv. At. Mol. Phys., **10**, 295 (1974).

^{e} J. E. Mentall, and P. M. Guyon, J. Chem. Phys., **67**, 3845 (1977).

^{f} J. B. Mitchell, J. L. Forand, C. T. Ng, D. P. Levac, R. E. Mitchell, P. M. Mul, W. Claeys, and A. Sen, Phys. Rev., **51**, 885 (1983); see D. Auerbach, R. Cacak, R. Caudano, T. D. Gaily, C. J. Keyser, J. McGowan, J. Mitchell, and S. F. Wilk, J. Phys. B: At. Mol. Phys., **18**, 3797 (1977); B. Peart, and K. T. Dolder, J. Phys. B: At. Mol. Phys., **7**, 1567 (1974); D. Mathur, S. Khan, and J. Hasted, J. Phys. B: At. Mol. Phys., **11**, 3615 (1978).

^{g} G. Carney, and R. Porter, J. Chem. Phys., **65**, 3547 (1976).

^{h} N. G. Adams, and D. Smith, In Astrochemistry, (IAU 120th symposium, D. Reidel., 1987); D. Smith, and N. G. Adams, Astrophys. J., **284**, L13 (1984); T. Amano, Astrophys. J., **329**, L121 (1988); H. Hus, F. Youssif, A. Sen, and J. B. A. Mitchell, Phys. Rev. A., **38**, 658 (1988).

^{i} R. L. Brown, and W. G. Mathews, Astrophys. J., **160**, 939 (1970).

^{j} R. Johnsen, and M. A. Bondi, Icarus, **23**, 139 (1974).

^{k} K. Hiraoka, and P. Kebarle, J. Chem. Phys., **62**, 2267 (1975).

The reaction rates for the vibrationally excited components are known approximately (although we do not have level by level discrimination) but in the ionosphere in particular, where densities are of the order of 10^8 cm^{-3} or 10^9 cm^{-3}, the H_3^+ can relax to the ground state through radiation with a time constant of 1. s or less. If the recombination rate of the H_3^+ remained about the same for the ground vibrational level the ionosphere would be dominated by H^+. However, there are strong indications that the recombination rate of $H_3^+(0)$ could be exceedingly slow and we are left with the problem of trying to establish the identity of the major ion. McConnell and Majeed[11] point out that if the much smaller rate for $H_3^+(0)$ is used in ionospheric calculations for Jupiter the predicted electron densities are far in excess of measured values. Thus if the argument for a slower recombination rate for $H_3^+(0)$ is correct, physical chemistry or atmospheric dynamics that has not been previously considered must be controlling the ion population. It is clear that this basic ionospheric problem is a major obstacle to progress in thermospheric physics. Table 1 contains a list of the main reactions involved in excited hydrogen atmospheres. Most of those reactions with known rates given in the list are referenced by Shemansky[12], with exceptions noted in Table 1. Thus the rate for the recombination reaction

$$e + H_3^+(0) \rightarrow H_2 + H, \tag{R7}$$

is a critical quantity for outer planet atmospheric science. The theoretical argument for a low rate for R7 has been presented by Michels and Hobbs[13]. R7 is endothermic with no apparent available curve crossings, and H_3^+ must be in at least the (1,0,0) level to allow fast recombination with cool electrons. A rate for R7 is difficult to establish experimentally because the reactions R3a,b populate the higher vibrational levels of H_3^+ and there is generally insufficient time in beam experiments to allow for relaxation to $H_3^+(0)$. More specialized techniques have been applied recently in attempts to determine the R7 rate, but the results are not in agreement and appear to be dependent on experimental method with no clear explanation for the differences. Flowing afterglow experiments by Smith and Adams[14] and Adams and Smith[15] have produced values

$$\kappa_{R7} \leq 2 \times 10^{-8} \text{ cm}^3 \text{ s}^{-1} \text{ and}$$
$$\kappa_{R7} \leq 1 \times 10^{-11} \text{ cm}^3 \text{ s}^{-1}$$

respectively. However, Amano[16] has measured the decay in a hollow cathode discharge cell using a laser absorption technique to obtain

$$\kappa_{R7} = 1.8 \times 10^{-7} \text{ cm}^3 \text{ s}^{-1},$$

and Hus et al.[17] have obtained

$$\kappa_{R7} = 2 \times 10^{-8} \text{ cm}^3 \text{ s}^{-1}$$

in a merged beam experiment. The reasons for this several orders of magnitude disagreement need to be explored and resolved. The invlovement of impurities and electric fields have been suggested. There is general agreement on the rates for vibrationally excited H_3^+ recombination (R4, R5, Table 1).

If the rate for R7 is as low as the Adams and Smith[15] and theoretical estimates[13] suggest then H_3^+ will very likely be the major ion in the outer planet ionospheres[11]. Other reactions that may be competative in influence on the H_3^+ population then must also be considered. Reaction with H_2 to form H_5^+ and subsequent physical chemistry will need to be quantified (R11 - R13, Table 1). The consequences of electron and collisional excitation of H_3^+ will also have to be considered. The evaluation of the competition between H_3^+ and H^+ as positive ions in the ionosphere also requires the determination of the rate coefficient for R10 (Table 1) as a sink for H^+. The rate processes that control the production of $H_2(v > 0)$ must also be quantified in order establish the importance of R10. Good progress has been made on low energy (< 5 eV) electron cross sections for ro-vibrational excitation of H_2X but there is a need for further work at higher energies[18]. The physical chemistry of vibrationally excited H_2 under the current circumstances needs a thorough review, in order to establish the best quantitative estimates of deactivation rates through atom-atom exchange[19], vibration- translation transfer (see for example Ref. 20), proton collisions [21], and dissociation reactions such as the electron induced transfer to the repulsive $^3\Sigma_u$ state from $H_2(*)$[22].

THE IO PLASMA TORUS

The three distinctive regions of the Io plasma torus shown in the schematic Figure 1 pose a serious problem in plasma dynamics. The inner region composed of cold plasma in near thermal equilibrium with electron and ion temperatures $T_e \sim T_i \sim 10^4$ K is separated from the non LTE hot outer region by a transition region at about 5.6 R_J with a width of $< .1$ R_J leading to temperatures of $T_e \sim 8. \times 10^4$ K and $T_i \sim 10^6$ K at 5.7 R_J. The energy and mass involved in this system, as discussed above is a major perturbation on Jupiter's magnetospheric dynamics, loading the entire region to the bow shock at ~ 100 R_J with heavy ions.

Diffusion across the transition region must be very slow because the structure is temporally stable and cannot be attributed to even a slow transient event. However, in order to explain the power dissipation in the system it has been argued that outward diffusion from the hot region must be fast, so that ions picked up in the rotating magnetic field may supply the required energy (see for example Ref 23,24). Recent calculations[6] have indicated that sulfur ions radiate too efficiently to make this a viable explanation. Furthermore the structure necessary for a high diffusion rate driven by convective turbulence appears not to be present according to recent analysis of in situ plasma data[25]. There are also theoretical arguments suggesting that flux tube interchange in the torus is not viable[26]. It appears that energy must be supplied from a source external to the plasma in order to maintain the system. However, model calculations[27] based on collisonal diffusion have encountered difficulty in matching observed plasma parameters, particularly with the introduction of recent improved calculations of charge exchange reaction rates[8]. A lack of definitive observational data, particularly for the distribution of neutrals in the plasma volume has prevented the determination of the source of the discrepancies. The errors in the modeling process most likely stem from one or more of four areas; deficient diffusion theory, inaccurate ion-ion reaction rates, inaccurate dielectronic recombination rates, or through the exclusion of unrecognized significant reactions.

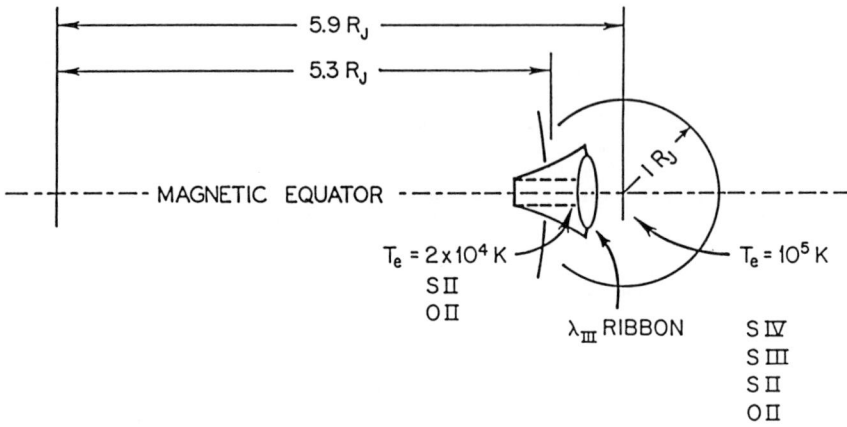

Figure 1: Io Plasma Torus

The Figure shows a schematic representation of observed radial and latitudinal distribution in the Io plasma torus surrounding Jupiter. The outer region centered at $\approx 5.75\ R_J$ is relatively hot with a large radiative emission rate maintaining a substantial separation between electron and ion temperatures. A field aligned feature in the hot torus described as a ribbon in the Figure has a substantially larger density in a very confined region in the radial direction and a limited extent in magnetic longitude. A cold inner region peaking in density near 5.3 R_J shows electrons in thermal equilibrium with singly ionized oxygen and sulfur.

Table 2 shows a list of the major reactions included in the modeling process. The basic tendency of the model calculations is to suppress the population of the lower ionized sulfur species relative to reduced observational data obtained from both spacecraft and ground based experiments. The model thus has the tendency to make the sulfur ion partitioning appear to be at a higher apparent temperature than the reality of the observations. The solution to this problem is to either invalidate the theory controlling the diffusion characteristics by postulating higher diffusion rates driven by turbulent processes, argue that the rates for reactions R25, R29, R31 - R33, and R36[8] are for some reason too large, artificially suppressing SII, or that dielectronic recombination rates for the sulfur ions are too low relative to reality.

TABLE 2
IO PLASMA TORUS REACTIONS

Reaction		$\overleftrightarrow{\kappa}$	$\overleftarrow{\kappa}$	$\overrightarrow{\kappa}$
			10^{-9} cm^3 s^{-1}	
R14	$O + e \leftrightarrow O^+ + 2e$	1.7 [a]	4.1×10^{-3} [b]	
R15	$O^+ + e \leftrightarrow O^{2+} + 2e$	2.1×10^{-2} [a]	5.0×10^{-3} [b]	
R16	$O^{2+} + e \leftrightarrow O^{3+} + 2e$	4.6×10^{-4} [a]	1.0×10^{-2} [b]	
R17	$O^{3+} + e \leftrightarrow O^{4+} + 2e$	1.4×10^{-6} [a]	3.1×10^{-2} [b]	
R18	$S + e \leftrightarrow S^+ + 2e$	12.0 [a]	3.9×10^{-2} [b]	
R19	$S^+ + e \leftrightarrow S^{2+} + 2e$	.43 [a]	4.1×10^{-2} [b]	
R20	$S^{2+} + e \leftrightarrow S^{3+} + 2e$	2.5×10^{-2} [a]	.11 [b]	
R21	$S^{3+} + e \leftrightarrow S^{4+} + 2e$	1.0×10^{-3} [a]	.19 [b]	
R22	$S^{4+} + e \leftrightarrow S^{5+} + 2e$	2.4×10^{-5} [a]	.20 [b]	
R23	$O + O^+ \rightarrow O^+ + O$	12.0 [d]		13.2 [e]
R24	$O + S^+ \rightarrow O^+ + S$	0.06 [d]		.09 [e]
R25	$S + O^+ \rightarrow S^+ + O$	8.0 [d]		3.0 [e]
R26	$S + S^+ \rightarrow S^+ + S$	16.0 [d]		24.0 [e]
R27	$O^{2+} + S \rightarrow S^+ + O^+$	11.0 [d]		23.4 [e]
R28	$O^{2+} + S \rightarrow S^{2+} + O^+ + e$	3.0 [d]		16.2 [e]
R29	$S^{2+} + O \rightarrow S^+ + O^+$	6.0 [d]		2.3 [e]
R30	$O^{2+} + O \rightarrow 2O^+$	0.2 [d]		.52 [e]
R31	$S^{2+} + S \rightarrow 2S^+$	0.6 [d]		0.3 [e]
R32	$S^{3+} + O \rightarrow S^{2+} + O^+$	4.4 [d]		19.2 [e]
R33	$S^{3+} + O \rightarrow S^+ + O^{2+}$	5.6 [d]		0.0 [e]
R34	$O^{2+} + O \rightarrow O + O^{2+}$	5.0 [d]		5.4 [e]
R35	$S^{2+} + S \rightarrow S + S^{2+}$	7.0 [d]		7.8 [e]
R36	$S^{3+} + S \rightarrow S^{2+} + S^+$	20.0 [d]		13.2 [e]
R37	$O^{2+} + S^+ \rightarrow O^+ + S^{2+}$	5.6 [d]		1.4 [e]
R38	$O^{2+} + S^{2+} \rightarrow O^+ + S^{3+}$	2.8 [d]		0.9 [e]

[a] $T_e = 6.5\times10^4$ K, [e] = 1500 cm^{-3}; R. A. Brown, D. E. Shemansky, and R. E. Johnson, Astrophys. J., **264**, 309 (1983).

[b] $T_e = 6.5\times10^4$ K, [e] = 1500 cm^{-3}; E. W. Davis, Diffusion in the Io Plasma Torus and it's Relation to the Torus Spatial Structure, (Ph.D. Thesis, Univ. Arizona, 1990).

[c] R. F. Stebbings, A. C. Smith, and H. Ehrhardt, In Atomic Collision Processes (North-Holland, 1964) 814.

[d] J. E. Johnson, and D. F. Strobel, J. Geophys. Res., **87**, 10385 (1982); R. A. Brown, D. E. Shemansky, and R. E. Johnson, Astrophys. J., **264**, 309 (1983).

[e] M. A. McGrath, and R. E. Johnson, J. Geophys. Res., **94**, 2677 (1989).

If the error in the model lies mostly with the calculated diffusion properties, our understanding of the physics of the system will be significantly affected. The determination of diffusion parameters in most plasma systems is notoriously difficult and generally requires the support of complete species partitioning and kinetic parameters. However, as discussed above the critical observational quantities, mainly the neutral population distribution, are not observationally well determined and we must rely on establishing an accurate atomic data base. Possibly the most critical of the atomic parameters is the set of charge exchange and charge changing reactions that significantly affect estimates of the rate of mass transfer through the the system, as well as ion partitioning. Apart from O^+ symmetric charge transfer[28], no experimental data is available for the sulfur/oxygen plasma considered here. Theoretical calculations by Johnson and Strobel[29] and more recently by McGrath and Johnson[8] are the only source of quantitative rates. The reactions and rate coefficients are given in Table 2 (R23 - R38). Both sets of values are based on variations of the impact parameter method[30]. The latest values[8] embodied improvement in the treatment of electron spin states and should on this basis provide more accurate rates. However, the changes in absolute quantities from the last set of values have uniformly tended to suppress S^+ ions[27] (Table 2, R25, R29, R31 - R33, R36). The changes in magnitude for the critical reactions between the two calculations are as large as a factor of 3, and in one case (R33) a formerly finite rate was suppressed to zero. It is admittedly difficult to estimate the accuracy of the calculated rates[29] and the best method of verification is calculation by other methods and experimental measurement. The energy involved in the ion-neutral reactions is determined by the bulk motion of the ions, ~ 500 eV and ~ 250 eV for sulfur and oxygen. The ion-ion reactions have more uncertain kinetics because the ion temperatures are not observationally separable, but rates are needed over the range $T_i \sim 10^4 - 2 \times 10^6$ K.

The ionization rates for the sulfur species shown in Table 2 are not as well established as for oxygen but it appears that uncertainties are not as large as for dielectronic recombination. The model calculations by Shemansky[6] have used dielectronic recombination rates calculated using the complete atomic structures, but through the application of the Burgess[31] semi empirical method. According to the extensive theoretical review by Bell and Seaton[32] the Burgess[31] method is adequate at relatively high electron temperatures. However, the Io plasma torus hot outer region shows temperatures in the range $T_e = 5 - 8 \times 10^4$ K. Storey[33] and Nussbaumer and Storey[34] have argued that theoretically low lying resonance states play a role in significantly enhancing recombination at temperatures in the $T_e \sim 10^4$ K range. Nussbaumer and Storey[34] have provided calculations of the rates for oxygen ions, but equivalent calculations for sulfur appear not to have been published. Table 2 gives the current values for these rates in the model. The effect of higher recombination rates for oxygen is not critical, but given the possibility of low diffusive loss rates, the partitioning of the sulfur species may be critically affected. Plasma conditions in this case[6] are such that the major impact of dielectronic recombination is to control ion partitioning, but emission from the process is not competative with direct electron excitation.

EXOSPHERIC BODIES: GAS - SURFACE INTERACTION

The solar system bodies such as Mercury, the Moon, and the icy outer planet satellites do not support what is ordinarily considered to be viable atmospheres, since the densities are significantly lower than any laboratory hard vacuum. However the state of the exospheric gas is important since it may contain information on the evolutionary history of the body, as well as current interaction with the solar or magnetospheric environment. Sodium and potassium were observed by Potter and Morgan[35,36] in 1985 and 1986 on Mercury and in 1987 on the Moon[9,37]. The remarkable aspect of these observations is the fact that the abundance of sodium on Mercury is comparable to that of oxygen, and on the Moon oxygen has never been observed[38]. Since oxygen is cosmically more abundant than the alkali metals, and source processes are expected to supply dissociated oxygen with about the same efficiency as the metals, it is surprising that the relative atmospheric concentrations are on the same scale[38]. The low relative abundance of oxygen therefore seems to require a fast loss process. Since homogeneous gas collisions do not occur the loss must be associated with a combination of gas-surface interaction, photoionization, and possible kinetics of the source mechanisms. According to experimental evidence (see Ref 38), the loss of oxygen through source kinetics is small. Comet and meteorite input does not produce gas at temperatures high enough for escape, and the dissociation of H_2O or OH does not release sufficient kinetic energy to the heavy product to allow escape. The processes of desorption from solar wind and photon deposition also do not produce energetic products (see Ref 39). Since the photoionization rate for oxygen is slow, the possible gas-surface interaction processes need to be explored for loss mechanisms.

The reactions produced by the impact of ions, electrons and EUV photons on insulator solids all have similar primary effects, given that impact primary energies are high enough to be in the region where the collision parameters can be represented by the first Born approximation. The simplified physics of the reaction is described by a scenario in which the primary particle or photon penetrates the solid surface, and loses energy through excitation of the electrons in the solid (see Ref 39). A significant branch of the process is excitation into the ionization continuum and the production of secondary electrons. The secondary electrons diffuse through the solid, and some of these pass into the space above the surface. The secondary electron energy is of the order of a few tens of eVs, and very effective in neutral dissociative excitation of the solid molecules. As a result the exothermicity in neutral dissociation of the solid molecules near the surface layer allows the desorption of the atomic product through the surface. Our needs in relation to these processes are to obtain more quantitative data on rates, particularly for interactions with amorphous silicates. The physical chemistry is complex because in astrophysical environments the surfaces are aged in the first few hundred Angstroms by the particle and photon bombardment. The subsequent processes of low energy atom-surface reaction of the desorbed products are of major interest. Relevant research in this subject appears to be mainly in work oriented toward semiconductor physics. The laboratory experimental environment is difficult, because of the limited ability to achieve the required ultra high vacuum. Notable work in this area on amorphous SiO_2 has been reported by Johannessen and Spicer[40] and Carrière and Lang[41]. Carrière and Lang[41] discuss the physical chemistry of the reactions including the effect of trace amounts of carbon, and show evidence of low energy chemistry of the interaction of oxygen with surfaces damaged by the bombardment process. Feuerbacher et al.[42] have measured photoelectron emission from Lunar surface fines, but direct quantitative measurement of the related atomic desorption properities is not available. Lord[43] describes the effects of hydrogen and helium ion implantation in solid olivine and enstatite, but no information was obtained on the impact induced desorption of the solid atoms. A possible loss process by charge transfer on interaction of the gas atom with a negatively charged surface has been suggested by Morgan and Shemansky[38], but there appears to be no relevant experimental evidence in the published literature.

In summary, research on the atmospheres of exospheric planetary bodies is in need of more quantitive information on the desorption properties of amorphous silicates induced by impact of protons(low keV range), other heavier ions such as O^+, electrons and photons, as a function of surface aging and contaminants such as carbon. The use of laser beam scattering techniques should allow definitive work in this experimental discipline. The physical chemistry of low energy gas-surface interaction is also of critical interest particularly for O, OH and H_2 in collision with neutral and negatively charged silicates at temperatures in the range 50 - 800K. H_2 is included here because of interest in accretion disk theory. Table 3 shows a schematic description of the reactions of interest.

TABLE 3
LOW DENSITY GAS-SURFACE REACTIONS OF INTEREST TO SOLAR SYSTEM SCIENCE

R39	$H^+ + (surf)^a$	$\rightarrow$ ⎡
R40	$e + (surf)$	$\rightarrow$ ⎢ dissociated desorbed products
R41	$h\nu + (surf)$	$\rightarrow$ ⎣
R42	$O + (surf)^-$	$\rightarrow O^- + (surf)$
R43	$O + (surf)$	$\rightarrow (surf)^b$
R44	$OH + (surf)^-$	$\rightarrow OH^- + (surf)$
R45	$OH + (surf)$	$\rightarrow (surf)^b$
R46	$H_2 + (surf)^-$	$\rightarrow H_2^- + (surf)$
R47	$H_2 + (surf)^-$	$\rightarrow H + H^- + (surf)$
R48	$H_2 + (surf)$	$\rightarrow (surf)^b$

[a] (surf) refers to solid state surface, generally composed of silicates.
[b] Reactions ending in a chemical bond at the surface.

Acknowledgments: The research for this review was supported by NASA grants NAGW-649 and NAGW-744 to the University of Arizona.

REFERENCES

[1] Science, December 15 issue (1989).
[2] R. G. French, and P. J. Gierasch, J. Atm. Sci., **31**, 1707 (1974).
[3] D. F. Strobel, and S. K. Atreya, In Physics of the Jovian Magnetosphere, edited by A. J. Dessler, (Cambridge Univ. Press., 1983), 51.
[4] D. M. Hunten, and A. J. Dessler, Planet. Space Sci., **25**, 817 (1977).
[5] Y. L. Yung, G. R. Gladstone, K. M. Chang, J. M. Ajello, and S. K. Srivastava, Astrophys. J. Lett., **254**, 65 (1982).
[6] D. E. Shemansky, J. Geophys. Res., **93**, 1773. (1988).
[7] F. Bagenal, In Time-variable phenomena in the Jovian system, edited by M. J. S. Belton, R. A. West, and J. Rahe, (NASA., 1989), 196.
[8] M. A. McGrath, and R. E. Johnson, J. Geophys. Res., **94**, 2677 (1989).
[9] A. E. Potter, and T. H. Morgan, Science, **241**, 675 (1988).
[10] P. Drossart, and et al., Nature, **340**, 539 (1989).
[11] J. C. McConnell, and T. Majeed, Planet. Space Sci., , in press. (1990).
[12] D. E. Shemansky, J. Geophys. Res., **90**, 2673 (1985).
[13] H. H. Michels, and R. H. Hobbs, Astrophys. J., **286**, L27 (1984).
[14] D. Smith, and N. G. Adams, Astrophys. J., **284**, L13 (1984).
[15] N. G. Adams, and D. Smith, In Astrochemistry, edited by M. S. Vardyaand S. P. Tarafdar, (IAU 120th symposium, D. Reidel., 1987), 1-18.
[16] T. Amano, Astrophys. J., **329**, L121 (1988).
[17] H. Hus, F. Youssif, A. Sen, and J. B. A. Mitchell, Phys. Rev. A., **38**, 658 (1988).
[18] J. W. McConkey, S. Trajmar, and G. C. M. King, Comments At. Mol. Phys., **22**, 17 (1988).
[19] M. Audibert, C. Joffrin, and J. Ducuing, Chem. Phys., **25**, 158 (1974).
[20] M. Alexander, Chem. Phys., **20**, 83 (1977).
[21] W. R. Gentry, and C. F. Giese, Phys. Rev., **11**, 90 (1975).
[22] M. Cacciatore, M. Capitelli, and M. Dilonardo, Chem. Phys., **34**, 193 (1978).
[23] A. Eviatar, and G. L. Siscoe, Geophys. Res. Lett., **7**, 1085 (1980).
[24] A. J. Dessler, Icarus, **44**, 291 (1980).
[25] J. D. Richardson, and R. L. McNutt, Geophys. Res. Lett., **14**, 64 (1987).
[26] D. J. Southwood, and M. G. Kivelson, J. Geophys. Res., **94**, 299 (1989).

[27] E. W. Davis, Diffusion in the Io Plasma Torus and it's Relation to the Torus Spatial Structure, (Ph.D. Thesis, Univ. Arizona., 1990).
[28] R. F. Stebbings, A. C. Smith, and H. Ehrhardt, In Atomic Collision Processes, edited by M. R. C. McDowell, (North-Holland., 1964), 814.
[29] J. E. Johnson, and D. F. Strobel, J. Geophys. Res., **87**, 10385 (1982).
[30] O. B. Firsov, Zh. Eksper. Teor. Fiz., **21**, 1001 (1951).
[31] A. Burgess, Astrophys. J. Lett., **139**, 776 (1964).
[32] R. H. Bell, and M. J. Seaton, J. Phys. B: At. Mol. Phys., **18**, 1589 (1985).
[33] P. J. Storey, Mon. Not. R. Astron. Soc., **195**, 27 (1981).
[34] N. Nussbaumer, and P. J. Storey, Astron. Astrophys., **126**, 75 (1983).
[35] A. Potter, and T. H. Morgan, Science, **229**, 651 (1985).
[36] A. Potter, and T. H. Morgan, Icarus, **67**, 336 (1986).
[37] A. L. Tyler, R. W. H. Kozlowski, and D. M. Hunten, Geophys. Res. Lett., **15**, 1141 (1988).
[38] T. H. Morgan, and D. E. Shemansky, J. Geophys. Res., In Press (1990).
[39] N. H. Tolk, R. G. Albridge, A. V. Barnes, and D. P. Russell, Inst. Phys. Conf. Ser. No. 94: Section, 385 (1988).
[40] J. S. Johannessen, W. E. Spicer, and Y. E. Strausser, J. Appl. Phys., **47**, 3028 (1976).
[41] B. Carrière, and B. Lang, Surface Sci., **64**, 209 (1977).
[42] B. Feuerbacher, M. Anderegg, B. Fitton, L. D. Laude, R. F. Willis, and R. J. L. Grard, Geochimica et Cosmochimica Acta , **3**, 2655 (1972).
[43] H. C. Lord, J. Geophys. Res., **73**, 5271 (1968).

Dense Plasmas

PROBLEMS IN LINE BROADENING AND IONIZATION LOWERING

J. Davis
Radiation Hydrodynamics Branch, Plasma Physics Division
Naval Research Laboratory, Washington, DC 20375

M. Blaha
Laboratory for Plasma Research, Univ. of Maryland
College Park, MD

ABSTRACT

The sensitivity of spectral line shift has been investigated in the impact approximation for a variety of assumptions using the distorted-wave with exchange approach. The formalism is applied to the Lyman-alpha line of hydrogen and ionized helium to determine the influence of various approximations on the line shift as well as identifying the differences between a neutral and ionized radiator. These various assumptions and their consequence will be discussed and our results will be compared and contrasted with other calculations. Some speculation on the differences between them will be offered. In addition, the ionization state Z of aluminum has been calculated in the average atom model and density functional approximations as a function of temperature and density. The values of Z were then used to determine the ionization lowering for several ionization stages. The lowering of ionization potential was calculated from the eigenvalues of the highest bound orbital energies. The important differences of results are compared with the ion sphere and Stewart-Pyatt models, and the different theories critiqued.

ELECTRON COLLISION SHIFT OF THE LYMAN-ALPHA LINE IN He+ AND H

The interaction of plasma microfields with radiating atoms and ions causes a line broadening that governs both the the half-width and shift of the spectral line, i.e., the intensity distribution in frequency is altered to reflect the presence of the plasma. The line shifts are of particular interest both for their intrinsic value in terms of providing insight into the fundamental nature of the underlying physical processes affecting the emission of radiation as well as its diagnostic value in the study of high density plasmas. In addition, exact wavelengths of spectral lines of highly charged ions are required in the determination of line opacities and level positions for x-ray laser transitions.

Since the investigation by Berg, et al.[1] on the effects of plasma interactions on the frequency shift of spectral lines it still remains an interesting and challenging problem. This is due primarily to its N-body nature for which only approximate solutions can be found. However, in low density plasmas the mutual interactions of perturbing particles may be neglected converting an intractable problem into a manageable one, making it possible to evaluate line profiles in the impact approximation via the Baranger formalism.[2] This theory is based on the knowledge of the scattering matrix for collisions of electrons and the radiator in the upper as well as in the lower state of the line. In the application of Baranger's theory, the initial perturber-radiator correlations are generally neglected, and the theory leads to negative (i.e. red) frequency shifts of Lyman-alpha lines of hydrogen-like ions[3,4], contrary to experimental results[5-8] which yield blue shifts. It has been suggested[9] that another effect which is associated with initial perturber-radiator correlations, namely plasma polarization, may cause blue shifts of hydrogen-like lines. Plasma polarization effects on He$^+$ lines have been calculated by several authors[5-8,10] with the assumption of charge neutrality outside the orbit of the bound electron with the result of blue line-shifts. However, a procedure similar to the quantum-mechanical approach of Volonte[10], but with inclusion of electric charges in the outer region, leads again to red shifts of Lyman lines.[11]

In the present investigation, we concentrate our efforts on a few special problems in the theory of the line-shift, mainly the sensitivity of calculated shifts to various effects taken into account in the evaluation of scattering matrices. For this reason we have chosen the Lyman-alpha lines of hydrogen and of ionized helium as two typical examples where the effects can be demonstrated and differences between a neutral and an ionized emitter may be seen. Our approach is fully quantum-mechanical, and the scattering matrices are obtained in a distorted-wave approximation with exchange. This approximation, although rather simple, is convenient for our purposes because it is very well suited for the calculation of various effects - addition or deletion of different terms in the scattering matrix can be performed easily. We calculated only the electron collision contributions to the shift. Effects of plasma ions will be neglected. Our basic assumption is that perturber-perturber interactions are weak and can be ignored, and therefore the applicability of the results for the Lyman-alpha line is limited to low-density plasmas, where the Debye shielding length is much larger than the dimensions of the radiator, i.e. for $N_e < 10^{18}$ cm^{-3}.

The shift of the Lyman-alpha line of hydrogen and ionized helium produced by electron collisions will be calculated with the assumption that the impact approximation is valid, which means that results will be applicable to low-density plasmas only. We will also assume that: 1) the perturber-perturber interactions are small and can be neglected, 2) the wave functions of the bound electron are not changed by plasma interactions, and 3) we will ignore the effect of positive ions. No uniform positive background will be included in the calculations, and the fine structure splitting of atomic levels will be neglected. All calculated shifts will be normalized to the electron density N_e=1 cm^{-3}. Unless indicated otherwise, all quantities throughout the paper will be given in atomic units (e=h/2π=m=1).

Within the validity of the impact approximation, the electron collision profile of a spectral line is determined by electron scattering on the upper and the lower level of the line. According to Baranger[2], the angular frequency shift $\Delta\omega$ of the radiative transition $n_p p \rightarrow n_s s$ is proportional to the electron density N_e and it can be expressed in terms of diagonal elements of the scattering matrix $S(\alpha_p, \alpha_p)$ and $S(\alpha_s, \alpha_s)$ by

$$\Delta\omega = \frac{1}{12} \pi N_e \int k^{-1} f(k) \sum_{\ell, L_p^T, S^T} (2L_p^T + 1) \qquad (1)$$

$$\times (2S^T + 1) \text{Im}\left[1 - S(\alpha_p, \alpha_p) S^*(\alpha_s, \alpha_s)\right] dk \quad .$$

The scattering channel α is characterized by the momentum k and the angular momentum ℓ of the colliding electron, and by the total orbital and spin angular momenta L^T and S^T which satisfy the conditions $S_p^T = S_s^T \equiv S^T$, $\ell_p = \ell_s \equiv \ell$, $k_p = k_s \equiv k$. $f(k)$ is the momentum distribution function of plasma electrons which is represented by a Maxwellian distribution in the present calculation.

It is convenient to rewrite (1) in terms of transmission matrices T. From the relation $T = 1 - S$ it follows

$$1 - S(\alpha_p, \alpha_p) S^*(\alpha_s, \alpha_s) = T(\alpha_p, \alpha_p) + T^*(\alpha_s, \alpha_s) - T(\alpha_p, \alpha_p) T^*(\alpha_s, \alpha_s) \qquad (2)$$

and the total line-shift is thus decomposed into two parts: contributions from direct terms $T(\alpha_p, \alpha_p) + T^*(\alpha_s, \alpha_s)$ and from the interference term $- T(\alpha_p, \alpha_p) T^*(\alpha_s, \alpha_s)$.

In the evaluation of diagonal elements of the transmission matrix T it is generally necessary to take into account coupling of several scattering channels. However, in the case of the Lyman-alpha line, the energy difference between the n=1 and n=2 levels in hydrogen and He$^+$ is much larger than the average kinetic energy of plasma electrons at the temperatures considered in our calculation, so that only a very small fraction of electrons scattered from the 1s level are affected by coupling to other channels. Therefore, we simplify our procedure by omitting all other levels in the evaluation of the matrix elements $T(\alpha_s, \alpha_s)$. On the other hand, we have included levels 1s, 2s, 2p, 3s, 3p, 3d in the T matrix for the evaluation of $T(\alpha_p, \alpha_p)$, but we have ignored all elements which do not involve the 2p level and consequently have only a small effect on $T(\alpha_p, \alpha_p)$.

All matrix elements $S(\alpha, \alpha')$ have been calculated in the distorted wave approximation with exchange (DWX), and the procedure is described in detail in Ref. 12.

First, we calculate the ρ matrix[13] with elements expressed in terms of direct and exchange integrals and coefficients f_λ and g_λ given by Percival and Seaton.[14]
The transmission matrix is then given by

$$T = 1 - e^{i\tau}(1 + i\rho)(1 - i\rho)^{-1} e^{i\tau} \quad . \qquad (3)$$

$e^{i\tau}$ are diagonal matrices with elements $e^{i\tau_\alpha}$ and τ_α is the phase shift of the continuum radial wave function representing the scattering in channel α.

In the present version of the distorted-wave approximation, the channels included in the ρ matrix are uncoupled. However, transformation (3) introduces a certain amount of coupling into the scattering matrix S, and consequently the line-shift calculated from eq. (1) depends on the number of levels included in the ϱ and S matrices.

One of the goals of this investigation was to establish the importance of different contributions to the total line-shift. Therefore we have performed a series of calculations in which certain terms were omitted in the expression for $\Delta\omega$. Results of the distorted-wave approximation are presented on Figs. 1 and 2 for He^+ and on Fig. 3 for hydrogen. In both cases we have found red line shifts.

Table 1

Results for the Lyman-alpha shifts shown on Figs. 1-3.
(DWX = distorted-wave approximation with exchange
 DW = distorted-wave approximation without exchange)

Curve	Levels included in the scattering matrix	Remark
A	1s2s2p3s3p3d	DWX
B	1s2p	DWX
C	1s2p	DW, only elastic terms included,
D	1s2s2p	DWX
J	1s2s2p3s3p3d	DWX, same as A without interference term
K	1s2p	DW, same as C without interference term
S	1s2s2p3s3p3d	DWX, with free electron functions generated using Yamamoto and Narumi's[4,17] potential
T	1s2s2p3s3p3d	DWX, same as S, without direct monopole terms

Curve A on Figs. 1-3 represents the most complete DWX calculation with all levels 1s-3d included in the S matrix.

For He^+ (Fig. 1), the omission of levels 3s, 3p, 3d leads to curve D, which indicates that the higher levels have only a very small effect on the frequency shift of Lyman-alpha line. If only 1s and 2p levels are included, the result is given by B, and if, in addition, all exchange terms are omitted and only elastic terms retained, the shift is represented by C. A glance at Fig. 1 and comparison of A and C shows that the pure elastic scattering is by far the most important contribution to the frequency shift. In this particular case $\Delta\omega$ depends only on phase shifts τ_α, because $\varrho = 0$, and consequently

$$\Delta\omega = -\pi N_e \int k^{-1} f(k) \sum_\ell (2\ell + 1) \sin 2\left(\tau_{p\ell} - \tau_{s\ell}\right) dk , \qquad (4)$$

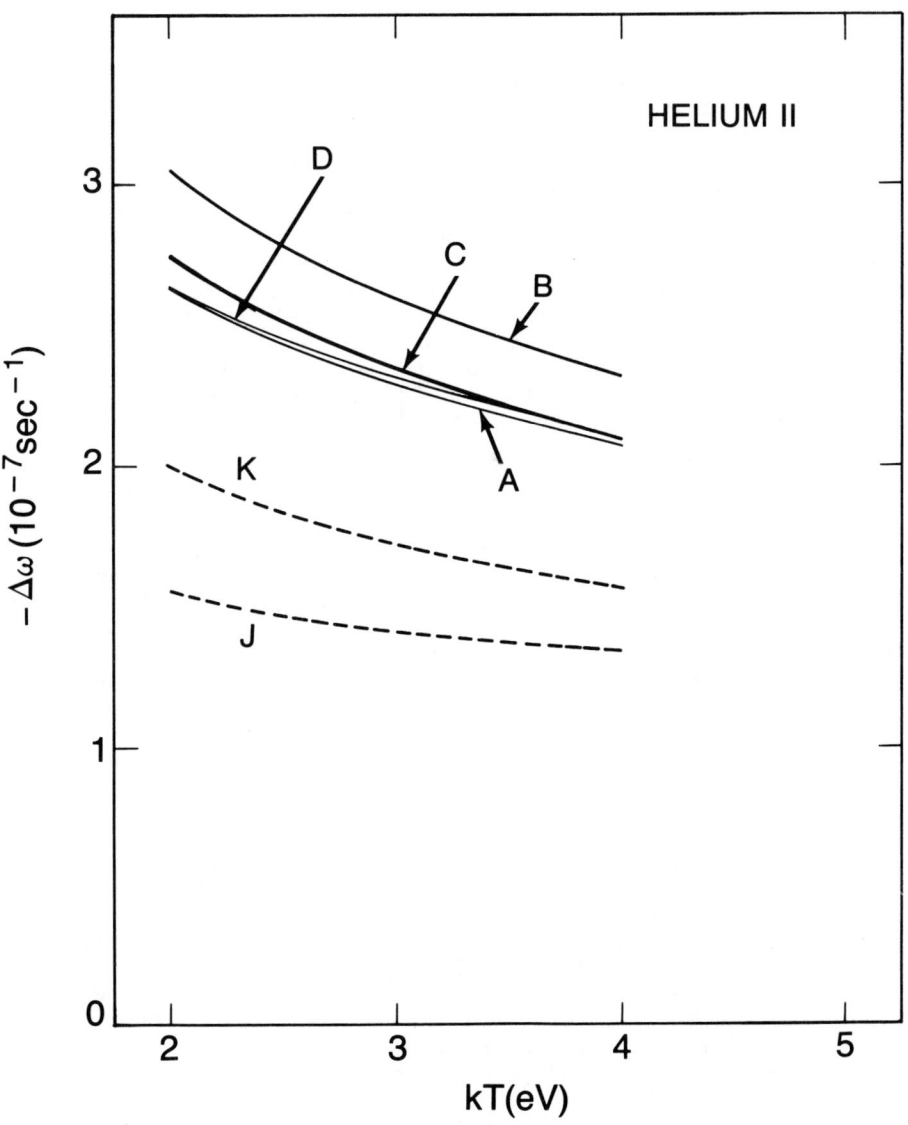

Fig. 1 Lyman-alpha shift of He$^+$ (N_e = 1 cm^{-3}). Symbols are explained in Table 1.

where $\tau_{p\ell}$ and $\tau_{s\ell}$ are phase shifts associated with the scattering on the 2p and 1s levels, respectively. Furthermore,

$$\sin 2\left(\tau_{p\ell} - \tau_{s\ell}\right) = \sin 2\tau_{p\ell} - \sin 2\tau_{s\ell} + 4\sin\tau_{p\ell}\sin\tau_{s\ell}\sin\left(\tau_{p\ell} - \tau_{s\ell}\right). \quad (5)$$

The first two terms on the right-hand side of (5) correspond to the direct terms, and the third term to the interference term according to (2).

The importance of the interference term for He$^+$ is shown on Fig. 1 by curves J and K. J was obtained in the same way as A, but the interference term was not included. K similarly corresponds to C. It can be seen that the inclusion of upper levels and of exchange contributions significantly changes the magnitude of the interference term even if the total frequency shift is affected minimally. The inclusion of exchange increases the value of the red shift as can be seen by comparing curves B and C.

In Fig. 2, our result for ionized helium (A) is compared to the semi-classical (SC) calculation of Griem[3] (circles) and to the R-matrix method of Yamamoto and Narumi.[4]

The semi-classical result is about 35% lower than the DWX method at kT = 2.5 eV, but the agreement improves with higher temperature. However, there are basic differences between the semi-classical procedure and our method. The semi-classical result is derived from the second-order terms in the S matrix elements, and in the expansion for the interaction potential only dipole and quadrupole terms have been taken into account. The monopole contributions have been approximated by the "strong collision term" and the SC results also contain contributions from perturbing levels higher than n=3 and from collisions with electrons whose energies are below excitation threshold. In the DWX method the main contribution to the shift comes from the monopole part of the potential and higher order contributions to the S matrix are due to the unitarizing relation (3). The ρ matrix has only first-order terms. In spite of the differences of the SC and DWX methods, the agreement of the results appears to be relatively good at higher temperature. In a recent calculation[15], Griem added to the previous results second order contributions from the 2p-2s interaction, obtained from the kinetic theory approach of Boercker and Iglesias[16]. Due to the degeneracy of the 2s and 2p levels, these contributions could not be properly evaluated in the original SC procedure. The new values of the Lyman-alpha shift in the He$^+$ are represented in Fig. 2 by triangles. The 2p-2s interaction is obviously very important. In our DWX approximation this contribution is not included because the ρ matrix contains no second-order terms.

The R-matrix calculation[4] yields results which are almost three times smaller than the DWX approximation. Yamamoto and Narumi included in their calculation the same number of levels as we did (1s-3d), but there are several other differences which may cause the disagreement. We suspect that the evaluation of elastic contributions in the R-matrix calculation is responsible for the discrepancy. We have performed two additional calculations to substantiate this suspicion: First we replaced the potential for colliding electrons (which is not the same for different target

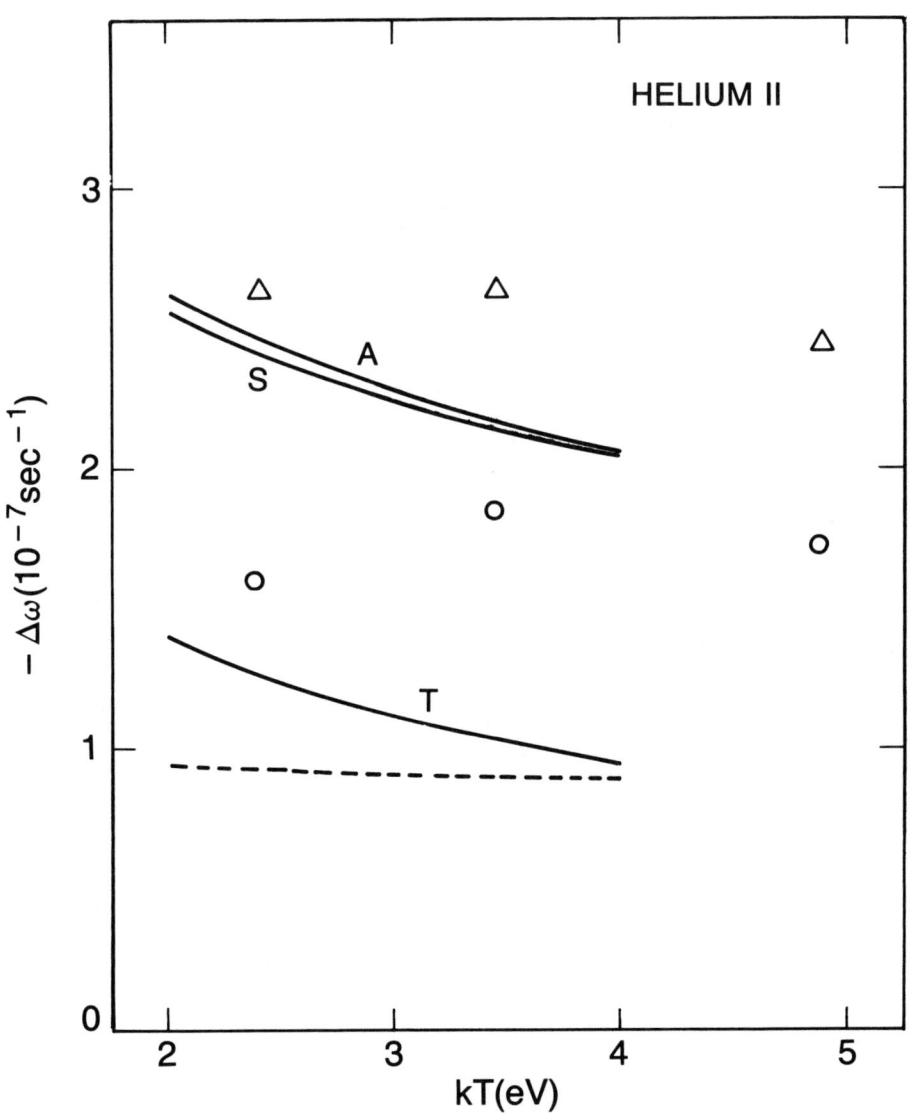

Fig. 2 Same as Fig. 1.
Circles: semi-classical calculation of Griem[3] (without contributions from 2p-2s interaction)
Triangles: semi-classical calculation of Griem[15] (with contributions from 2p-2s interaction)
Dashed line: R-matrix method of Yamamoto and Narumi[4].

levels in our procedure) by the potential used in the work of Yamamoto and Narumi[17]. This potential has the same form for all levels and therefore additional monopole terms representing different elastic scattering from different levels should be added to the ϱ matrix. The result is represented by curve S and it indicates that the difference of potentials is of small importance. However, if we ignore the monopole terms, the result reduces to curve T and the agreement with Yamamoto and Narumi is better, but still not satisfactory over the whole temperature range. A possible inadequacy of diagonal R-matrix elements in ref. 4 has already been pointed out by Griem.[3]

Observational evidence[5-8] concerning the Lyman-alpha shift in ionized helium is to a certain degree contradictory due to experimental difficulties. In contrast to our calculation which yields negative shifts, experimentally found shifts are positive (i.e. blue) and of much larger magnitude. At the present time it is not even clear if this discrepancy is caused by experimental inaccuracy or if it represents incompleteness of the theory.

For hydrogen (Fig. 3), the sensitivity of frequency shift to the inclusion of atomic levels in the $\underline{S}$ matrix is much larger than for He^+.

The smallest shift is obtained with pure elastic scattering without exchange, (i.e. with 1s and 2p levels included and $\varrho=0$, curve C on Fig. 3). The shift even reverses its sign and becomes blue at low temperature. Addition of exchange and quadrupole terms leads to curve B, and calculation with 1s2s2p levels included is represented by D. Further addition of n=3 levels reduces the magnitude of the red shift (curve A). In contrast to ionized helium, contributions from levels higher than n=3 may not be negligible. The effect of exchange on the shift calculated with all levels 1s-3d is opposite to that found for He^+. In hydrogen, the exchange substantially reduces the value of the red shift. Calculations without the interference term are represented by J and K (Fig. 3). J corresponds to A and K to C.

The SC results of Griem[18] for the Lyman-alpha shift of hydrogen are by a factor of 5 larger than the DWX results, and the inclusion of the 2p-2s interactions following the theory of Boercker and Iglesias[16] leads to a further increase of the shift[15]. It is therefore obvious that second-order terms in the scattering matrix which were omitted in our calculation, have a dominant effect on the result.

IONIZATION LOWERING

The lowering of the ionization potential in plasmas has been of academic interest for many years because of its fundamental importance to understanding how the plasma microfields influence atomic structure and processes. However, in more recent times because of the advent of drivers powerful enough to generate above solid density plasmas, knowledge of ionization lowering is of considerable practical importance because of its relationship with the spectroscopy of plasmas. In an effort to better understand this phenomena a number of theoretical studies have been carried out over the years with varying degrees of success. Among the generally accepted and more commonly used results are due to the Debye, Ion

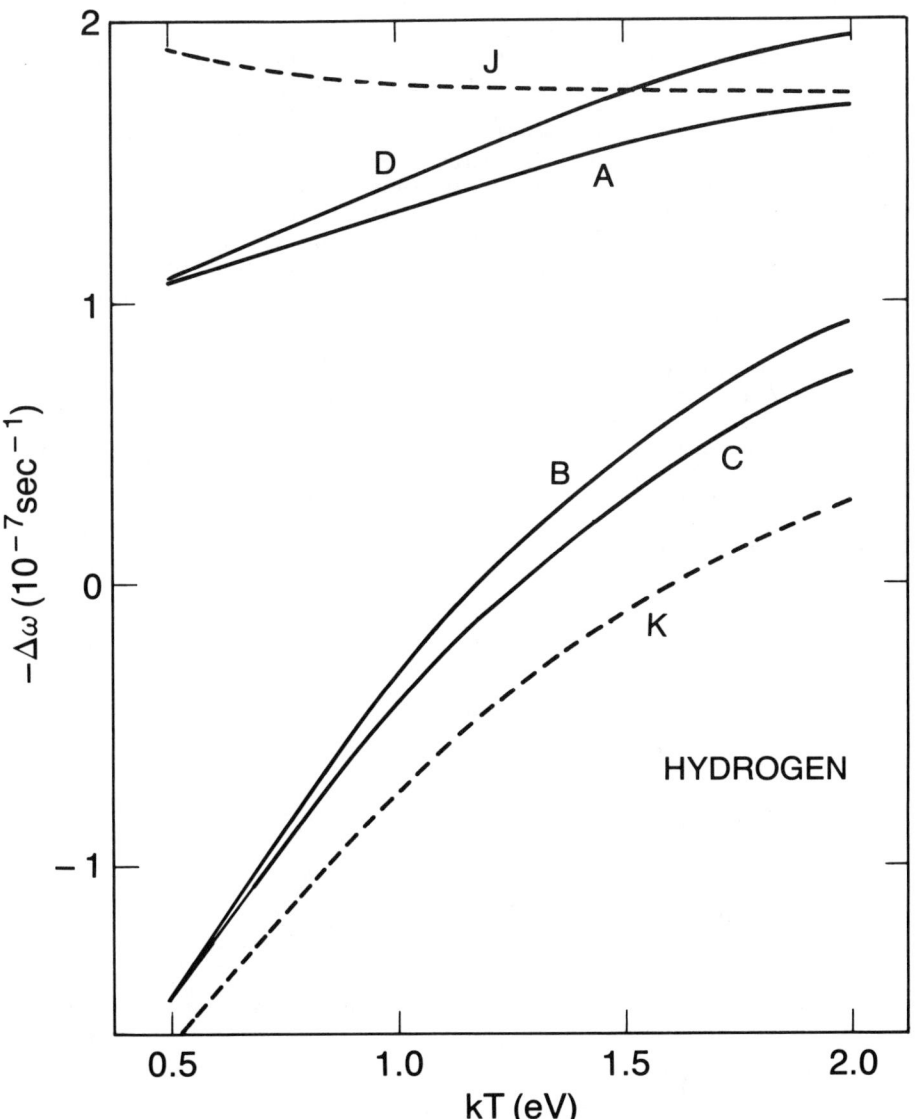

Fig. 3 Lyman-alpha shift of hydrogen ($N_e = 1 \text{cm}^{-3}$). Symbols are explained in Table 1.

Sphere and Stewart-Pyatt models. More recently, a number of Average Atom Models (AAM) have been proposed to deal with ionization lowering. The latest, and possibly the most general AAM is due to Crowley.[19] On the other hand, we have adopted a different approach based on a selfconsistent density functional method.[20] This method has the advantage of generating selfconsistent potentials as well as treating specific atomic states making it very suitable for dealing with the analysis of experimental data. Also, it is not our intent here to review the existing theories but instead to provide quantitative results for comparison and benchmarks with the standard models mentioned above. To this end, we have evaluated the ionization lowering of the aluminum ions for several plasma densities and temperatures of experimental interest.

The average charge Z of aluminum ions was calculated for ion densities N_i from 10^{21} to 10^{24} cm^{-3} and for temperatures from 20 to 100 ev. The calculation was based on the spherically symmetric non-relativistic AAM with infinite boundary. It was assumed that the local density of plasma ions $n_i(r)$ with charge Z is given by Boltzmann statistics as a function of distance from the central ion, and the local electron density $n_e(r)$ was found by solving the corresponding equations of the density functional theory. The equations were solved self-consistently and the ionization state Z was found from the difference of the nuclear charge and the number of bound electrons which is determined by the solution of the equations. An example of the electron and ion density profiles from the AAM is shown on Fig. 4.

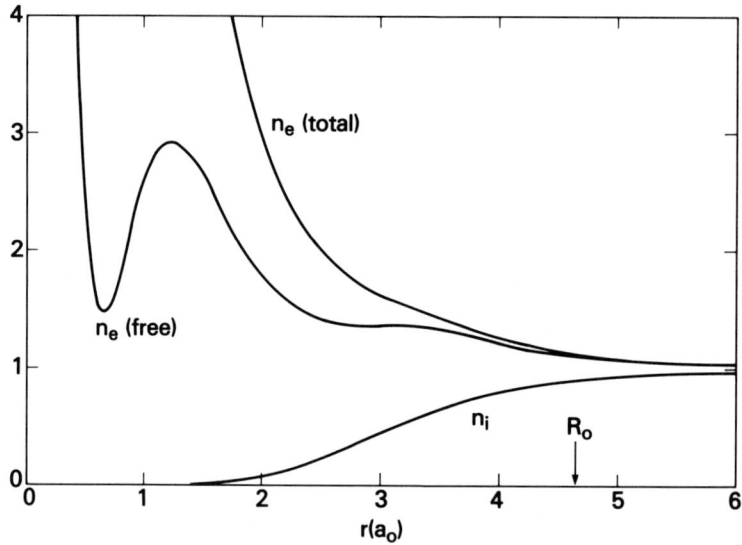

Fig. 4. Electron and ion density profiles (normalized to unity at infinity[22]) in the average atom model for aluminum plasma at N_i = 1.6×10^{22} cm^{-3}, kT = 20 eV. R_o indicates the ion sphere radius. (Number of bound electrons = 9.885, Debye length = 1.38 a_o, Γ=2.84.)

On Fig. 5, the average ion charge $\bar{Z}$ is plotted as a function of ion density N_i for three values of temperature. On Fig. 6, $\bar{Z}$ is shown as a function of temperature for a constant ion density $N_i=10^{22} \text{cm}^{-3}$ together with the result of a collisional-radiative model[21] in which no continuum lowering was taken into account.

Using the value of $\bar{Z}$ from the AAM for plasma ions, and keeping the number of bound electrons in each $n\ell$ orbital constant so that they would correspond to the ground configuration of a particular aluminum ion with charge Z, the same system of equations for $n_i(r)$ and $n_e(r)$ was self-consistently solved and energy eigenvalues $\varepsilon_{n\ell}$ for bound electron were obtained. The continuum lowering ΔI for a Z ion was determined from

$$\Delta I = \varepsilon^{\circ}_{n\ell} - \varepsilon_{n\ell} \, ,$$

where $\varepsilon^{\circ}_{n\ell}$ is the eigenvalue corresponding to an isolated ion, and $n\ell$ is the bound orbital of the ground configuration with the highest value of n and ℓ.

The results for aluminum ions from Al III to Al XI for the ion density $N_i = 10^{22} \text{cm}^{-3}$ are displayed on Fig. 7. The value of ΔI is little sensitive to temperature for any given Z ion, but the lowering for different ions at the same temperature may be different. The most striking example is the value of ΔI for the 3s electron in Al III. The value of ε_{3s} at these plasma conditions is small and the 3s electron is very loosely bound. However, the value of $|\Delta I/I|$ monotonically decreases with Z, and this behavior is partly responsible for the difference of $\bar{Z}$ values on Fig. 6 at low temperature, where the omission of continuum lowering in the collisional-radiative model produces the largest effect.

Fig. 8 shows the lowering derived from the Debye-Hückel model (ΔI_D), from the ion-sphere model with a uniform electron distribution (ΔI_{IS}), and with the simplified Stewart-Pyatt formula (ΔI_{SP}). These results should be compared with our values on Fig. 7. The lowering in these approximations is given by (in atomic units)

$$\Delta I_D = -\frac{Z+1}{D} \, , \quad \Delta I_{IS} = -\frac{3}{2}\frac{Z}{R_o} \, ,$$

$$\Delta I_{SP} = -\frac{3}{2}\frac{Z}{R_o} \left\{ \left[1 + (D/R_o)^3\right]^{2/3} - (D/R_o)^2 \right\} \, ,$$

where D = Debye screening length, R_o = ion-sphere radius. At plasma conditions corresponding to Figs. 7 and 8, D<R_o (see Fig. 4) and the Debye-Hückel theory is unacceptable. Fig. 4 also indicates that the free electron density n_e inside the ion sphere is substantially greater than the average density N_e so that any model based on a uniform electron density necessarily leads to a much smaller value of $|\Delta I|$ as shown on Fig. 8 for ΔI_{IS} and ΔI_{SP}.

The results presented here are meant to remind the reader that these (olde) problems are still with us and remain, for the most part, unresolved. Much more remains to be done both experimentally

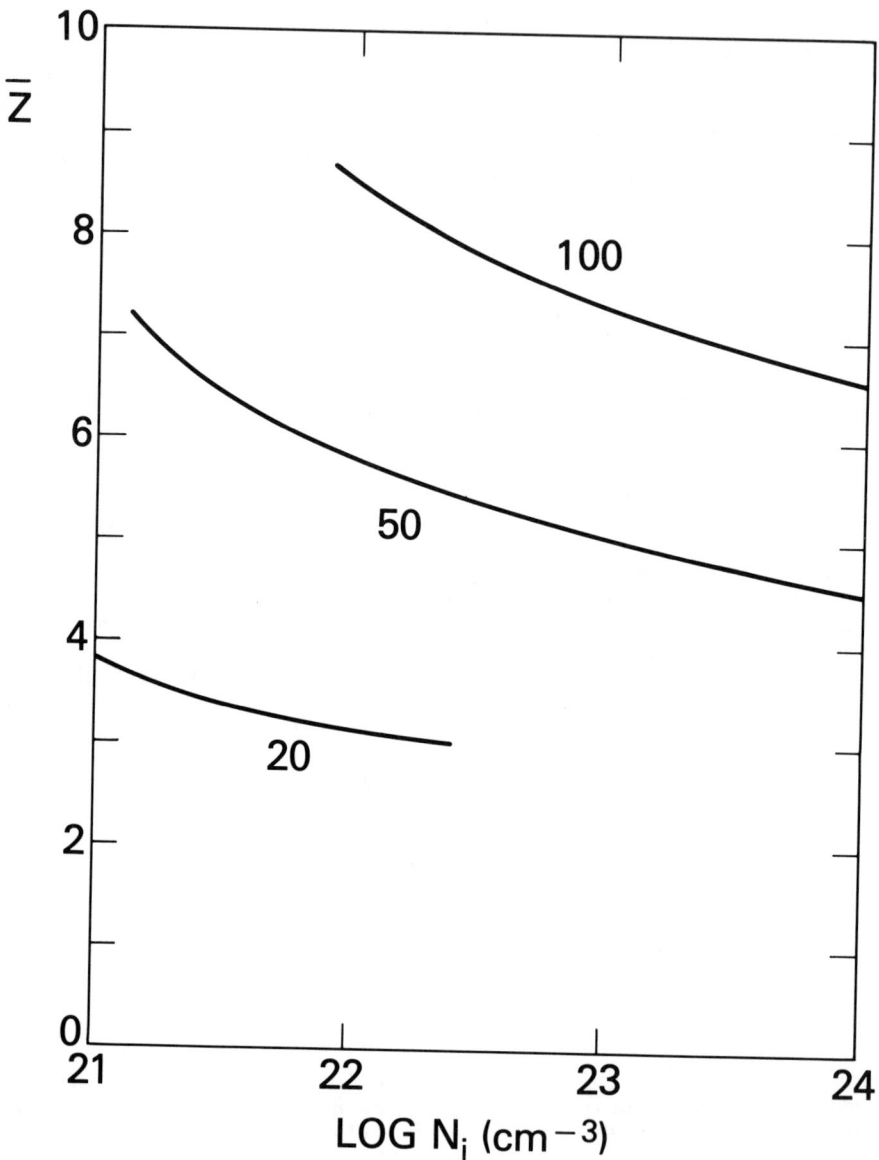

Fig. 5 Average ion charge $\bar{Z}$ in aluminum plasma as function of ion density N_i for kT = 20, 50 and 100 eV.

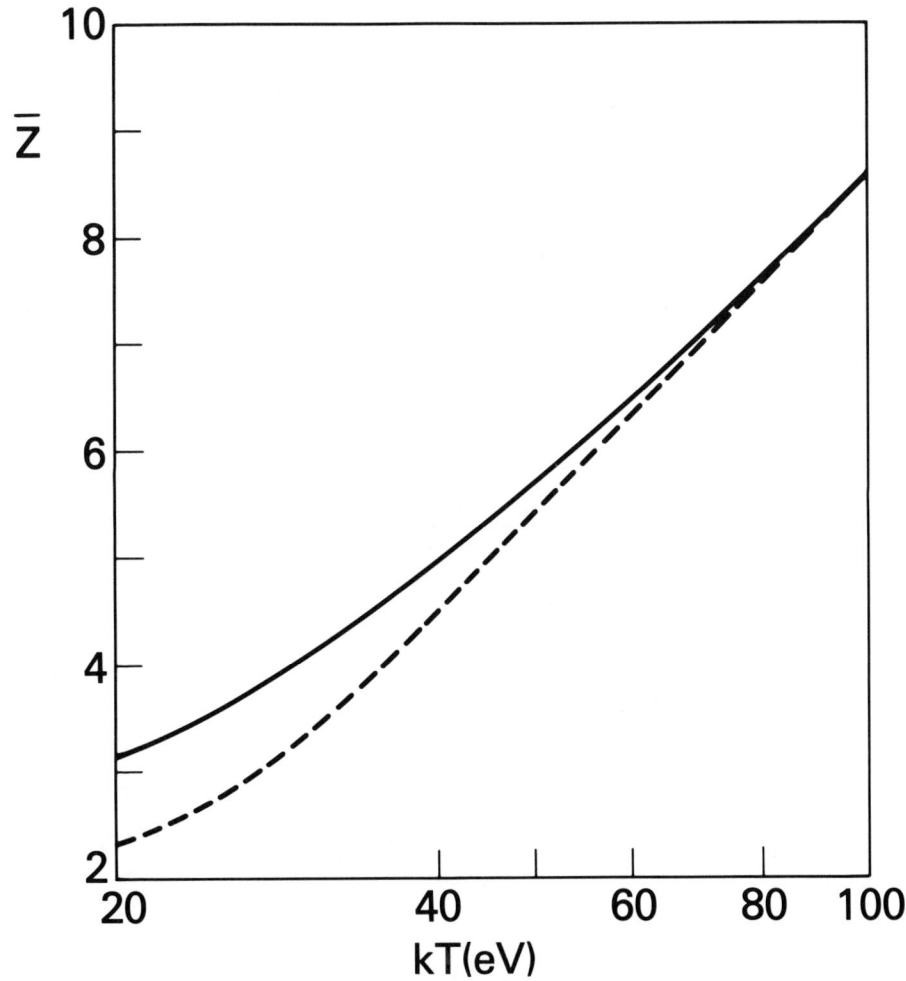

Fig. 6 Average ion charge $\bar{Z}$ in aluminum plasma as function of temperature for ion density $N_i = 10^{22}$ cm^{-3}.
Full line: present average atom model
Broken line: collisional-radiative model[21]

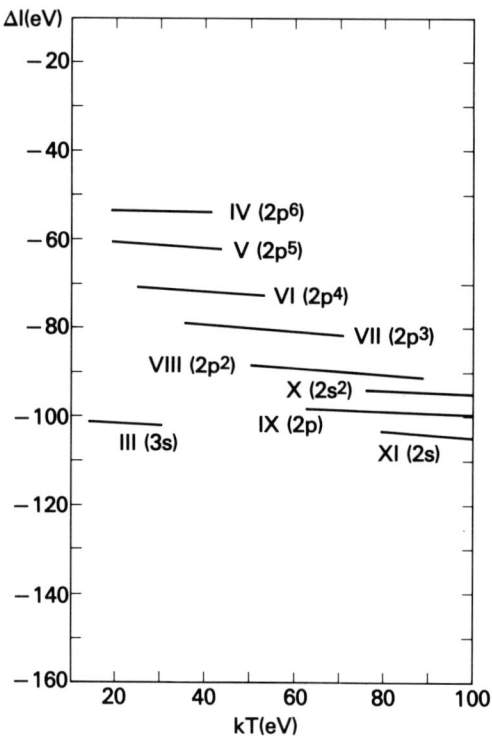

Fig. 7 Continuum lowering ΔI for ground configuration of aluminum ions at ion density $N_i = 10^{22}$ cm^{-3} (present results).

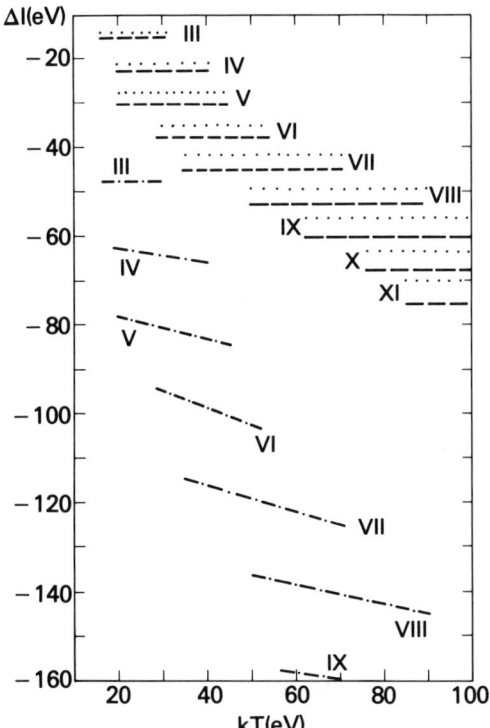

Fig. 8 Continuum lowering ΔI for ground configuration of aluminum ions at ion density $N_i = 10^{22}$ cm^{-3}.
Dashed lines: ΔI_{IS} (ion sphere model with uniform electron distribution)
Dash-dotted lines: ΔI_D (Debye–Hückel model)
Dotted lines: ΔI_{SP} (Stewart–Pyatt theory)

as well as theoretically. However, if any real progress is to be made in these areas, it is essential that well conceived and diagnosed experiments be performed.

ACKNOWLEDGEMENTS

This work was supported by the Office of Naval Research.
The authors are indebted to Prof. Hans R. Griem for valuable discussions and comments.

REFERENCES

1. H.F. Berg, A.W. Ali, R. Lincke, and H.R. Griem, Phys. Rev. 125, 199 (1962).
2. M. Baranger, Phys. Rev. 111, 481 (1958).
3. H.R. Griem, Phys. Rev. A 27, 2566 (1983).
4. K. Yamamoto and H. Narumi, Progr. Theor. Physics 64, 436 (1980).
5. J.R. Greig, H.R. Griem, L.A. Jones, and T. Oda, Phys. Rev. Lett. 24, 3 (1970).
6. A.H. Gabriel and S. Volonté, J. Phys. B:Atom. Molec. Phys. 6, 2684 (1973).
7. T. Goto and D.D. Burgess, J. Phys. B:Atom. Molec. Phys. 7, 857 (1974).
8. M. Neiger and H.R. Griem, Phys. Rev. A 14, 291 (1976).
9. O. Theimer and P.C. Kepple, Phys. Rev. A 1, 957 (1970).
10. S. Volonté, J. Phys. B:Atom. Molec. Phys. 8, 1170 (1975).
11. R. Cauble, J. Quant. Spectrosc. Rad. Transfer 28, 41 (1982).
12. M. Blaha and J. Davis, NRL Memorandum Rep. No. 6294 (1988).
13. H.E. Saraph, M.J. Seaton, and J. Shemming, Phil. Trans. Roy. Soc. London A 264, 77 (1969).
14. I.C. Percival and M.J. Seaton, Proc. Camb. Phil. Soc. 53, 654 (1957).
15. H.R. Griem, Phys. Rev. A 38, 2943 (1988).
16. D.B. Boercker and C.A. Iglesias, Phys. Rev. A 30, 2771 (1984).
17. K. Yamamoto, J. Phys. Soc. Japan, 49, 730 (1980).
18. H.R. Griem, Phys. Rev. A 28, 1596 (1983).
19. B.J.B. Crowley, to be published.
20. J. Davis, M. Blaha, and U. Gupta, "Radiative Properties of Hot Dense Matter" (J. Davis, et al., editors), World Scientific 1985, p. 261.
21. D. Duston and J. Davis, Phys. Rev. A 21, 1664 (1980).

LARGE SCALE CALCULATIONS OF THE STRUCTURE AND DYNAMICS OF DENSE MATTER

Stephen M. Younger
X-Division, Los Alamos National Laboratory, Los Alamos, NM 87545

ABSTRACT

The physics of matter at high density has been found to be strongly influenced by transient quasi-molecular phenomena, the distribution of electron charge density among several nearby atoms. Calculations performed using the self-consistent field molecular dynamics approximation, a synthesis of techniques derived from quantum chemistry and molecular dynamics, have shown that such effects can result in more than a 30% increase in the diffusion coefficient of helium at 1 g/cc. The effect of finite electron temperature is to weaken many-electron collective phenomena, but significant effects are expected to occur even in fully ionized dense plasmas.

INTRODUCTION

An understanding of the structure and dynamics of dense plasmas is important for several applications. The recent observations of the gas giant planets by the space probe Voyager II revealed anomalies in the atmospheric temperature distributions which could be affected by the energy transport rates in dense warm hydrogen-helium mixtures. At higher temperatures, models of stellar structure depend both on the equation of state of the plasma as well as the detailed radiative and collisional properties of matter under extreme conditions of temperature and density. On Earth, the inertial confinement fusion program routinely produces plasmas of many grams per cubic centimeter at temperatures of up to several kilovolts. In materials science, shock compression is being explored as a route to synthesize new materials with potential commercial applications.

Traditional methods for computing the properties of dense matter can be divided into two major categories: those based on some form of average atom approximation and statistical theories of many particles interacting via predefined potentials. Average atom approximations attempt to describe the physics of the entire plasma in terms of a single effective structure.[1] The electron distribution for the atom is constrained to occupy a fixed volume, often taken as the ion sphere volume. Boundary conditions on the wavefunction and its derivative mimic the effect of the external plasma. Finite electron temperature can be included in the calculation by distributing the electronic population over excited states according to a Fermi distribution.

In statistical theories of plasma, effective interaction potentials are employed to simulate the interaction of a large number of atoms.[2] Such theories include molecular dynamics, which tracks the classical motion of a large number of atoms, and Monte Carlo approximations, which are based on generating the most probable configurations for an ensemble.

Both the average atom and the statistical theories are expected to break down at high density where the internuclear distance is comparable to an atomic diameter. When several atomic wavefunctions overlap quasi-molecular behavior can occur, leading to significant changes in interatomic interactions as well as collisional and radiative rates. It is not necessary for such structures to result in permanent or even metastable bound configurations. Random collisions resulting from thermal motion will stochastically lead to close approaches in which strong atomic orbital overlap occurs. In order to properly treat such phenomena it is essential to have a methodology which explicitly treats the self-consistent redistribution of electronic charge density during the time evolution of a many-atom ensemble.

SELF-CONSISTENT FIELD MOLECULAR DYNAMICS

Self-consistent field molecular dynamics (SCFMD) is a new computational tool for the study of dynamic processes in many-atom ensembles based on a synthesis of methods used in quantum chemistry and classical molecular dynamics. SCFMD models the time evolution of complex many-atom clusters through the repetition of three steps. First, the electronic structure of the entire ensemble is computed using a Hartree-Fock Gaussian orbital molecular structure code. Second, this charge distribution is used to calculate the electrostatic force acting on each nucleus. Third, the positions and velocities of each atom are adjusted using classical equations and the computed forces. By repeating these steps one can generate the time histories of all of the atoms in the sample.[4]

The Hartree-Fock Gaussian orbital approximation employed here is based on the HONDO code of Dupuis and King.[5] The total wavefunction of the sample is expanded in molecular orbitals, which are in term described by expansions over simple Gaussian functions centered on the nuclei.

Interatomic forces are computed by means of the Hellmann-Feynman approximation. This method ignores small corrections to the force due to the explicit dependence of the basis set on the nuclear coordinates. These corrections were found to be small for binary interactions involving the simple atomic structures discussed here.

Atomic motion is computed classically using the quantum forces as input. The timestep of the calculation is automatically controlled to avoid large changes in interatomic separation during a single move. Positions, velocities, accelerations, and other properties of the calculation are written to a storage file to allow later post-processing of time-averaged quantities.

The effect of finite electron temperature is included in the calculations by allocating the electronic population over the virtual orbital spectrum according to a Fermi distribution. Note that this distribution takes place within the self-consistent field algorithm itself.

SCFMD STUDIES OF DENSE HELIUM

Helium was chosen for the initial application of SCFMD to dense matter for several reasons. It has a simple $1s^2$ ground state which can be approximately described by very small basis sets consisting of only two or three Gaussian functions. Helium has a high first excitation energy, so that the electrons will be in their ground state even at relatively high ion temperatures. High ion temperatures are desirable since it is in close atomic collisions that the simple interatomic force algorithm employed here is expected to be most accurate. A high excitation energy also implies a small polarizability so that simple basis sets are adequate to describe interactions. Finally, helium has been extensively studied using a variety of approximations in statistical mechanics and has also been observed in two-stage gas gun experiments at densities up to 0.68 g/cm^3.[6]

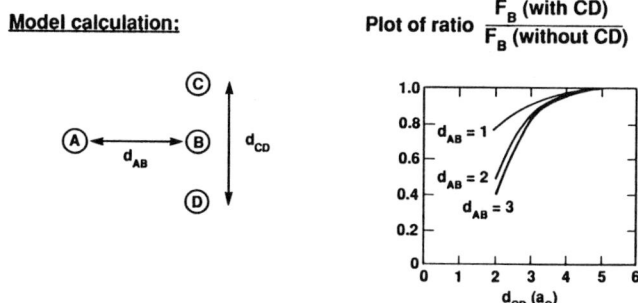

Figure 1. Ratio of the force on a test atom B due to another atom A with the presence of neighbor atoms C and D to the force computed from atom A alone.

Before turning to the results of dynamic calculations, it is instructive to examine the effect of neighboring atoms on the interaction of two sample atoms in a static configuration. Figure 1 shows the ratio of the force on atom B due to atom A when neighbor atoms C and D are present to that where only A and B are considered. The symmetry of the sample configuration was chosen such that if interatomic forces added in a pairwise fashion, the neighbors would have no effect on the force on atom B and the ratio would be one. Departures from unity signify additional screening of the force of atom A on atom B due to the self-consistent redistribution of electronic charge density among atoms B,C, and D. The nuclear potentials of atoms B,C, and D add such that the deepest part of the potential well is at position B. Thus the largest concentration of electronic charge density will be near atom B, in just the position to screen that atom from the effect of atom A.

In order to assess the effects of such screening in dynamic ensembles we applied SCFMD to the study of dense helium at densities between 0.1-3.0 g/cm^3 and at temperatures of 1 eV and 5 eV. We studied clusters containing 20-30 atoms contained in a three dimensional box. Approximate periodic boundary conditions were employed to mimic an infinite medium. The electronic wavefunction of the ensemble was described by two or three Gaussian basis sets located on each nucleus. Using the output from these calculations, we computed pair correlation functions, velocity autocorrelation functions, and other dynamic properties of dense helium.

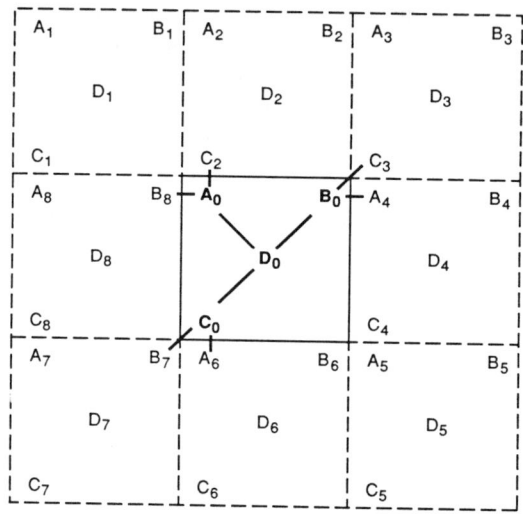

Figure 2. Periodic boundary conditions and the effective range approximation.

The computational labor associated with obtaining the Hartree-Fock solution for a large ensemble of atoms increases roughly as the third power of the size of the total basis set employed. Most of this time is spent in the calculation of the two-electron matrix elements. In our application of periodic boundary conditions we employed an effective range approximation wherein each atom interacts only with the nearest image of each of the other atoms, as illustrated in Figure 2. When an atom passes through a boundary of the box it reappears at the opposite edge with the same velocity. Although such an algorithm is straightforward for pairwise additive potentials, complications arise in finite basis expansions since the spatial extent of the orbitals is fixed and the integral package employed here is not equipped to truncate the matrix element integration at the box boundary. As an approximation to true numerical periodic boundary conditions, we employed an algorithm wherein for any matrix element the orbitals are chosen so that the closest images interact at any given time. This insures that the strongest interactions are correctly modeled. Residual errors associated with the neglect of small distant interactions are expected to be small when averaged over many hundreds of cycles.

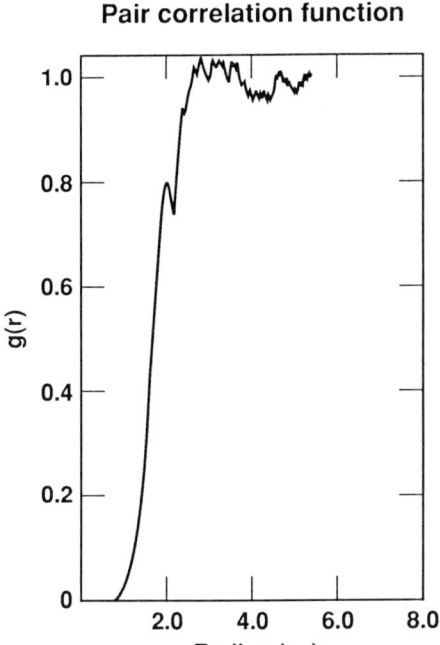

Figure 3. Pair correlation function for helium at 1 g/cm^3 and 5 eV.

Figure 3 shows the pair correlation function g(r)

$$g(r) = n(r) / (4\pi r^2 \Delta r n_o) \quad (1)$$

computed for a 23 atom sample of helium at 1 g/cm^3 and 5 eV. Here n(r) is the number of atoms between r and r+Δr from a test atom in a plasma with average number density n_o. The pair correlation function thus gives the probability for finding another atom at a given distance from a sample atom relative to a random distribution. Peaks in g(r) correspond to preferred spatial orientations such as molecular bonds or crystal lattices. The lack of structure in Figure 3 is typical of high temperature material. The position of the shoulder at small radii is indicative of the distance of closest approach and the width of the shoulder is a reflection of the convolution of the temperature distribution of the sample and the steepness of the repulsive interaction potential between atoms at short ranges. At large radii g(r) approaches unity, equivalent to a random distribution of atoms in a disordered medium.

The velocity autocorrelation function

$$Z(t) = <v(t)v(0)>/<v(0)v(0)> \quad (2)$$

is a fundamental measure of the persistence of dynamic correlations within the

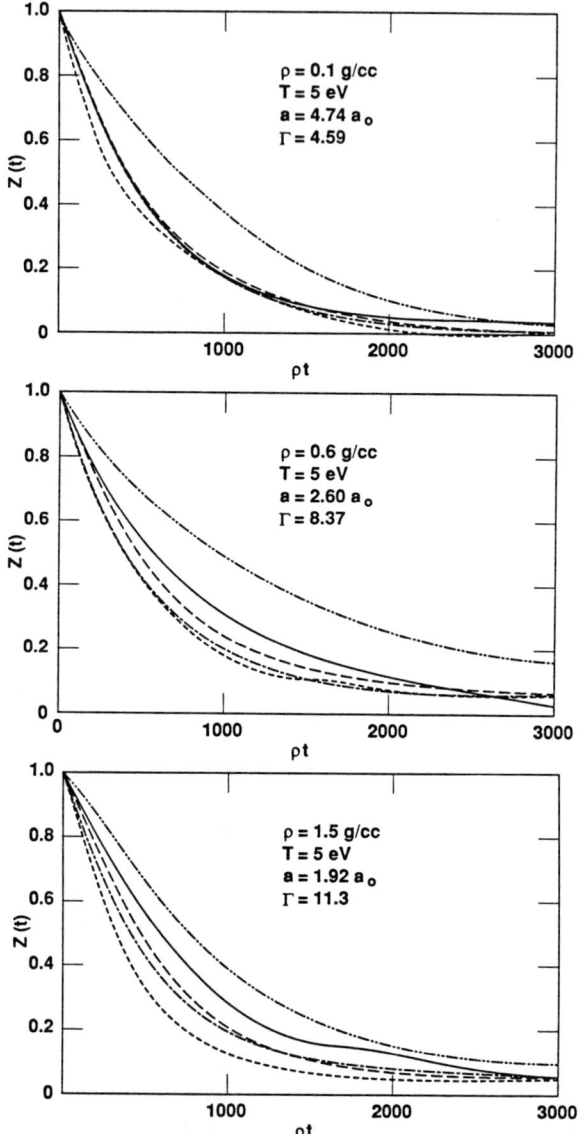

Figure 4. Velocity autocorrelation function for helium at 5 eV as a function of density. Solid curve: SCFMD; dot-dashed curve: molecular dynamics using Hartree-Fock pair potential derived from same basis set as was used in the SCFMD calculation; dotted curve: pair potential due to Aziz et al., based on an accurate Hartree-Fock calculation at short range[7]; dashed curve: pair potential due to Aziz et al. based on quantum Monte Carlo calculations at short range[8]; dot-dot-dashed curve: pair potential of Young et al. based on high density linear muffin tin orbital calculations[9].

sample. The denominator in Z(t) provides unit normalization at t=0. At high temperatures one expects the velocity autocorrelation function to decay with a rate related to the mean collision time. Figure 4 shows Z(t) for helium at 5 eV and densities of 0.1, 0.6, and 1 g/cm^3. In addition to the SCFMD curve, we show the results of classical molecular dynamics calculations using the same number of atoms and the same kinematics algorithms but with pair interaction potentials corresponding to a number of different theoretical and semi-empirical approximations for the binary interaction of helium atoms. The dotted curves in Figure 4 were derived from the pair potential of Aziz et al.[7] based on accurate Hartree-Fock calculations for small separations. The long dashed curve corresponds to a quantum Monte Carlo short range potential in a revised potential due to Aziz et al.[8] The dot-dot-dashed curve was calculated using a linear muffin tin orbital (LMTO) approximation for the short range potential due to Young et al.[9] This approximation is expected to be valid at very high density and low temperature where many-atom screening is significant. It will be less accurate at low density and high temperature where atomic kinetic energy can result in the close approach of two atoms without the presence of the nearest neighbors assumed in the LMTO calculations. Finally, the dot-dashed curves in Figure 4 correspond to classical molecular dynamics calculations using a pair potential derived from the same basis set used in the full SCFMD calculations. Comparison of the solid (SCFMD) and dot-dashed curves is a direct indication of the departure of the interaction from a superposition of pair potentials.

At 0.1 g/cm^3 the SCFMD results are in good agreement with the pair potential calculations except for the LMTO case. This is expected, since at low density binary collisions dominate and there is little dynamic screening. The failure of the LMTO calculation at low density and high temperature was expected, based on the arguments presented above. At 0.6 g/cm^3 the SCFMD results begin to diverge from the pair potential curves, indicating a softer effective interaction. This effect is more pronounced at 1 g/cm^3, where differences in the pair potentials are also apparent.

The physical mechanism for the softening of the effective atomic interaction evident from the velocity autocorrelation function is as follows. At high temperature, significant momentum transfer occurs only during close collisions. During such a collision, the addition of the potentials of the colliding nuclei results in a potential well much deeper and wider than that associated with a single atom. This well attracts not only the charge density associated with the collision partners, but also part of the electronic charge density of surrounding atoms. The concentration of negative charge in the collision region results in additional screening of the nuclear repulsion, allowing a closer approach and an associated slower decay in Z(t).

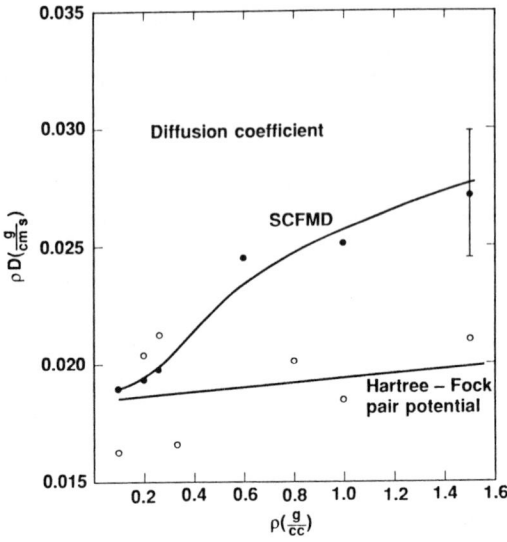

Figure 5. Coefficient of self-diffusion as a function of density. The departure of the SCFMD results from the Hartree-Fock pair potential curve is an indication of dynamic screening at high densities.

The coefficient of self-diffusion D is proportional to the time integral of the velocity autocorrelation function. Figure 5 shows D as a function of density for helium at 5 eV. As the density increases to a value such that there is significant overlap among atomic orbitals D experiences a rapid rise compared to results from Hartree-Fock pair potential calculations.

A limited number of calculations have been performed using a finite electron temperature version of SCFMD. These calculations employed three-Gaussian basis sets for ten atoms. Although such basis sets cannot possibly describe diffuse excited state orbitals for isolated atoms, at high density the proximity of neighbor atoms provides additional basis flexibility which will allow for an approximate description of excitations. Continuum functions will appear as orbitals distributed over many different centers. The diagonalization of the electronic Hamiltonian will attempt the span the eigenspace within the confines of the primitive basis set given.

To describe the ionization balance of the plasma we plot in Figure 6 the number of electrons occupying negative energy states vs. temperature at densities of 0.6 and 1 g/cm^3, demonstrating two competing phenomena in dense plasmas. First, one notes that there is slightly more ionization (fewer bound electrons) for the higher density case, consistent with the theory of pressure ionization. Second, as the temperature is increased at constant density there is less ionization than one would expect based on an isolated atom model of the plasma. This decrease in ionization has nothing at all to do with recombination processes (radiative, dielectronic, three-body, etc.) which are not included in the SCFMD calculation, but is a reflection of a change in the plasma ground state. During collisions the superposition of nuclear potentials which results in dynamic screening also results in very tightly bound orbitals. This is a familiar result from elementary quantum chemistry and corresponds to the approach to the united atom limit. At high temperature the ionic collision rate is high, so that at any given time there are an appreciable number of hybrid orbitals with energies more negative than those of isolated atoms.

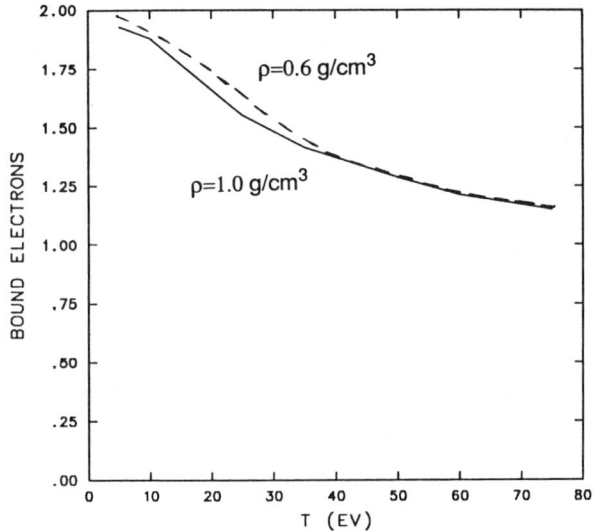

Figure 6. Average number of bound electrons on a helium atom vs. temperature and density, illustrating the competition between pressure ionization and the increase in ionization energy due to the formation of quasi-molecular orbitals.

DISCUSSION

The above calculations demonstrate that at high density many-electron interactions are important in determining both the structure and dynamics of dense matter. The formation of quasi-molecular structures during close collisions not only results in increased screening of the interatomic potential, but an increase in the ionization energies as well.

There are several pieces of experimental information which support these findings. First, the interpretation of the results of a two stage gas gun experiment performed on helium which achieved a density of 0.68 g/cm^3 at an estimated temperature of 2 eV required much softer effective pair interaction potentials than are deduced from binary scattering experiments or accurate theoretical calculations.[6,9] Since the statistical calculations included the effect of many-atom interactions, such a softening must imply a change in the potential itself rather than simply a many-body effect involving the sum of pair interactions.

A second very interesting piece of experimental data relating to energy level lowering in dense plasma is the recent observation of high energy wings on spectral lines occurring at very high density and temperature during laser fusion implosions.[10] Although many phenomena can be invoked to explain low energy features in the spectrum (screening, satellite lines, etc.) an increase in transition energy requires the invocation of a more attractive potential, such as would be obtained during atomic collisions in the plasma.

Although we have concentrated on ionic transport phenomena, we expect the formation of transient quasi-molecular structures in dense matter to have a profound effect on electron transport processes as well. We are examining methods for extracting electron transport quantities from our SCFMD calculations.

Finally, we note that residual effects of orbital hybridization are expected to occur even at temperatures high enough that the plasma is completely ionized. As an illustration, consider the problem of charged particle stopping in dense matter. Traditional theory treats this process by calculating the polarization of a uniform electron gas by the particle, and then calculating the reactive force and energy loss due to this polarization. In a real plasma, however, the electron distribution is not uniform but is polarized by the plasma ions. This additional polarization mechanism reduces the ability of the test particle to polarize the plasma and hence lose energy. The result is that many-body plasma effects should decrease the net energy loss rate and hence increase the effective range of ions in dense plasmas.

ACKNOWLEDGMENTS

This work was performed under the auspices of the United States Department of Energy by the Los Alamos National Laboratory. Part of the work was done while the author was at the Lawrence Livermore National Laboratory. I would like to thank Gayle Sugiyama and Alan Harrison of Livermore for many helpful conversations during the course of this work.

REFERENCES

1. B.F. Rozsnyai, Phys. Rev. A **5**, 1137 (1972).
2. M.P. Allen and D.J. Tildesley, **Computer Simulation of Liquids**, (Oxford University Press, Oxford, 1988).
3. S.M. Younger, A.H. Harrison, K. Fujima, and D. Griswold, Phys. Rev. Lett. **61**, 962 (1988).
4. S.M. Younger, G. Sugiyama, and A.H. Harrison, Phys. Rev. A **40**, 5256 (1989).
5. M. Dupuis, J. Rys, and H.F. King, J. Chem. Phys. **65**, 111 (1976).
6. W.J. Nellis, N.C. Holmes, A.C. Mitchell, R.J. Trainor, G.K. Governo, M. Ross, and D.A. Young, Phys. Rev. Lett. **53**, 1248 (1984).

7. R.A. Aziz, V.P.S. Nain, J.S. Carley, W.L. Taylor, and G.T. McConville, J. Chem. Phys. **70**, 4330 (1979).
8. R.A. Aziz, F.R.W. Mc Court, and C.C.K. Wong, Mol. Phys. **61**, 1487 (1987).
9. D.A. Young, A.K. McMahan, and M. Ross, Phys. Rev. B **24**, 5119 (1981).
10. C.F. Hooper, D.P. Kilcrease, R.C. Mancini, L.A. Woltz, D.K. Bradley, P.A. Jaanimagi, and M.C. Richardson, Phys. Rev. Lett. **63**, 267 (1989).

SPECTROSCOPIC ANALYSIS OF HOT DENSE LASER PRODUCED PLASMAS

C. F. Hooper, Jr., R. C. Mancini, D. P. Kilcrease
Department of Physics
University of Florida
Gainesville, FL 32611

L. A. Woltz
Atomic and Plasma Radiation Division
National Institute of Standards and Technology
Gaithesburg, MD 20899

ABSTRACT

We present time-integrated and time-resolved x-ray line spectra recorded during laser driven implosion experiments. These spectra are compared with theoretical Stark broadened line profiles in order to infer the plasma temperature and density at the time of line emission.

I. Introduction

During the last decade high-power lasers have been employed to implode microballoons filled with gases such as neon, argon, deuterium, or a mixture of deuterium and argon. These implosions have generated high-temperature (~ 1 keV), high electron density ($\sim 10^{23}$/cc $- 10^{25}$/cc) plasmas.

Two centers currently performing such experiments are the Lawrence Livermore National Laboratory and the University of Rochester Laboratory for Laser Energetics.

Diagnostics for these experiments have included a variety of instruments, but we will concentrate on x-ray spectrometer results, both time-integrated and time-resolved. Analysis of the x-ray line spectra resulting from such implosions has required increasingly sophisticated theoretical interpretation.

In section II, we will discuss three sets of experiments and the corresponding theoretical analysis. In section III, we will draw some conclusions and mention some outstanding problems that remain.

II. Theory and Experiment

A. Preliminary Theoretical Considerations

Standard theory of plasma line broadening uses the static ion approximation and treats dipole radiation. Hence, the electrons are treated as dynamic and the ions stationary during the radiative lifetime.[1]

The expression for the Stark broadened line shape is given by

$$I(\omega) = \int_0^\infty P(\varepsilon) J(\omega, \varepsilon) d\varepsilon, \tag{1}$$

where $P(\varepsilon)$ is the microfield probability distribution and

$$J(\omega,\varepsilon) = \frac{1}{\pi}\text{Re Tr}_{re}\int_0^\infty e^{i\omega t}\vec{d}\cdot\rho_{re}e^{-iH(\varepsilon)t/\hbar}\vec{d}e^{iH(\varepsilon)t/\hbar}\,dt \qquad (2)$$

is the electron-broadened line profile for a radiating ion in the presence of the ion microfield $\vec{\varepsilon}$.

Tr_{re} is a trace over radiator and electron states, $\vec{d}$ is the radiator dipole operator, ρ_{re} is the density operator for the radiator and perturbing electrons, and $H(\varepsilon)$ is given by

$$\begin{aligned}H(\varepsilon) &= H_r + e\varepsilon z_r + H_e + V_{er}\\ &= H(r) + H_e + V_{er}\end{aligned} \qquad (3)$$

Here,

H_r is the unperturbed Hamiltonian for the radiator,

z_r is the z coordinate of the radiating electron,

H_e is the Hamiltonian for the plasma electrons,

V_{er} is the radiator-electron interaction, and

$H(r)$ is the Hamiltonian for the radiator in the presence of the static ion field.

For convenience we have chosen the z axis to be in the direction of the static ion field and the radiator nucleus to be the origin of our coordinates.

Employing well-known transform techniques and neglecting any plasma-electron generated level shifts, which are believed to be small, we can write Eq. (2) as

$$J(\omega,\varepsilon) = -\frac{1}{\pi}\text{Im Tr}_r\vec{d}\cdot[\omega - L(r) - M(\omega)]^{-1}f(r)\vec{d}, \qquad (4)$$

where:

$f(r)$ is the radiator density matrix,

$L(r)$ is the Liouville operator for the radiator in the presence of a static ion field, and

$M(\omega)$ is the electron width and shift operator which includes the dynamic electron broadening.

This formulation has recently been extended to allow line broadening calculations to be performed for multi-electron radiators.[2] Further generalizations have also allowed a treatment of the ion quadrupole effect that includes ion correlations. In these cases atomic physics information for the radiator is generated by a separate code such as that developed by Cowan at LANL.

Doppler broadening is included by convolution:

$$I(\omega) = \int_{-\infty}^{+\infty} I_{\text{Doppler}}(\omega - \omega')I_{\text{Stark}}(\omega')d\omega' \qquad (5)$$

Finally, opacity broadening is included with the following approximate expression (uniform slab model):

$$\Psi(\omega) \propto \left[1 - e^{-\tau_o(I(\omega)/I(\omega_o))}\right]$$
$$\tau_o \propto f\, N_1 I(\omega_o) \ell$$
$$f = \text{oscillator strength}$$
$$N_1 = \text{ion ground state density} \quad (6)$$
$$I(\omega_o) = \text{intensity at line center}$$
$$\ell = \text{characteristic geometric length}$$

Here, $I(\omega)$ is the Stark + Doppler profile.

Changes from the standard theory will be discussed in the context of our experimental analysis.

B. Experiments and Analysis

1. **Rochester (LLE) Experiments (1985)**

   a) Primary purpose: study feasibility of using L-shell x-ray emission as a density diagnostic.

   b) Laser driver was 24-beam Ω system with $\lambda = 3500$ Å and energy $\sim 2000 J$ delivered in ~ 600 ps.

   c) Target: Ar/Kr filled (50/50) plastic shell (350 μm x 5 μm), 500 Å Al layer.

   d) Spectra and analysis: time integrated spectra, multi-electron radiator line-broadening theory.

The analysis and main results for this series of experiments are presented in Figures 1 to 6. Figure 1 shows the Ar K-shell line emission and the Kr L-shell $n = 4$ to 2 transitions, while Figure 2 displays an example of the fitting of the Ar K-shell lines that allowed us to infer the plasma conditions. Figure 3 displays the emission from $n = 3$ to 2 transitions in Li- and Be-ℓ Kr; a comparison between theoretical and experimental spectra is presented in Figure 4. These lines do not have a significant electron density sensitivity in the $1 \times 10^{23} - 1 \times 10^{24}$ cm^{-3} range. On the other hand, we found that the $n = 4$ to 2 transitions are sufficiently density sensitive to make them useful for diagnostic purposes; this fact is demonstrated in Figure 5 while Figure 6 shows the best theory-experiment comparison which was found at an electron density of 2×10^{23}, in good agreement with the results of Ar K-shell analysis.

We conclude that Stark line-broadening analysis of L-shell x-ray spectra does offer promise as a plasma-density diagnostic.

2. **Rochester (LLE) Experiments (1987)**

   a) Purpose: follow-up to experiments described in 1.

   b) Driver was again 24-beam Ω-laser system with $\lambda = 3500$ Å and energy $\sim 2000 J$ delivered in ~ 600 ps; new laser beam smoothing techniques were employed.

c) Targets:
   1. Shot number 15761: 450 μm x 6 μm with 1000 Å Al layer; 2 atm Ar fill.
   2. Shot number 15771: 450 μm x 6 μm with 1000 Å Al layer; 10 atm Ar fill.
   3. Ar/Kr filled and empty microballoons with nominally the same physical characteristics.

d) Spectra and Analysis:
   1. For shots 15761 and 15771, we obtained time-resolved x-ray spectra. The He_α line with its Li-like satellites was observed as was the Ly_α line plus He-like satellites (see Figures 7 and 8).
   2. There was little or no line emission observed from microballoons with Kr/Ar (50/50) fills.

e) Analysis
   1. For shots number 15761 and 15771, Stark broadening analysis was employed to evaluate the T and N_e dependence of the Ly_α line (and satellites) and the He_α line (and satellites). An illustration is shown in Figure 9. As a result of our analysis thus far, we have assigned the following values of T, N_e to spectra from 15761.[3]

TABLE 1. Density and temperature results for the analysis of data in Figs. 7 and 8. Time is measured with respect to the peak of the laser pulse (1×10^{24} cm$^{-3} \simeq 4$ g cm^{-3}).

Figure, Spectrum	Time, interval (ps)	Electron density (cm^{-3})	Temperature (eV)
7,(a)	109–171	$2. - 6. \times 10^{24}$	600–900
7,(b)	171–233	$4. - 6. \times 10^{24}$	600–900
7,(c)	202–264	$4. - 7. \times 10^{24}$	600–800
7,(d)	233–295	$6. - 8. \times 10^{24}$	500–600
7,(e)	295–357	$2.5 - 6. \times 10^{24}$	450–600
7,(f)	357–419	$2.5 - 4. \times 10^{24}$	450–600
8,(a)	312–375	$3. - 6. \times 10^{24}$	600–900
8,(b)	348–405	$5. - 7. \times 10^{24}$	600–800
8,(c)	375–427	$6. - 8. \times 10^{24}$	500–800

   2. The blue satellite features are unexplained as yet, but may be indicative of a cooperative phenomenon.
   3. The observation of K_α radiation from Be-, B-, and C-like argon is consistent with the picture obtained by the model: hot compact core, cooler outer layer.
   4. Relatively low opacities are a surprise but are consistent with ionization-balance calculations.

3. LLNL Experiments (1988)

Microballoons filled with D_2 and a fraction of a percent of Ar were imploded using the Nova laser system.[4] Time-resolved Ar spectra were obtained. Samples of spectra are illustrated in Figures 10 and 11. These spectra show the line emission from the He_β (3684 eV) and He_γ (3874 eV) of He-like Ar and the Ly_β (3935 eV) of Hy-like Ar. From the three line emissions recorded in this energy range, the He_β provided the best quality signal and was therefore selected for diagnostic purposes.

As an initial step in analyzing these spectra we have calculated the He_β line of Ar^{+16} (not yet including the observed Li-like satellite structure). Our calculations have included

a) field mixing of the $n = 3$ and $n = 4$ levels.

b) inclusion of the ion quadrupole effect.

Figures 12, 13, and 14 show some of the results of our theoretical calculations, while Figure 15 displays the best fit to the He_β line of Figure 11 indicating an electron density of 1×10^{24} cm^{-3}.

III. Conclusions

A. With increased laser power and improved beam quality, even higher densities are possible.

B. It is highly probable that plasmas created at these densities will lead to new physics.

C. Plasma spectroscopy is, and will continue to be, an indispensable tool in understanding these plasma phenomena.

D. The resulting plasmas from implosions are still not well characterized.

1. We need more and better designed experiments

2. We need continued improvement in diagnostic techniques.

E. Outstanding problems still exist.

1. Can we observe the nucleation of the plasma-solid-state phase transition?

2. Treatments of continuum lowering are still open to question.

3. More study of non-LTE effects is needed, both theoretical and experimental.

4. To what degree are higher order field-effects important at high densities?

5. Ion-dynamic effects in line broadening can be accurately simulated, but other theoretical approaches are still not adequate.

6. Claims persist that at high electron densities very large line shifts occur. To what extent are these claims valid?

7. Where is the inclusion of higher multipole radiation important?

Much has been done, much remains to be done.

REFERENCES

1. H. Griem, Spectral Line Broadening by Plasmas (Academic, New York, 1974).
2. L.A. Woltz and C.F. Hooper, Jr., Phys. Rev. **A38**, 4766 (1988).
3. C.F. Hooper, Jr., D.P. Kilcrease, R.C. Mancini, L.A. Woltz, D.K. Bradley, P.A. Jaanimagi and M.C. Richardson, Phys. Rev. Lett. **63**, 267 (1989).
4. R.E. Turner, P. Bell, S.G. Glendinning, S. Hatchett, J.D. Kilkenny, G. Power, L. Suter, R.S. Thoe and R. Uphdye, Bull. Am. Phys. Soc. **33**, 1945 (1988).
5. R.W. Lee, B.L. Whitten and R.E. Stout, II, J. Quant. Spec. Rad. Transfer **32**, 91 (1984).

FIGURES

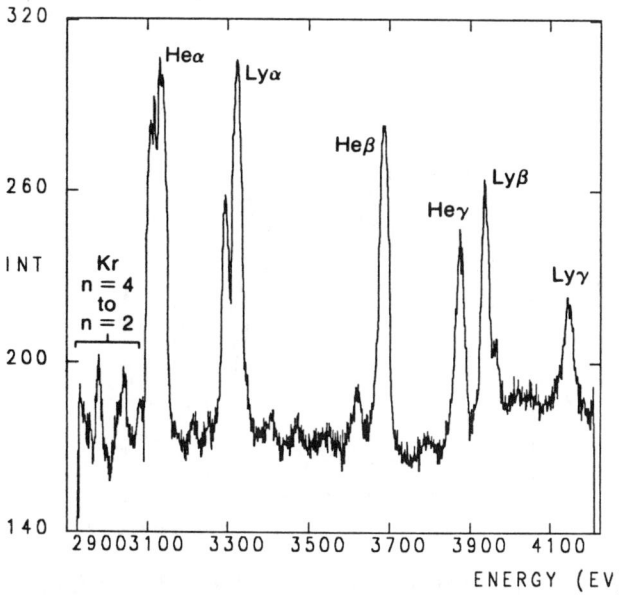

Fig. 1: Time integrated spectra in the range 2900–4200 eV. This spectrum shows the Ar K-shell and the Kr L-shell $n = 4 \to 2$ lines.

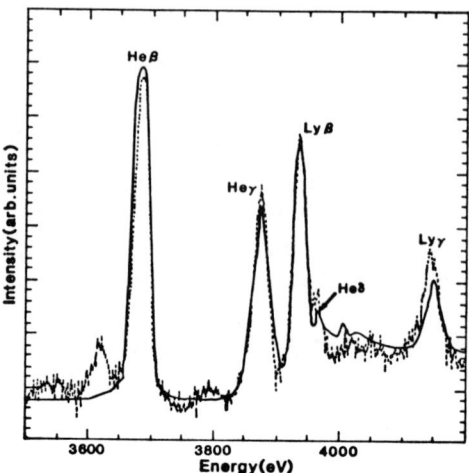

Fig. 2: Calculations using the RATION code[5] are shown in comparison to the Ar spectra of shot 11582. Plasma conditions for this shot are estimated to be: $T = 1100$ eV, $N_e = 1.5 \times 10^{23}$ cm^{-3}.

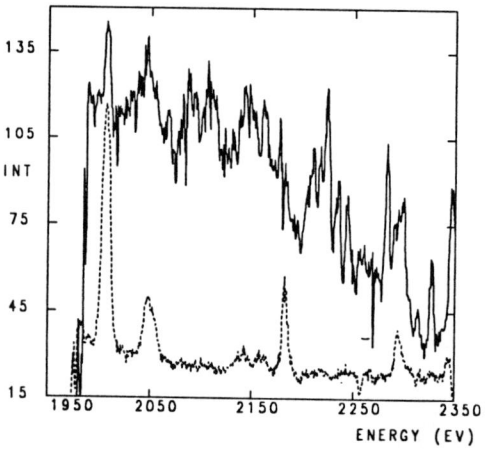

Fig. 3: Time integrated spectra in the range 1900–2400 eV. Lower spectrum, shot 11581: an unfilled polymer microballoon coated with 500 Å of Al. The most prominent lines corresponds to Si^{+13}Ly$_\alpha$ (2005 eV), Si^{+12}He$_\beta$ (2183 eV), Si^{+12}He$_\gamma$ (2294 eV), and Al^{+12}Ly$_\beta$ (2048 eV). Upper spectrum, shot 11582: a polymer microballoon coated with 500 Å of Al and filled with 50% of Ar and 50% of Kr. In addition to the lines seen in shot 11581, we now see emission from Kr L-shell lines. (These spectra have been multiplied by arbitrary factors in order to make them fit in the same figure.)

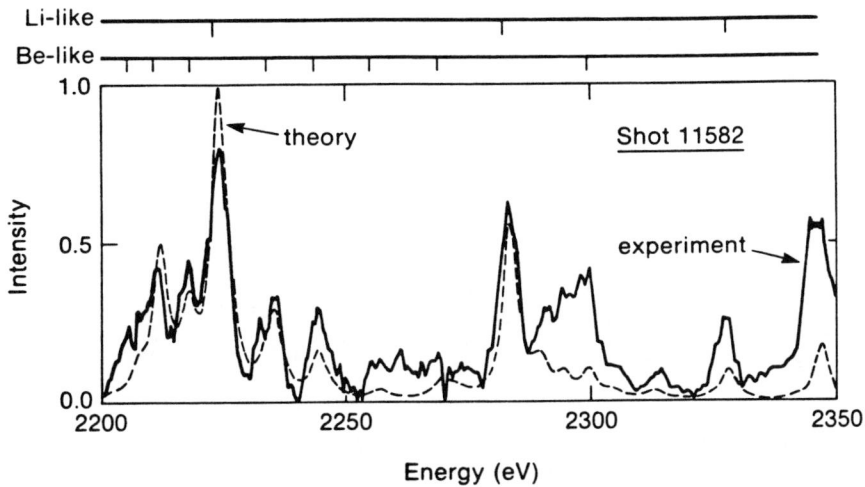

Fig. 4: Comparison of the combined Li-like and Be-like Kr theoretical spectra (– – –) with experimental spectra (——) in the range 2200–2350 eV for shot 11582. Silicon He$_\gamma$ (2294 eV) and He$_\delta$ (2346 eV) lines contribute to the experimental spectrum.

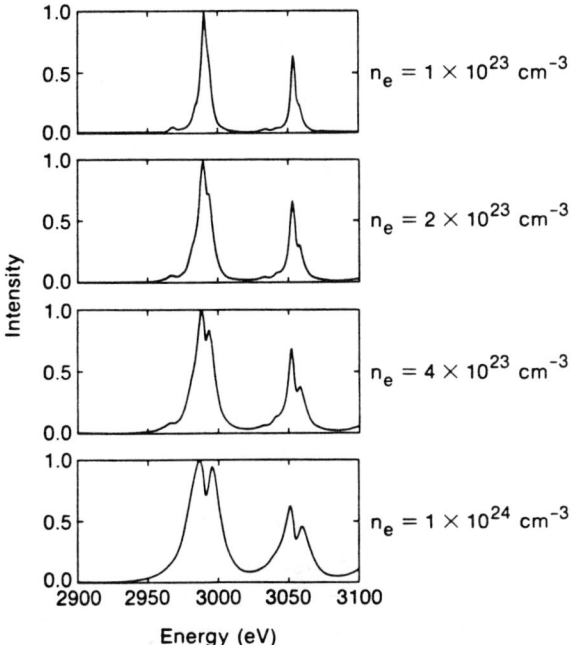

Fig. 5: Electron density sensitivity for the Li-like Kr spectra, $n=4$ to $n=2$ transitions in the 2900–3100 eV range for a temperature of 1100 eV.

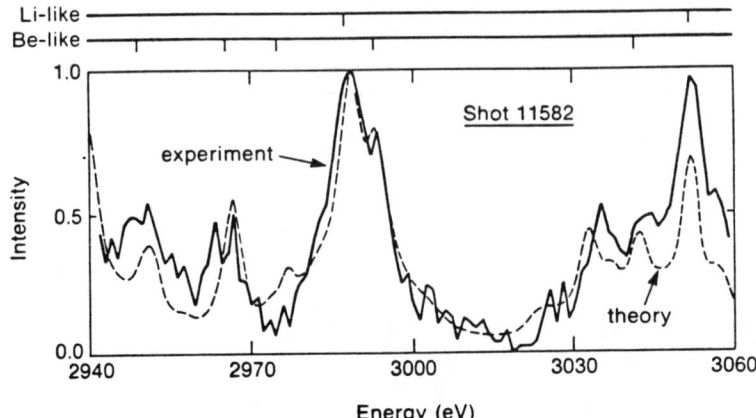

Fig. 6: Comparison of combined Li-like and Be-like Kr theoretical spectra (– – –) with LLE shot 11582 (———), 2940–3060 eV.

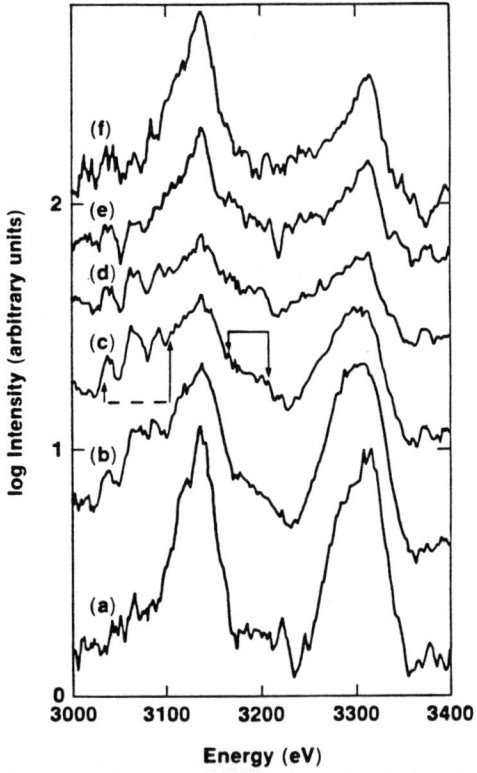

Fig. 7: Series of time-resolved spectra for the 2 atm case. (c) and (d) correspond to implosion stagnation; (– – –) inner-shell transitions in C-, B- and Be-like Ar; (———) broad "blue satellite" feature. These spectra have been shifted vertically by arbitrary amounts for the purpose of suitable display in the same picture.

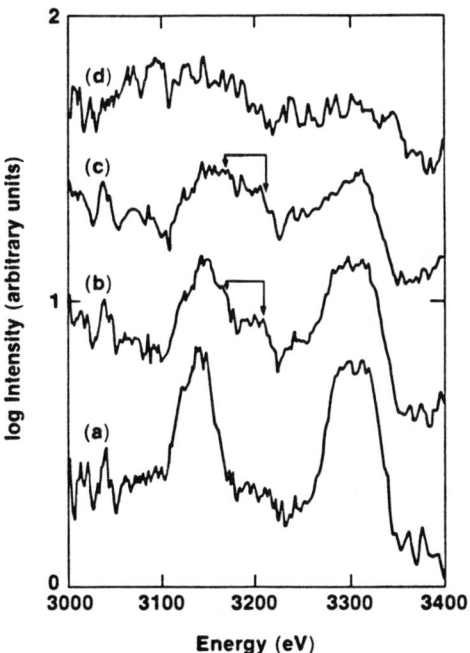

Fig. 8: Same as previous figure, but for the 10 atm case. In this case spectrum (c) is characteristic of the stagnation. Note that after the stagnation (d) only continuum emission is seen.

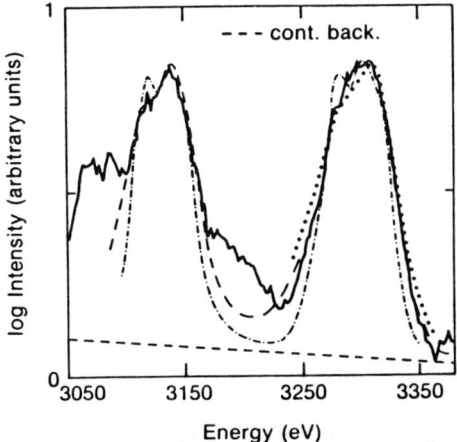

Fig. 9: An example of the fittings obtained. This case corresponds to spectrum b in Figure 7. Experimental spectrum (———); theoretical fittings: $N_e = 1 \times 10^{24}$ (—·—·—·), 5×10^{24} (- - -) and 8×10^{24} (· ··), in cm^{-3}.

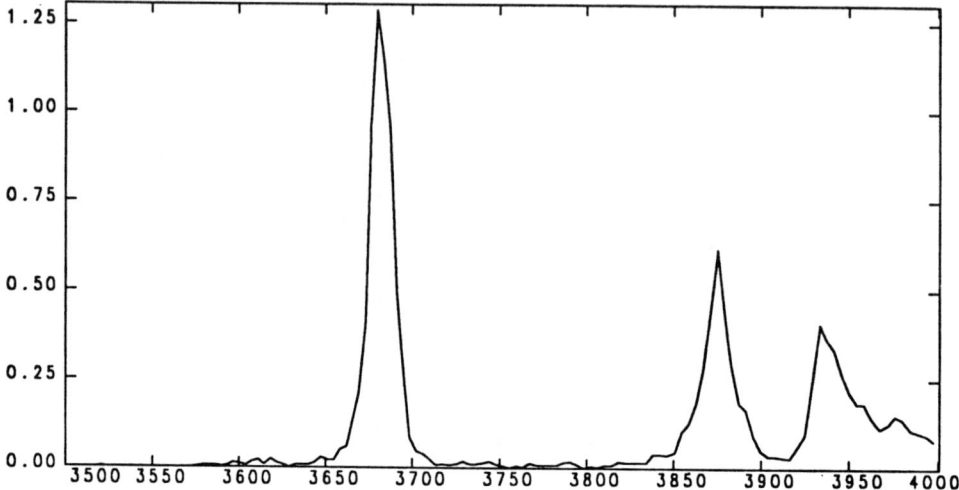

Fig. 10: Time-resolved argon spectra (He$_\beta$, He$_\gamma$, Ly$_\beta$) from an Ar doped DD implosion.

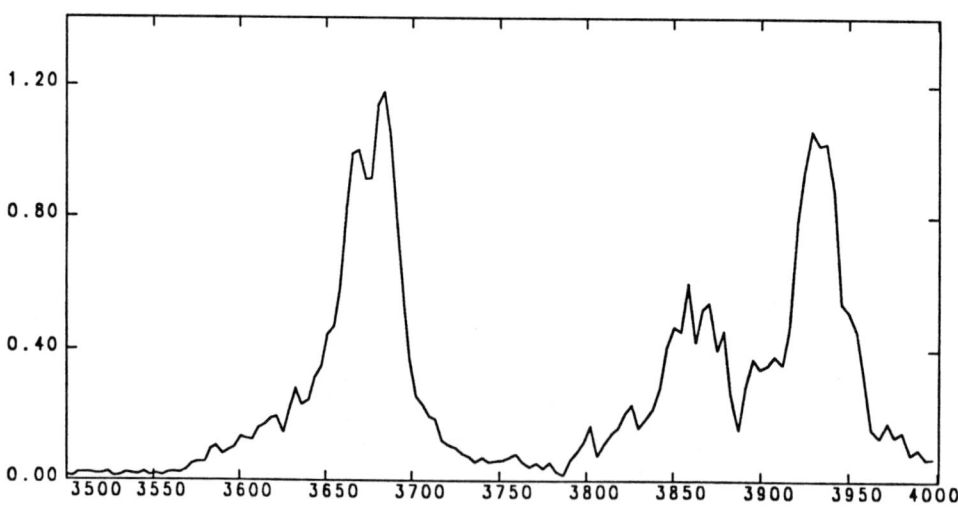

Fig. 11: Time-resolved argon spectra (He$_\beta$, He$_\gamma$, Ly$_\beta$) from an Ar doped DD implosion.

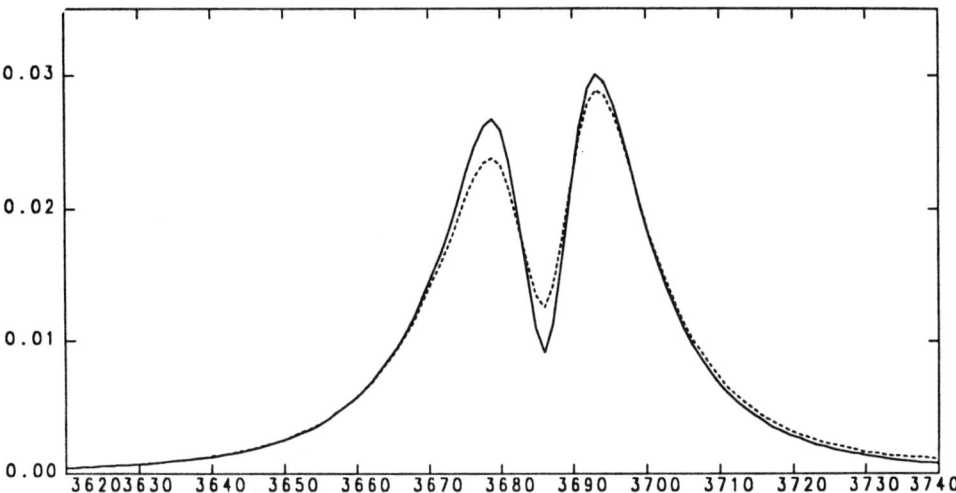

Fig. 12: He$_\beta$ line profile for He-like Ar at a density $N_e = 1 \times 10^{24}$ cm^{-3} and temperature $T = 1000$ eV. (———): only the 1s3ℓ states included in the calculation; (— — —): 1s3ℓ and 1s4ℓ states included.

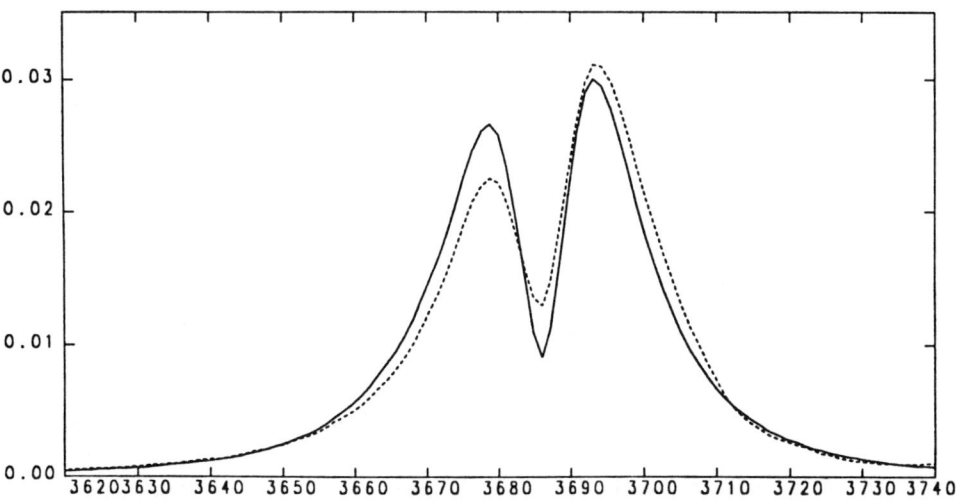

Fig. 13: He$_\beta$ line profile for He-like Ar at $N_e = 1 \times 10^{24}$ cm^{-3} and $T = 1$ keV. (———): only the 1s3ℓ states included in the calculation; (— — —) indicates the result obtained when considering the 1s3ℓ and 1s4ℓ states together with the ion quadrupole effect.

216 Hot Dense Laser Produced Plasmas

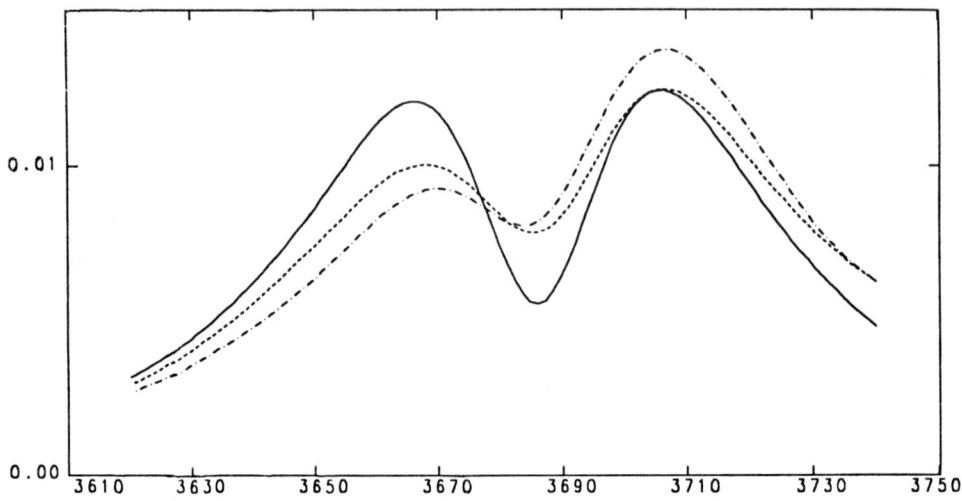

Fig. 14: He$_\beta$ line profile for He-like Ar at a density $N_e = 5 \times 10^{24}$ cm^{-3} and temperature $T = 1000$ eV. (———): result including 1s3ℓ states; (– – –): result considering 1s3ℓ and 1s4ℓ states; (– · – · –): result with 1s3ℓ and 1s4ℓ states, and the ion quadrupole effect.

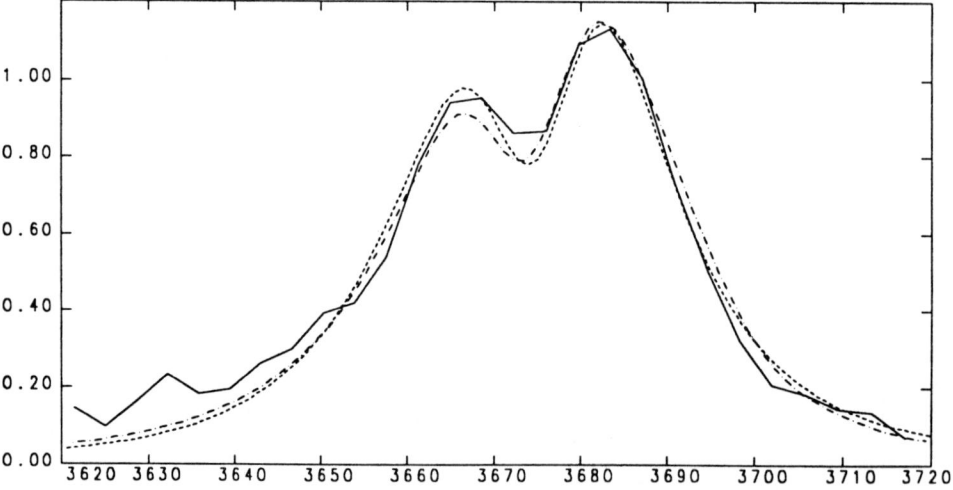

Fig. 15: Theoretical fit to the He$_\beta$ line profile. (———): experimental spectra; (– – –): theoretical data for $N_e = 1 \times 10^{24}$ cm^{-3} and $T = 1000$ eV, calculation included 1s3ℓ and 1s4ℓ states; profile has been corrected for opacity with $\tau_o = 0.3$; (– · – · –): the theoretical calculations included 1s3ℓ and 1s4ℓ states as well as ion quadrupole effect; the opacity correction was done with $\tau_o = 0.8$.

ATOMIC PROCESSES IN PLASMAS UNDER ULTRA-INTENSE LASER IRRADIATION[*]

G. T. Schappert, D. E. Casperson, J. A. Cobble, J. C. Comly, L. A. Jones, G. A. Kyrala, K. J. LaGattuta, P. H. Y. Lee, G. L. Olson, and A. J. Taylor

Los Alamos National Laboratory
Los Alamos, New Mexico, 87545

ABSTRACT

Lasers delivering subpicosecond pulses with energies of a fraction of a Joule have made it possible to generate irradiance levels approaching 10^{20} W/cm^2. We presently operate two such systems, a KrF based excimer laser capable of producing a few 10^{17} W/cm^2 at 248 nm with a repetition rate of 3-5 Hz and a XeCl based excimer laser capable of producing mid 10^{19} W/cm^2 at 308 nm and 1 Hz. We will discuss some experimental results and the theory and modeling of the interaction of such intense laser pulses with aluminum. Because of a small ASE prepulse the high intensity interaction is not at the solid surface but rather at the $n_e = 2 \times 10^{22}$ cm^{-3} (KrF) laser critical density of the blowoff plasma generated by the ASE. The transient behavior of the plasma following the energy deposition by the intense subpicosecond pulse can be viewed as the energy-impulse response of the plasma. Experimental results and modeling of the x-ray emission from this plasma are presented.

I. INTRODUCTION

Recent new developments in excimer laser technology have made it possible to achieve irradiances on target exceeding 10^{19} W/cm^2. The developments responsible for this are oscillator designs capable of generating pulses of the order of femtosecond duration and large aperture amplifiers capable of amplifying these pulses without significant pulse stretching or wavefront distortion. Hence modest energies, e.g one joule in a 200 fs pulse, focused to a nearly diffraction-limited spot a few microns in diameter results in an irradiance of nearly 10^{20} W/cm^2. Such laser systems typically operate at a repetition rate of several Hz, ideal for experimental programs.

Laser matter interaction experiments and modeling of these experiments are in progress at several laboratories. Such experiments are of great interest because in this intensity range the radiation electric fields can exceed the atomic binding fields, leading to an area of atomic physics which has never been available for experimental study. At 10^{19} W/cm^2 the peak laser electric field is nearly 10^{11} V/cm, a factor of twenty larger than the basic atomic electric field unit e/a_o^2. The classical quiver energy of a free electron in such a laser field is 180 keV for the XeCl wavelength at 308 nm. Clearly, laser matter interactions under such conditions lead to ultrahigh energy density conditions in the material. In this paper we will describe the two high irradiance laser systems at Los Alamos and report on high energy density experiments in an aluminum plasma.

II. LASER FACILITY

Subpicosecond, high-brightness excimer laser systems are being used to explore the interaction of coherent ultraviolet radiation with matter. The vast majority of such systems, based on the amplification of subpicosecond pulses in small aperture (1 cm^2) XeCl or KrF amplifiers, deliver[1-4] focal spot irradiances of approximately 10^{17} W/cm^2. Scaling to higher irradiances requires an additional large aperture amplifier[5,6,7] which preserves nearly diffraction-limited beam quality and subpicosecond pulse duration. We describe both a small aperture KrF system which routinely provides irradiances >10^{17} W/cm^2 to the experiments discussed here, as well as a large aperture XeCl system which produces 0.25-J, subpicosecond pulses and yields irradiances on target (with f/1 optics) of 6.4×10^{19} W/cm^2. The output parameters of both systems are summarized in Table I.

	LABS I (KrF)	LABS II (XeCl)
Repetition rate (Hz)	5	1
Wavelength (nm)	248	308
Energy (mJ)	30 ± 1	250 ± 10
Pulsewidth (ps)	.67	.335
Intensity (W/cm^2)	4.4×10^{17} (f/3)	4.6×10^{18} (f/3.7)
Intensity (W/cm^2)	4.0×10^{18} (f/1)	6.4×10^{19} (f/1)

Table I. Operating conditions of the Los Alamos Bright Source (LABS) laser systems LABS I and LABS II.

The small aperture KrF system consists of a "front-end" which produces 248-nm seed pulses, followed by two KrF amplifiers. The system is sketched in Figure 1. The

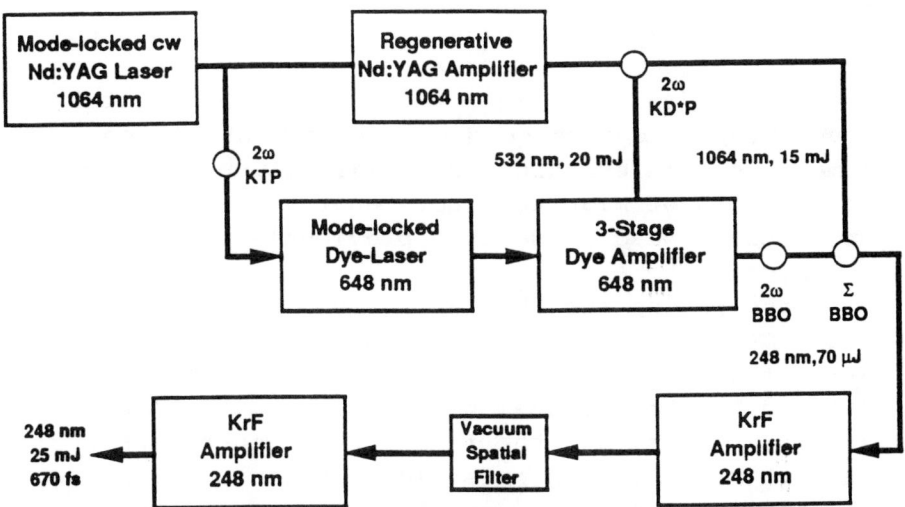

Fig 1. Small aperture high-brightness KrF laser system

output pulse train from a mode-locked Nd:YAG laser is split into two beams. One beam is frequency-doubled in KTP and used to synchronously pump a mode-locked dye laser with separate gain and absorber jets. The gain medium is DCM dissolved in benzyl alcohol and ethylene glycol, while the absorber is DTDCI in methanol and ethylene glycol. A pulse from the remaining train is selected at a repetition rate of 5 Hz and amplified by a regenerative amplifier up to the 40-mJ level. This 100-ps pulse is frequency doubled in KD*P with 50% conversion efficiency and used to longitudinally pump a three stage dye amplifier seeded by the 650-fs, 648-nm pulses from the dye laser. The amplified pulse energy is 2.0 mJ and the beam quality is better than two times diffraction-limited. We observe no temporal broadening of the 650-fs pulses through the amplifier. Advantages of this synchronous amplification scheme include low spontaneous emission, nearly diffraction-limited amplified beam quality, elimination of timing jitter between the pump pulses and the seed pulses to the amplifier and the availability of the unconverted 100-ps, 1064-nm pulses for mixing purposes. The amplified 648-nm pulses are then frequency-doubled in a 2-mm long BBO crystal. The resulting pulses at 324 nm are finally sum-frequency mixed with unconverted 1064-nm pulses from the regenerative amplifier in a second 2-mm BBO crystal to produce 70-μJ subpicosecond seed pulses at 248 nm. This scheme, based on synchronous amplification, produces one to two orders of magnitude more energy in the 248-nm subpicosecond seed pulse than previously reported in other systems[2-6].

These pulses are then amplified by two Lambda Physik EMG 200 Series KrF amplifiers, separated by a vacuum spatial filter to suppress amplified spontaneous emission (ASE) and to improve beam quality. The output beam diameter is 25 mm and the final output energy is 30±1 mJ with <0.5 mJ ASE. The pulsewidth, measured using two-photon ionization in NO, is 670 fs. The focused spot size achievable with this system has been determined indirectly by measuring the confocal parameter of a beam focused by f/3 optics. The inferred focal spot diameter is 3.6 μm, which implies an irradiance at the focal plane of 4.4×10^{17} W/cm^2. For most experiments parabolic mirrors are used as the focusing optics to preserve pulsewidth, minimize aberrations and avoid nonlinear absorption and refraction.

The large aperture XeCl system[7] is sketched in Figure 2. Pulses of 165 fs duration at 616 nm are initially generated in a linear-cavity, dispersion-

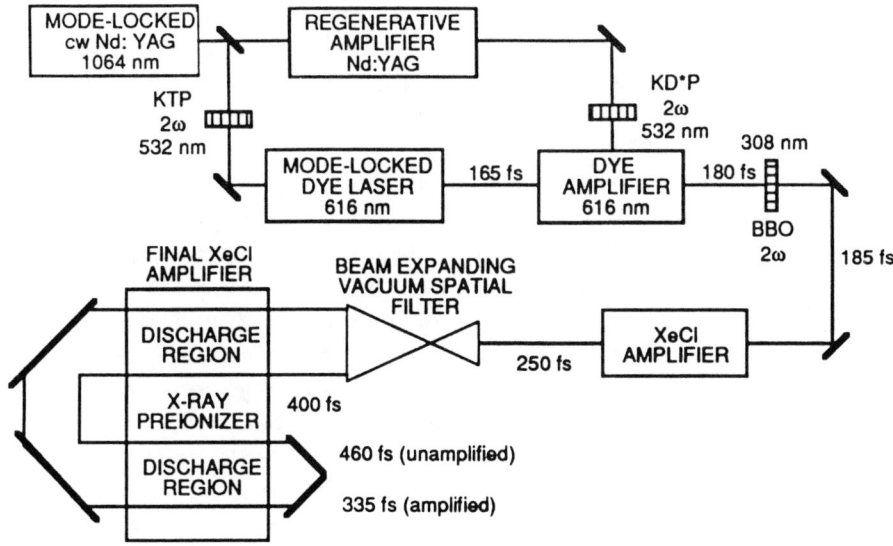

Fig. 2 Large aperture high-brightness XeCl laser system

compensated dye laser (Kiton Red/ DODCI) that is synchronously pumped by the frequency-doubled output of a cw mode-locked Nd:YAG laser. A synchronous amplification scheme, similar to that described for the KrF system, is used to amplify these pulses to the 1.5 mJ level[8]. The amplified 616-nm pulses are then frequency-doubled in a BBO crystal. Preamplification of these 0.15-mJ pulses to the 6-mJ level is accomplished with a single small aperture XeCl discharge laser. The pulsewidth at this stage is 250 fs. The beam is then expanded in a vacuum spatial filter before entering the final amplifier.

The 10x10 cm^2 aperture final amplifier consists of two independently pumped discharge gain regions, each sharing a common 130-keV x-ray preionizer. The discharges are pumped by low-jitter, thyratron-switched pulse modulators with two-stage magnetic compression. The resultant jitter is 5 ns over a 20-ns gain time. The small-signal gain length, g_oL, for each discharge region is 8.0 and useful gain is obtained over roughly one half of the aperture area. In order to maintain nearly diffraction-limited beam quality at a 1-Hz repetition rate, a gas flow system is employed which maintains nearly laminar flow transverse to the discharge electrodes. After hot gas is cleared from the discharge volume, this flow system establishes less than $\lambda/20$ wavefront distortion over the clear aperture before the next shot is fired.

The output energy of the system is 250 ±10 mJ in a 335 fs pulse. To illustrate the evolution of the pulse shape as the pulse propagates through the system, the pulse duration at each stage has been indicated in Figure 2. All values given in the figure are measured using autocorrelation techniques (two photon ionization in DABCO), with the exception of the value given after the BBO doubling crystal, which was calculated. The ultrashort pulse lies on a 20-ns, 50-mJ ASE pedestal, but only half of the ASE is included in the solid angle of the output beam. The beam quality of the fully amplified pulse is determined by measuring the transmission through a calibrated pinhole using an f/3.7 focusing optic. The measured transmission of 82% through a 10 μm-diameter pinhole implies FWHM dimensions of the focal spot of 3.4±0.7 by 4.1±0.8 μm^2. These dimensions are about 3 times larger than would be obtained with a diffraction limited beam. With these focal spot dimensions, the mean focal irradiance is 4.6×10^{18} W/cm^2. To our knowledge, this represents the highest demonstrated optical irradiance ever obtained with a laser source. With f/1 focusing optics, this system will produce an irradiance of 6.4×10^{19} W/cm^2, which corresponds to an electric field of about 30 atomic field units and results in a relativistic electron quiver motion.

Initial experiments on this system have recently started. The first was a test of multiphoton absorption in Ne at an irradiance of several 10^{18} W/cm^2. The laser was focused into low pressure Ne (10^{-7}-10^{-6} Torr to avoid space-charge effects) in an ion time of flight spectrometer which separates the ions according to their charge to mass ratio. A typical time of flight spectrum is shown in Figure 3 indicating eight charge states of Ne, including the two isotopes ^{20}Ne and ^{22}Ne for most of them. To generate Ne^{+8} requires a net investment of more than 1 KeV of energy or the absorption of over 250 4-eV photons.

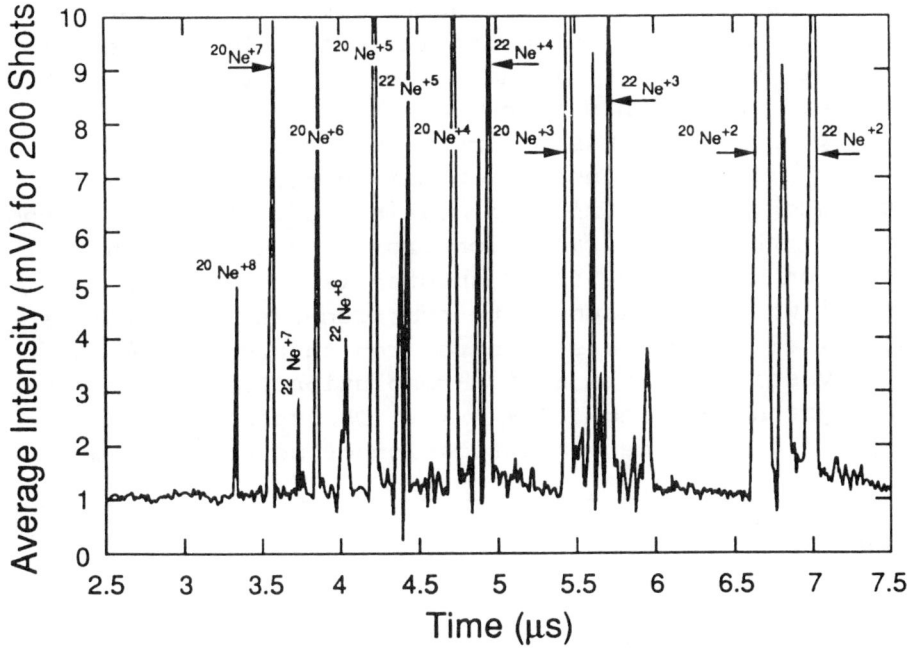

Fig. 3 Neon ions time of flight spectrum

II. SOLID TARGET EXPERIMENTS

The experiments discussed in this section are investigations of the interaction of high-irradiance pulses with solid targets performed on LABS-I. The purpose of this work is to study the resulting state of the hot, high-density plasma and possible applications.

A 20-mJ, 0.7-ps pulse is focused onto an aluminum target resulting in an irradiance of several 10^{17} W/cm^2 at a repetition rate of 3 Hz. The target slowly translates so that each pulse interacts with a new surface. The 0.7-ps main pulse is superimposed on a roughly 15-ns pedestal of amplified spontaneous emission (ASE) which contains about 5% of the total energy. Hence the main pulse interacts with a small blowoff plasma generated by this prepulse ASE. This raises the question whether the main pulse ever reaches the solid target surface since its propagation in the prepulse plasma is limited to electron densities below the 2×10^{22} cm^{-3}

critical density. Since we cannot eliminate the ASE prepulse, we have investigated its effect by varying the amount of ASE and the timing of the main pulse relative to the ASE onset. This gives rise to different scale-length prepulse plasmas in front of the solid.

The primary diagnostic on this experiment is x-ray spectroscopy. A transmission grating spectrograph is used to survey the plasma radiation from 5 to 100 Å with a resolution of 2 Å. We found substantial x-ray emission near 7 Å which corresponds to hydrogen- and helium-like aluminum emission. This spectral region around 7 Å is then investigated with a high-resolution flat PET crystal spectrograph covering 1.5 - 2 keV. The spectrum is shown in Figure 4 with the identification of several hydrogen-

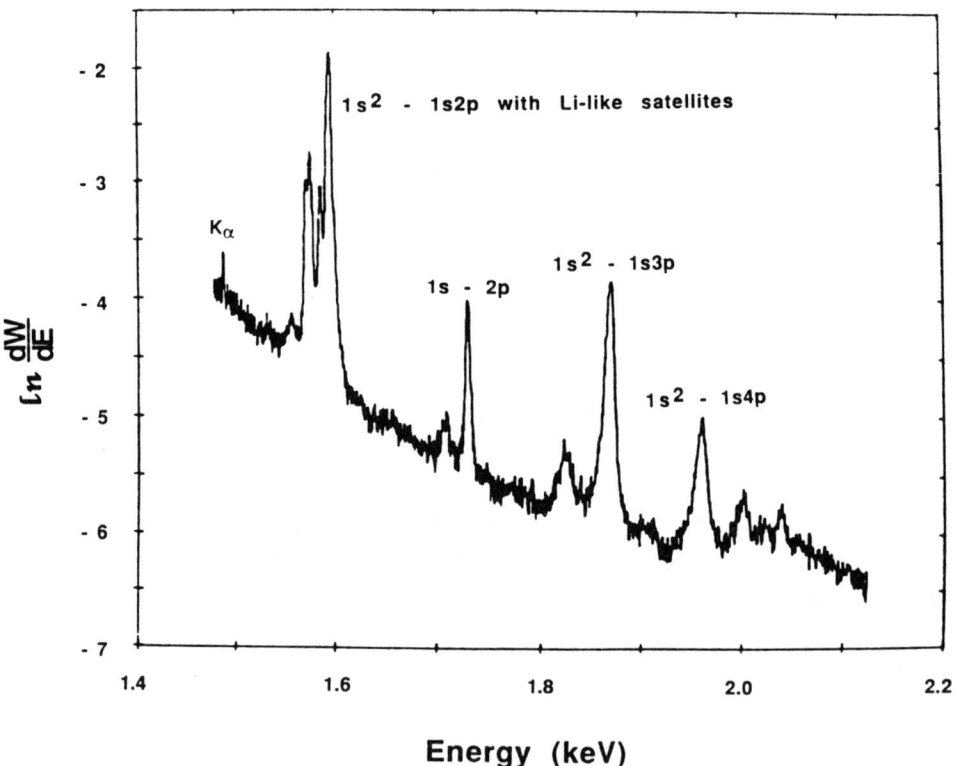

Fig.4 Natural log of x-ray flux from PET crystal spectrum of aluminum integrated over 100 shots

and helium-like aluminum lines. The spectrum is time
integrated, requiring about one hundred shots. The
temporal evolution of the 1.5-2 keV radiation is measured
with a filtered x-ray streak camera which indicated an
emission time of less than 10 ps. Filtered pinhole camera
pictures show the emission region of this radiation to be
a spot of <10 μm in diameter.

A simple analysis[9] of this highly-transient nonequilibrium plasma based on the dominant collisional and
radiative atomic rates resulted in a peak electron
temperature of about 1.5 keV and an electron density near
the critical density of $2 \times 10^{22}/cm^3$. Preliminary
estimates of x-ray conversion efficiency into keV line
radiation approach 0.5 % of the laser energy.

We now turn to the investigation of the role of the
ASE prepulse. By varying the ASE, we have subjected the
aluminum surface to a prepulse intensity between 10 -
2000 GW/cm^2, resulting in various scale-length blow-off
plasmas. Figure 5 depicts the typical situation, showing

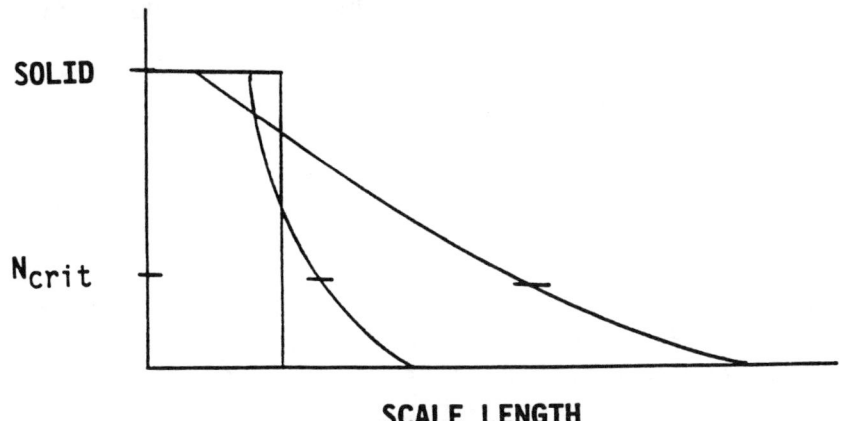

Fig.5 Blow-off plasma from ASE prepulse

the position of the initial solid aluminum boundary, a
short scale length blow-off plasma from a weak prepulse,
and a long scale length plasma from a stronger prepulse.
Since the lowest prepulse intensity (10 Gw/cm^2) is just
about the damage threshold for aluminum, one would expect
little if any blow-off and the interaction with the main
pulse takes place close to solid density aluminum.

The PET crystal spectroscopy data were analyzed as function of the prepulse irradiance. We concentrate on the He-like Al 3-1 and 4-1 line intensities and the line widths. The instrumental width of the spectrometer is about 1 eV.

The intensities of the He-like Aluminum 3-1 and 4-1 lines as a function of the prepulse irradiance are shown on Figures 6 a and 6 b. In both cases a threshold

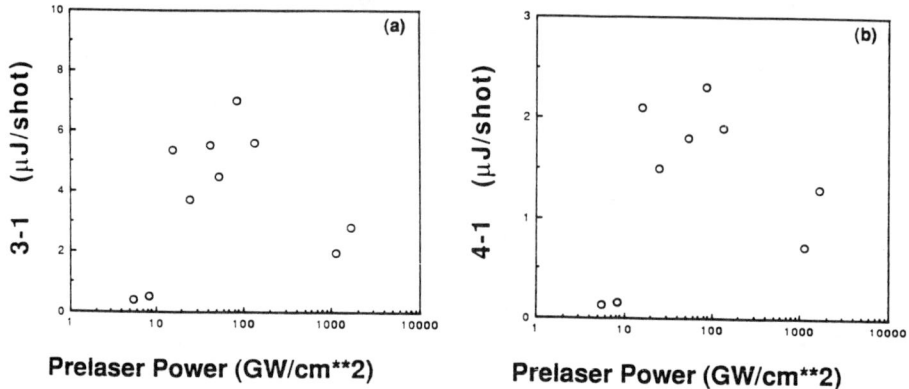

Fig.6 He-like Al (a) 3-1 line (b) 4-1 line intensity as a function of prepulse irradiance

prepulse is required to produce significant intensities, the intensities peak at a moderate level of prepulse, and decline at high prepulse levels. A qualitative explanation of this dependence is that at low prepulse the energy is deposited in a steep density gradient at critical and transported by electron thermal conduction to the higher density region behind critical. The resulting lower electron temperature is insufficient to excite the He-like aluminum. At the other extreme, the main pulse is dissipated in the long density scale length plasma without reaching the critical density. Again, the lower temperatures cannot excite the emission. A similar effect has been observed [10,11] using short pulse dye lasers. The blow-off plasma was generated by a first pulse followed by a second pulse with adjustable delay. Neither paper studied individual spectral lines but broad band regions, the above 1 keV region in aluminum in reference (10) and the 35 - 70 eV region in tantalum in reference (11). From these results it is clear that there is an optimum prepulse plasma for maximum laser to x-ray conversion.

Figures 7 a and 7 b show the linewidths (FWHM eV) of the 3-1 and 4-1 He-like lines as a function of the prepulse irradiance. These lines have nearly the same linewidth, which appears to be inconsistent with Stark broadening theory with roughly n^2 scaling. An analysis of the widths based on scaling a He-like argon

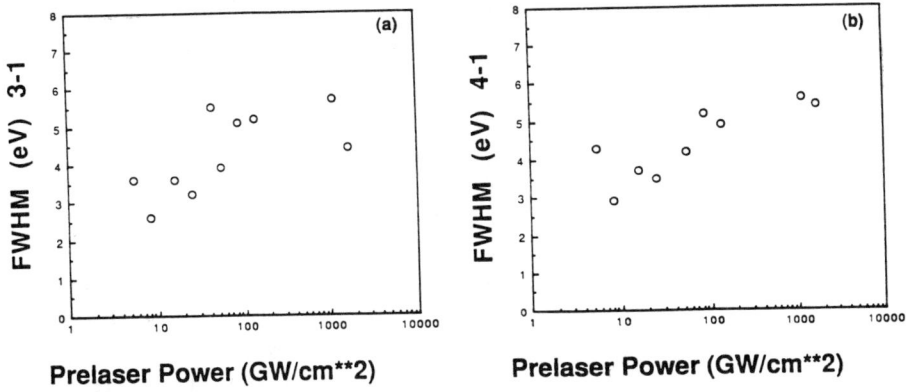

Fig.7 He-like Al (a) 3-1 line (b) 4-1 line width
       FWHM(eV) as a function of prepulse irradiance

calculation[12] which includes electron broadening and ion dynamical effects results in an electron density range of $6 \times 10^{21} - 1.5 \times 10^{22}$ cm^{-3} for the 3 - 1 line in Fig. 7 a, and somewhat lower, $4.5 \times 10^{21} - 10^{22}$ cm^{-3} for the 4-1 line in Fig. 7 b. This implies that the two lines may not be emitted from the same region, or that opacity broadening is involved in the 3-1 line emitted from a lower density region.

Next consider the factor of two increase in the linewidths as we increase the prepulse irradiance by over two orders of magnitude. Under the assumption of Stark broadening this implies an emission region of 2-3 times the density at the highest prepulse irradiance compared to the lowest. At our lowest prepulse intensity we generate little or no plasma, hence the main pulse interacts at the critical density in the steep density gradient very near the solid aluminum. For a longer prepulse plasma scale length, the critical density region will be at a greater distance from the high density region resulting in less Stark broadening. A similar effect has been observed[13,14] in the soft x-ray emission from short pulse irradiated silicon. These experiments used a dye laser which was stated to have no prepulse. By introducing a prepulse, the plasma could then be investigated over a wide range of density gradients. Although the density diagnostic was not line width but resonance - intercombination line ratios, the experimental results clearly indicated a narrower line width emission from what was interpreted as the higher density plasma, i. e. the no-prepulse case.

The above discussion questions whether the Stark effect is the dominant broadening mechanism for our He-like Al lines. The fact that the lines have nearly equal width suggests a motional effect. A simple Doppler interpretation would require ion temperatures of 15 - 60 KeV, which is unrealistic. Another possibility is the generation of large velocity fields. When the main pulse propagates up the prepulse generated plasma gradient a region of maximum energy deposition occurs. This is usually at critical if the pulse does not expend its energy earlier. The blow-off from this region then leads to a velocity field as shown in a one-dimensional ZAP-code simulation on Figure 8. In this simulation a Gaussian prepulse of 10 ns FWHM (truncated at ± 10 ns) with peak intensity of 10^{11} W/cm^2 is incident on an Aluminum surface located at 0.005 cm. The ensuing absorption, hydro-blowoff, and heating are calculated. At 3 ns the .7 ps, 10^{17} W/cm^2 main pulse reaches the original solid interface after propagating through the blowoff plasma established by the prepulse. The plasma is further heated and ionized. Figure 8 shows the velocity fields at three different times after the main pulse. Note the large velocities established within a few ps of the main pulse near the solid interface and that the outgoing components are several times larger than those propagating into the high density material. The 1-2 keV radiation emitted from this region would be Doppler-broadened and blue-shifted by about 4 eV. Consider now the same problem with a long scale length prepulse plasma such that the critical density is far from the original solid interface. A more symmetric velocity field, with shocks running both up and down the gentle density gradient is established resulting in a Doppler spread (and no shift) of twice that of the steep gradient case. Hence, the results of this simulation show that hydro

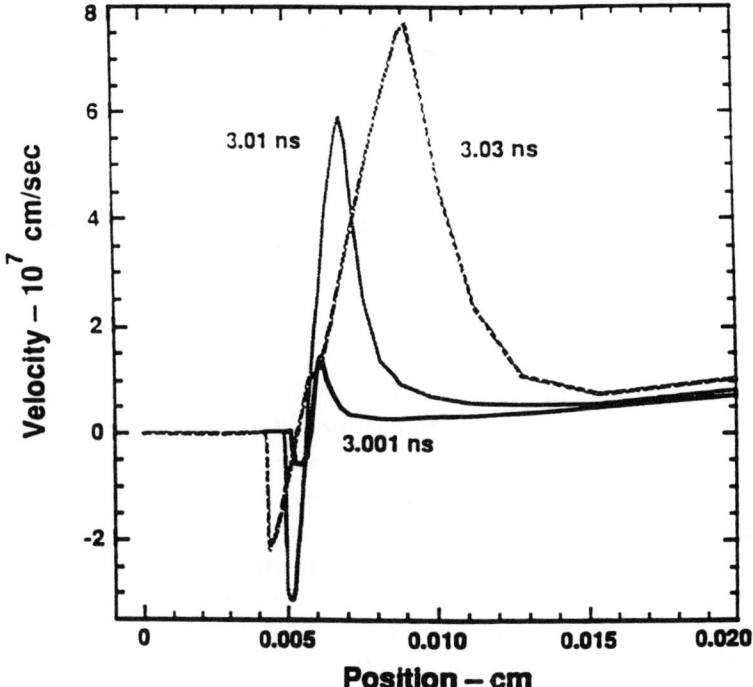

Fig.8 Calculated one dimensional velocity profile as a function of position 1, 10, and 30 ps after the main pulse

Doppler effects could explain the observed line width behavior provided they dominate the Stark broadening. This would require the emitting region be substantially below the critical density. Perhaps a Doppler-Stark coupling calculation[15] for He-like Al ions in a velocity field could resolve this problem.

More modeling and experimental work is in progress to resolve the above questions and lead to a better understanding of the role of prepulse plasmas in high irradiance laser matter interaction studies.

The role of multiphoton processes in the high density plasma interaction experiments is not clear. One might argue that if the collision frequency exceeds the optical frequency, the coherent interaction with the laser field is disrupted and multiphoton, or laser electric field dependent processes are diminished.

In general however, these results look promising for converting ps laser pulses into ps x-ray line radiation. We are presently investigating higher Z materials where one expects even better conversion efficiency. Such narrowband pulsed x-ray sources are valuable for applications such as x-ray laser research and time-resolved x-ray imaging.

IV. ACKNOWLEDGMENT

We wish to thank J. Grosso, S. E. Harper, T. R. Hurry, C. S. Lester, M. D. Maestas, J. P. Roberts, and K. A. Stetler for their technical support in laser operations and experiments and C. F. Fenstermacher, R. B. Gibson and R. J. Jensen for their help in the design and procurement of the large aperture amplifier for LABS II from Beta Development Corporation.

* This work was supported by the U.S. Department of Energy under contract W-7405-ENG-36 with the University of California.

REFERENCES

1. J. H. Glownia, J. Misewich, and P. P. Sorokin, J. Opt. Soc. Am. B **4**, 1061 (1987).
2. A. P. Schwarzenbach, T. S. Luk, I. A. McIntyre, U. Johann, A. McPherson, K. Boyer, and C. K. Rhodes, Opt. Lett. **11**, 499 (1986).
3. J. R. Roberts, A. J. Taylor, P. H. Y. Lee, and R. B. Gibson, Opt. Lett. **13**, 734 (1988).
4. W. Tighe, C. H. Nam, J. Robinson, and S. Suckewer, Rev. Sci. Instrum. **59**, 2235 (1988).

5. J. R. M. Barr, N. J. Everall, C. J. Hooker, I. N. Ross, M. F. Shaw, and W. T. Toner, Optics Commun. **66**, 127 (1988).
6. A. Endoh, M. Watanabe, N. Sarukura, and S. Watanabe, Opt. Lett. **14**, 353 (1989).
7. A. J. Taylor, J. P. Roberts, T. R. Gosnell, and C. S. Lester, Opt. Lett. **14**, 444 (1989).
8. A. J. Taylor, C. R. Tallman, J. P. Roberts, C.S. Lester, T. R. Gosnell, P. H. Y. Lee, and G. A. Kyrala to be published in Optics Letters, January 1990
9. J. A. Cobble, G. A. Kyrala,A. A. Hauer,A. J. Taylor, C. C. Gomez, N. D. Delamater, and G. T. Schappert Phys. Rev. A. **39**, 454, (1989)
10. D. Kuhlke, U. Herpers, and D. von der Linde Appl. Phys. Lett. **50**, 1785, 1987
11. H. W. K. Tom and O. R. Wood,II Appl. Phys. Lett. **54**, 517, 1989
12. H. R. Griem, M. Blaha, and P. C. Kepple Plasma Preprint UMLPR 89-029
13. M. M. Murnane, H. C. Kapteyn, and R. W. Falcone Phys. Rev. Lett. **62**,155, 1989
14. H. M. Milchberg and H. R. Griem Phys. Rev. Lett. **63** ,338,(1989)
15. R. Stamm, B. Talin, E. L. Pollock and C. A. Iglesias Phys. Rev. A. **34**, 4144, 1986

SPECTROSCOPY OF DYNAMIC, NONEQUILIBRIUM, AND TURBULENT PLASMAS IN PULSED-POWER SYSTEMS

Y. Maron

Physics Department, Weizmann Institute of Science, 76100 Rehovot, Israel.

ABSTRACT

High resolution spectroscopic diagnostic methods have been developed to reliably measure the temporal and spatial distributions of physical quantities in the strong-electric-field region and in the plasmas in high-power high-voltage devices. Spontaneous emission, laser-light absorption, and laser-induced-fluorescence were utilized, together with time-dependent collisional-radiative calculations. These methods were employed to investigate the properties of the acceleration gap and the behavior of the highly dynamic nonequilibrium anode plasma in intense ion diodes. Conclusions on the electron density and current density in the diode gap, ion lateral velocities in the gap, the magnetic field induced by the current flow, the plasma formation, the plasma conductivity, plasma heating, plasma expansion, particle fluxes from the anode surface into the plasma, particle density and velocity distributions in the plasma as well as near the anode surface, turbulent electric fields, electron energy distribution, and possible use in other pulsed-power configurations are discussed.

1. INTRODUCTION

Understanding the complicated phenomena that take place in a high power device can be improved only if high-resolution non-perturbing diagnostic methods are used to observe many physical quantities inside the device. Such measurements are difficult to perform because of the strong fields, high current densities, small size of the devices, and the experimental irreproducibility. In the last years we developed spectroscopic techniques capable to probe into both the acceleration gap (the region in which strong electric fields prevail) and the plasmas in pulsed power systems[1-10]. These techniques were employed to investigate the charge flow and the anode-plasma behavior in magnetically insulated ion diodes.

This research programme was initiated by the suggestion[1] that the electric field distribution in the diode acceleration gap can be measured by the observation of the Stark shift of line emission from ions transversing the gap. Also, the line Doppler broadening parallel to the electrodes can give the transverse velocity distribution of the accelerated ions. For these measurements the ions can be excited by a tunable laser[1] or spontaneous line emission can be utilized. These measurements yielded[2,3] the ion velocities in the diode acceleration gap, the charge and current densities, and the *actual* gap width, i.e., the time-dependent distance between the electric-field-excluding electrode plasmas. Prior to these investigations these quantities had been estimated from

1-D theoretical treatments or from simulations. However, the actual charge flow is severely affected by the finite size of the diodes, the spatial variations of the magnetic fields, and the plasma expansion and nonuniformities.

Later, we developed spectroscopic methods to investigate the plasmas in such systems[4-7]. For these measurements spectral profiles of two lines were observed as a function of time in a single discharge with spectral and temporal resolutions ≤ 0.1 Å and 5 ns, respectively. Line shifts were determined to within 0.02 Å and the spatial resolution could be varied between 30 μm and 1 mm. We obtained the particle velocity distributions in the anode plasma from Doppler broadening and shift and the time dependent magnetic field from Zeeman splitting. The electron temperature and temperature gradient were determined from line intensities and collisional radiative calculations. These calculations had to be *time dependent* since such plasmas are far from being in a steady state because they undergo rapid ionization during the ≤ 100-ns-long pulses[11], and also because they are usually being continuously replenished throughout the pulse by material flowing into them from the plasma source. The calculations accounted for the continuous injection of neutral particles and multiply charged ions into the plasma, thus allowing the time dependent absolute fluxes of the various species injected into the plasma and those drawn from the plasma into the gap to be obtained. Recently, we also investigated the particle velocity and density distributions at the immediate vicinity of the anode surface using laser spectroscopy[8], we studied turbulent electric fields from anisotropic Stark broadening and plasma satellites[9], and probed into the electron energy distribution by observing level population ratios for various atomic systems[10]. In this paper we describe the measurements and the main conclusions on the diode acceleration gap on the anode plasma.

II. EXPERIMENTAL ARRANGEMENT

We used planar magnetically insulated diodes[2,4] as shown in Fig.1(a). The magnetic field B_z applied parallel to the anode, which inhibits the electron flow, was 5 to 9 kG. The anode plasma was formed as a result of a flashover of the epoxy-anode surface. Various elements were mixed with the epoxy in small areas of the anode in order to seed the plasma with the species required for the various spectroscopic observations. The diode waveforms are shown in Fig.1(b).

The diagnostic system is shown in Fig.2 as described in Ref.4. Laser light could be directed into the diode in the y-direction (parallel to the anode) or in the x-direction (perpendicular to the anode). Light from the diode could be collected in the y and z-directions and at an up-to-50° angle with the anode surface in the $x - z$ plane. We used one-meter spectrographs equipped with 2400 grooves/mm gratings, the output of which was further optically dispersed and projected onto rectangular fibre-photomultiplier-tube arrays.

For the laser-spectroscopy[8] we used a high-power pulsed ($\simeq$ 6 ns) dye laser equipped with a wave extending unit to allow for tuning in the range

2160-9000Å. For the various measurements the laser light bandwidth could be changed from 1 to several cm^{-1}.

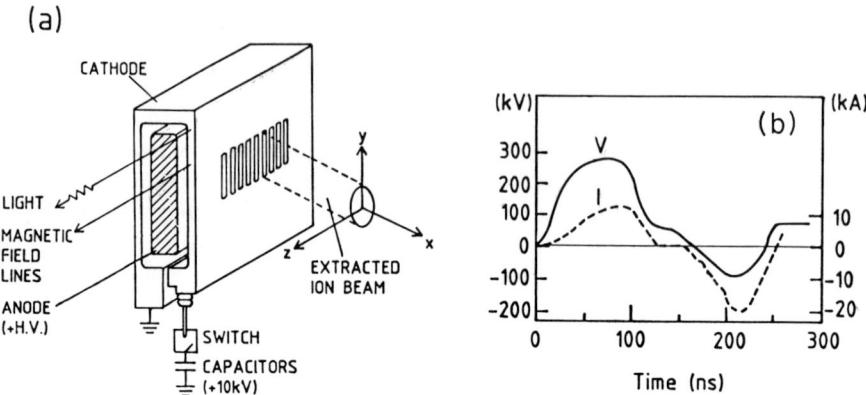

Fig. 1: (a) The planar magnetically insulated ion diode. The strong electric field E_x is produced in the 8-mm-wide anode-cathode gap by connecting the anode to the $\simeq +300$ kV voltage pulse. The insulating magnetic field B_z is produced by driving $\sim$300 kA current through the cathode that serves as a single-turn coil. The ion beam is extracted through slots in the cathode. (b) The diode voltage and total current waveforms for B_z=7.0 kG.

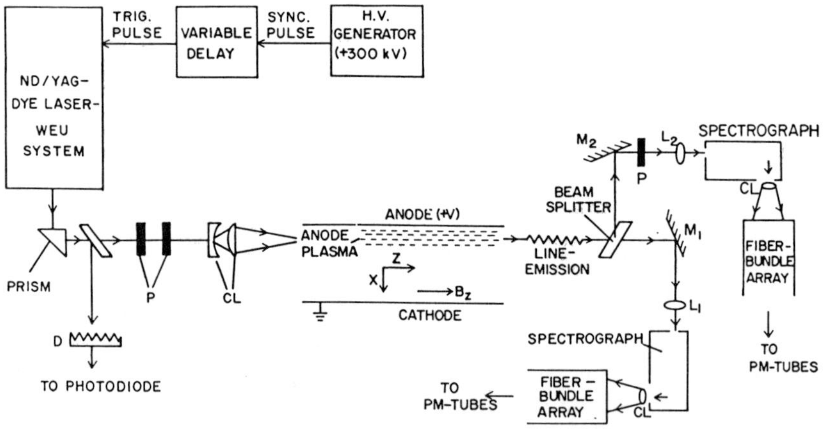

Fig. 2: The laser system and the diagnostic arrangement. WEU, D, P, CL, and M denote a wave extending unit, a diffuser, a polarizer, a cylindrical lens and a mirror, respectively. The laser light, synchronized with the diode voltage pulse, can be directed into the diode in the x and y directions. Laser light,

induced fluorescence, and spontaneous emission can be collected in various directions by the two spectroscopic systems.

III. MEASUREMENTS OF THE ELECTRIC FIELD IN THE DIODE ACCELERATION GAP

For these measurements, using spontaneous emission, the light-emitting ions have to be excited at the anode plasma to a level that is appropriately Stark shifted during the ion acceleration in the diode gap. The shift has to be larger than the Doppler broadening that results from the ion velocities parallel to the electrodes (See Sec. IV), unless Doppler-free spectroscopy is employed[1]. The level has to radiatively decay in the visible or U.V. regions in time comparable to the ion transit time (a few ns) in the few-mm-wide gap. Finally, the line emission from the desired ions, that usually constitute a small fraction of the ion beam of a density $\simeq 10^{12}$ cm^{-3}, has to be intense enough to allow for an observation of the spectral line profile in a single discharge with the required temporal and spatial resolutions. The electric field for various positions across the gap is obtained in separate discharges.

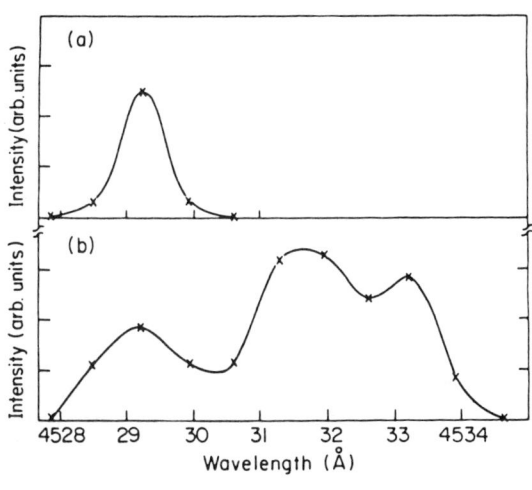

Fig. 3: (a) Spectral profile of the $4d_{5/2} \rightarrow 4p_{3/2}$ AlIII emission from the anode plasma ($x=0.15$ cm from the anode surface) at $t=70$ ns after the start of the diode voltage pulse. The zero-field wavelength is 4529.2 Å. The initial anode-cathode gap was 0.65.
(b) Profile of the emission from the acceleration gap: $x=0.375$ cm and $t=65$ ns. The uncertainty for each point is $\pm 10\%$. The signal at 4529.2 Å results from scattered plasma light. The lines in (a) and (b) indicate the trend.

We used the $4d \rightarrow 4p$ line emission from AlIII ions drawn from the plasma. The emission spectral profile in the plasma and in the gap are shown in Fig. 3. The electric field is unfolded from the Stark shifted emission using a calculation of the emission pattern under the effects of the electric field, the magnetic field, and the Doppler broadening. An example of a measured electric field distribution is shown in Fig.4. This distribution gave the distributions of the electric potential, of the ion velocity, of the ion density (using the total ion current density measured outside the diode), the electron density (using poission equation), and the electron current density (using one-dimensional analysis[12]. Furthermore, the zero-field positions gave the time dependent *actual* width of the acceleration gap unambiguously. This allowed comparisons with theoretical treatments[13], which are rather sensitive to the diode gap, to be made and also yielded the plasma expansion rate.

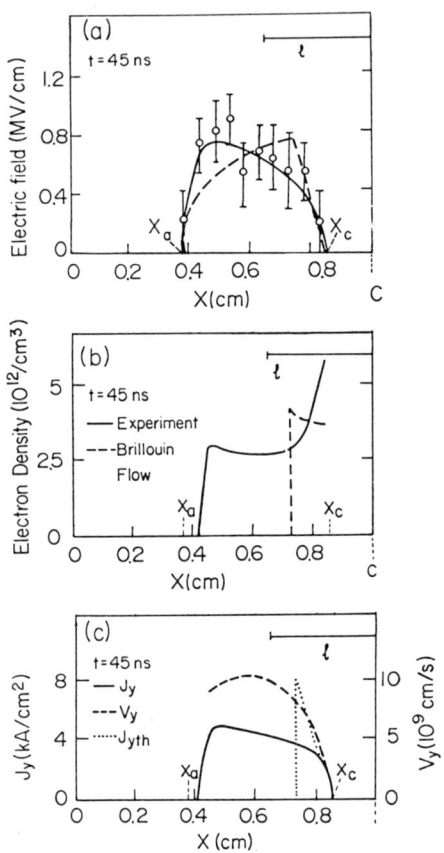

Fig. 4: (a) Measured $E(x)$ distribution using the Stark shift method. The voltage on the diode $V=280$ kV and the applied magnetic field $B_z=6.6$ kG. The solid anode surface is at $x=0$ and the cathode is at $x=1$ cm (denoted by C). In this experiment a cathode vane projected 0.35 cm off the cathode surface. The regions between $x=0$ and $x=x_a$ and between $x=x_c$ and $x=1$ cm are of the field-excluding anode and cathode plasmas, respectively.
(b) Electron density distribution inferred from $E(x)$ in (a) and that derived from the one-dimensional Brillouin-flow model (Antonsen and Ott in Ref. 13) using the actual diode gap $d = x_c - x_a$.
(c) The distribution of the electron current density ($J_y(x)$) inferred from $E(x)$ in (a) and that derived from the Brillouin-flow model ($J_{yth}(x)$) for the actual diode gap.

The distributions of the electron density and the electron current density were observed to spread towards the anode beyond the region of the electron sheath predicted by the 1-D solutions[13], see Fig. 4(b). This, together with the increased total number of electrons in the gap explained the enhancement of the measured ion current density over the calculated one. Knowledge of the electron density distribution in the diode gap is very important since the diode operation is sensitive to this parameter as demonstrated in the various theoretical solutions[13,14].

The distribution of the electron current density inferred (Fig.4(c)) showed a significant electron flow close to the anode plasma, which enhances the induced magnetic field on the anode side, as verified in later measurements[5]. In addition, these measurements showed a rapid gap closure early in the pulse resulting from expansion of the electrode plasmas (see Sec. V).

IV. MEASUREMENTS OF ION TRANSVERSE VELOCITIES IN THE DIODE GAP

The transverse velocity distributions of CIII and AlIII ions parallel to the electrodes in the diode acceleration gap were obtained from Doppler broadening of spontaneous emission as shown in Fig. 5[3]. These velocities were found to be significantly smaller than those previously observed for protons outside the diode, suggesting that the ions acquire considerable transverse velocities either upon their extraction through the cathode slots or during their drift outside the diode.

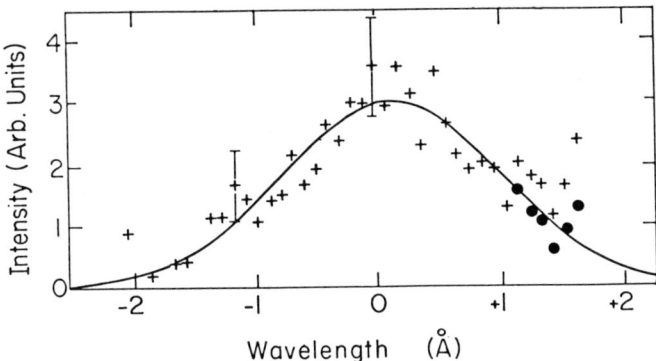

Fig. 5: Spectral profile of the CIII 4647.4-Å line emission from the acceleration gap at $x=5.5$ mm. The emission intensity is averaged over $t=30$ to $t=70$ ns. Spatial and spectral resolutions are 1.2 mm and 0.85 Å, respectively. The wavelength is given with respect to the line center at the anode plasma. The round points are data corrected for the contribution of the 4650.16-Å line of the triplet [2s(^{2}S)3p→2s3s]. The curve is a Gaussian fit to the data.

The comparison between the CIII and AlIII transverse velocities (the latter was obtained the AlIII 3601-Å line) suggested that the ion beam divergence angle inside the diode is independent of the ion mass. This would happen if the divergence is caused by transverse electric fields of a frequency much smaller than the inverse ion transit time in the gap.

V. PLASMA FORMATION AND EARLY FAST EXPANSION

Line emission was first observed at $t \simeq 20$ ns, and at $t \simeq 55$ ns the plasma was seen to occupy an about 1.5-mm-wide region near the anode surface, giving an expansion velocity $\gtrsim 3$ $cm/\mu s^4$, consistent with previous observations[2]. Early fast plasma expansion against the magnetic filed due to low-temperature (collisional) electrons is ruled out because of fast electron heating due to elastic collisions with the hotter ions[15]. We therefore, suggested a mechanism of plasma expansion based on the formation of a fast-neutral layer by charge exchange processes and the subsequent ionization of this layer[15]. The calculated expansion and ionization of this layer were in agreement with the plasma thickness and density observed early in the pulse[15]. This suggestion is consistent with the observed insensitivity of the plasma expansion to the applied magnetic field and the early presence of the magnetic field in the plasma (see Sec. VII).

VI. PARTICLE VELOCITY DISTRIBUTIONS IN THE ANODE PLASMA

The particle velocity distributions in the anode plasma were determined by the observation of line Doppler broadenings and shifts [4]. The hydrogen velocities were estimated by studying the shape of the dip in the Stark broadened profile of the H_β line[9]. For these measurement the effects of other broadening mechanisms were assessed and the high spectral resolution helped in discriminating against impurity-line effects. The velocity distributions parallel to the anode were found to be nearly Maxwellian[4]. The kinetic energies of neutral particles and ions of various charge states were found to be about 8 eV and 20-80 eV, respectively. The ion temperature is significantly higher than what has been commonly believed. The line Doppler broadening and shift, parallel to and at an angle with the anode (see Fig. 6), showed that the ions move away from the anode with nearly isotropic velocity distributions. Particularly smaller ionic Doppler shifts perpendicular to B_z were also observed in order to study, through the ionic motion, the electric fields resulting from the magnetic field penetration into the plasma[8].

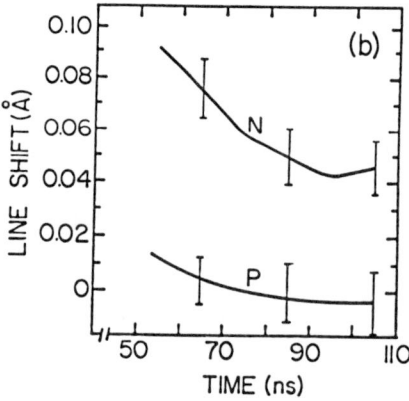

Fig. 6: The blue shift of the CIII 2297-Å line measured at 53° with the normal to the anode (N) giving the average CIII velocity away from the anode surface. Also shown is the zero shift measured parallel to the anode (P).

Ion elastic collisions with the colder electrons cause substantial electron heating. Also, the relatively large ion kinetic energies result in a high plasma pressure and thus a large plasma pressure-gradient. The latter enhances the plasma expansion against the magnetic field and also possibly affects the plasma stability[4] (see Secs. XI and XII).

VII. THE TIME DEPENDENT MAGNETIC FIELD IN THE PLASMA

The magnetic field in the anode plasma was measured by the observation of the Zeeman splitting of emission from BaII ions seeded in the plasma (see Fig.7(a)). The magnetic field applied in the diode was found to be present in the plasma a few nanoseconds after the plasma appeared.

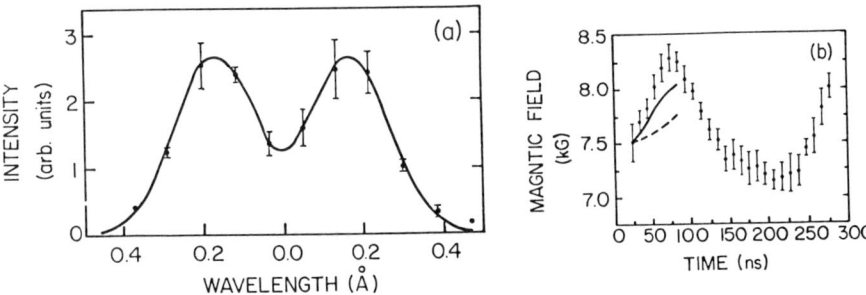

Fig. 7: (a) Spectral profile of the BaII 6142 Å line observed parallel to B_z with an applied magnetic field $B_z=7.6$ kG. The curve is a best fit of convolution of Doppler broadened Gaussian profiles assumed for each of the emission components.
(b) Magnetic field as a function of time within 0.1 mm near the anode surface. Also given is the calculated magnetic-field rise at the anode surface for classical plasma conductivity (dashed line) and for an anomalous conductivity 10× lower than classical (solid line), see Sec. X.

The rise of the magnetic field induced on the anode side by the electron flow in the diode gap was seen to be $\simeq 0.8$ kG (see Fig. 7(b)), which is $\simeq 3$ times larger the prediction of the 1-D model[13]. However, it is consistent with the rise estimated[5] using the pressure balance formula for the diode gap[12], the measured ion current density, and the electric field distribution across the diode acceleration gap[2], which showed a significant electron flow close to the anode plasma (see Sec. III).

VIII. ELECTRON DENSITY IN THE PLASMA

The electron density was obtained as a function of time and distance from the anode surface from the H_β Stark broadening, as shown in Fig. 8[4]. For this determination the H_β Doppler broadening obtained from the H_α spectral profile, was used. For the latter, the influence of ion dynamic effects[16] on the H_α Stark broadening in our $\simeq$20 eV proton plasma, was also considered[9].

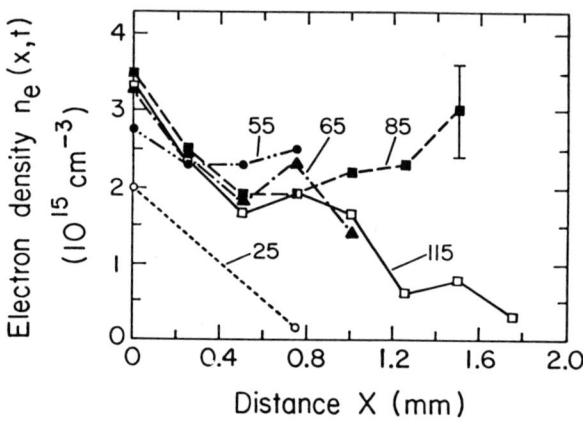

Fig. 8: Electron density axial distribution $n_e(x,t)$ obtained from the H_β Stark broadening, for five times (given in nanoseconds) throughout the pulse. The spatial resolution is $\simeq$0.3 mm. The data points are connected by straight lines.

The use of the observed electron density in studying the plasma formation, expansion, plasma heating, and level-population densities is discussed in Secs. V, XI, XII, and XV.

IX. ELECTRON TEMPERATURE AND TEMPERATURE-GRADIENT

Determination of the electron temperature in plasmas in short-pulse devices is difficult because of the time dependence of the excited-level populations and of the level-population ratios in such plasmas. Assuming a steady state for the level populations is shown[6] to lead to misleading conclusions on the electron temperature and on the charge state distributions.

We determined the electron temperature in our anode plasma by observing line intensities as a function of time and comparing them with *time-dependent* collisional-radiative calculations of the atomic level populations[6-7]. Important features of these calculations are that they use the observed electron density and that they also account for the continuous material flow into

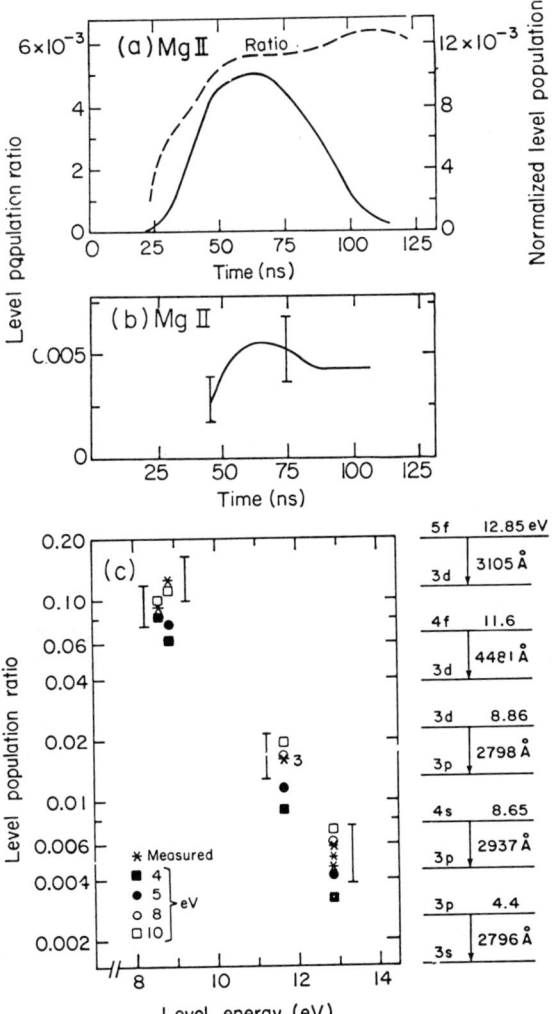

Fig. 9: (a) The calculated MgII 3p $^2P_{3/2}$ population and the ratio of the 5f ^2F population to the 3p $^2P_{3/2}$ population, using a continuous parabolic-in-time supply of MgII at the ground state (see Sec. XV). Populations are divided by the degeneracy and $T_e=7$ eV is assumed; (b) The ratio of the MgII 5f ^2F population to the 3p $^2P_{3/2}$ population measured in a single discharge; (c) Measured MgII level populations normalized to the population of the 3p level, averaged between $t=60$ to 95 ns. The observed lines and the level energies are given in the level scheme. Also shown are the calculated normalized populations for $T_e=4$, 5, 8, and 10 eV at $t \simeq 80$ ns, using the continuous particle source for the plasma.

the plasma. In these calculations atomic levels are coupled through electron collisional excitations and deexcitations, and through spontaneous radiative decays. Each two adjacent atomic/ionic species are coupled through ionization and recombination (radiative, three-body, and dielectronic). Our modeling includes sufficient number of atomic levels. Higher-n levels (up to the reduced continuum limit[11]) are assumed to be hydrogenic. A detailed description of the code and its uses in the investigations of transient phenomena in short-duration plasmas will be described elsewhere. The electron temperature was determined for times (the second half of the pulse, i.e., $t \gtrsim 60$ ns) in which the line intensity ratios were seen to vary slowly both in the calculations (see Fig. 9(a)) and in the observations (see Fig. 9(b)), and were relatively insensitive to details of the continuous material flow into the plasma.

The electron temperature was determined independently from line intensities of three species: MgII, AlIII, and CIII (triplet levels). For each species all important visible and U.V. transitions were observed. The results for each species yielded T_e between 5 to 8 eV as shown in (c) for MgII. For the calculations in this Figure the observed spatially-averaged time-dependent electron density (see Fig. 8) was used.

The dependence of the electron temperature on the distance x from the anode surface was obtained by observing the x-dependence of CIII line-intensity ratios. This is justified since the CIII propagation distance during the time required for equilibration between the excited levels (a few nanoseconds) is small with respect to the plasma thickness. An example of such a measurement is presented in Fig.10, showing, no x-dependence of the level population ratio. This yielded a variation <1 eV for the electron temperature across the plasma. The implication of this result will be discussed in Sec. XII.

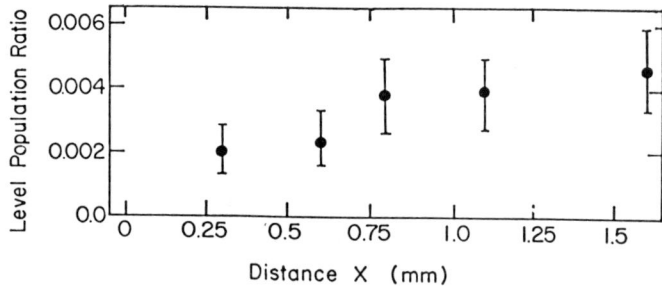

Fig. 10: Ratio between the populations of the CIII 2p3d ^{1}D and 2p^2 ^{1}D levels determined for each position in a single discharge using the 4326-Å and 2297-Å line intensities, respectively. The ratio is averaged over $t=75$ to 95 ns and the spatial resolution is $\simeq 0.5$ mm.

X. ANOMALOUS PLASMA CONDUCTIVITY

Using the time-dependent magnetic field rise on the anode side (see Sec. VII) we now calculate the penetration of the time dependent component of the magnetic field into the anode plasma. Assuming 1-D geometry and a constant plasma conductivity we calculate the rise of the magnetic field at the anode surface for a classical plasma conductivity, as shown in Fig. 7(b). The calculated rise of the magnetic field is significantly less than that observed. However, assuming an anomalous conductivity $\simeq$ 10× lower than classical, the calculated rise is much closer to the observed one (see Fig. 7(b)). In Secs. XI and XII we present three additional observations consistent with this anomalous conductivity.

XI. PLASMA EXPANSION AGAINST THE MAGNETIC FIELD

The measurement of the electric field distribution in the acceleration gap[2] indicated that after its early fast expansion (see Sec. III) the anode plasma expanded at a rate $\simeq$ 1 cm/µs. This rate agrees with later observations[4] and it is much higher than what had been expected from the plasma diffusion against the magnetic field.

Based on the observations described above we propose an explanation for this relatively fast plasma expansion. Since collisions with the neutral particles in the plasma are found to be insignificant and since the proton Larmor radius is smaller than the plasma thickness, we use the fluid equation for the expansion velocity of a fully ionized plasma $V_x = \frac{-c^2}{\sigma B^2}\partial_x P$, where P is the plasma pressure and σ is the plasma conductivity. Assuming $P=n(T_i+T_e)$ and using $T_i=25$ eV, $T_e=7$ eV, a plasma density $n=2.2\times 10^{15} cm^{-3}$, a pressure-gradient scale of 1 mm, and B=7.2 kG as observed in the plasma[4-6], gives for a classical conductivity $V_x \lesssim 0.1$ cm/µs. This is much smaller than the observed velocity ($\simeq$1 cm/µs). However, the use of an anomalous plasma conductivity that is $\simeq$10× lower than the classical conductivity (see Sec. X) explains the observed V_x.

XII. ELECTRON HEATING IN THE ANODE PLASMA

The electrons in the plasma can be Ohmically heated by currents in the plasma resulting from the magnetic field induced in the plasma (by the electrons drifting in the diode gap) and the currents resulting from the plasma pressure-gradient. The latter, $J_y = \frac{c}{B}\partial_x P$, is important due to the relatively high ion temperature (see Sec. VI). If a classical conductivity is assumed, the electron heating would be dominated by the induced magnetic field. Using the measured electron density[4], it is shown[6] that this heating should result in an electron temperature gradient in the plasma of about 5 eV/mm. This temperature gradient cannot be reduced by thermal convection or conduction because of the strong electron magnetization ($\omega_{ce}/\nu_e \simeq 40$). This predicted large temperature-gradient is in clear contradiction to the rather uniform temperature observed in plasma (see Sec. IX).

The temperature uniformity is, however, consistent with the plasma anomalous conductivity assumed (see Secs. X and XI). With this anomalous conductivity the induced currents in the plasma are much smaller and more uniform and the electron heating is dominated by the currents due to the plasma pressure gradient. This is likely to result in a uniform temperature in the plasma, especially since in the case of an anomalous conductivity the electron thermal convection against the magnetic field is relatively fast[6].

The electron Ohmic heating due to the total pressure-driven current in the plasma is estimated[6] assuming an anomalous plasma conductivity. Together with the electron heating due to elastic collisions with the hotter ions (see Sec. VI), this heating, averaged over the plasma, is shown to about balance the average electron cooling due to inelastic collisions with the plasma particles and due to thermal convection to the anode surface (using the collisional-radiative calculations and the known plasma composition the main inelastic collisions were found to be hydrogen ionization and impact excitations of CII and CIII ions[7]). This balance is consistent with the observed temporal constancy of the electron temperature[6], further supporting the anomalous plasma conductivity determined.

We note that no evidences for the specific plasma instability responsible for the anomalous conductivity are available as yet although a possible mechanism was discussed in Ref.4.

XIII. OBSERVATIONS OF TURBULENT ELECTRIC FIELDS IN THE PLASMA

Following the observations of the plasma anomalous conductivity we searched for turbulent electric fields in the plasma[9]. For these measurements we utilize the effect of such fields on the line broadening. The H_α spectral profile was observed for various directions of lines of sight and of emission polarizations. Differences in the Stark broadenings for different polarizations suggest the existence of anisotropic turbulent electric fields in the plasma[17].

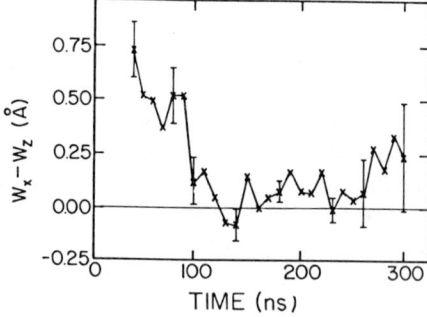

Fig.11: The difference between the FWHM W_x of the H_α line observed in the y-direction for the x-polarization and that for the z-polarization W_z, integrated over the plasma. Here, $W_x \simeq 2.2$ Å.

Fig. 11 shows the difference in the H_α widths observed in the y-direction for the x and z polarizations. It is seen that during the voltage pulse on the diode W_x is larger than W_z, consistent with turbulent fields, pointing mainly in the x-direction. The peak width-difference occurs at the beginning of the pulse and it is about 30% of the width. Using Stark broadening calculations based on a convolution of the Holtsmark field with quasistatic anisotropic fields of various distributions, the anisotropic field amplitude is estimated to be $\simeq 10$ kV/cm[9]. The dependence of the field amplitude with the distance x from the surface was also studied.

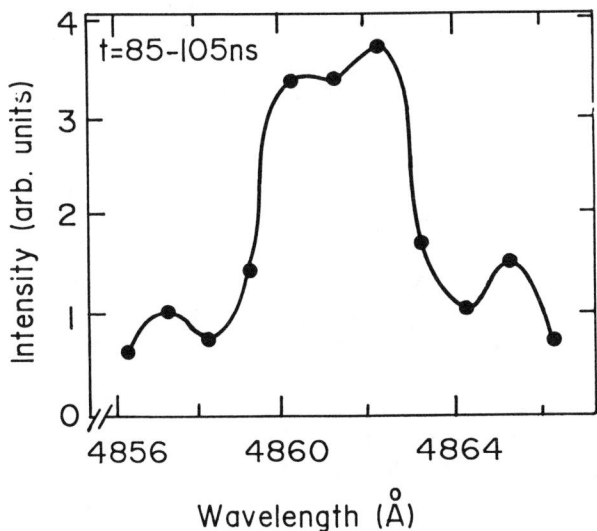

Fig.12: The spectral profile of the H_β line observed in the z direction at $x \simeq 0$ averaged over $t=$ 85 to 105 ns. The uncertainty in the data points is $\lesssim 10\%$. The satellite frequency shift from the line center is 3.3×10^{12} sec^{-1} and w_{pe} (obtained from n_e that is inferred from the FWHM of the main H_β line) is 2.9×10^{12} sec^{-1}. The line indicates the trend.

Turbulent fields were also investigated by searching for plasma satellites near the H_β line, as shown in Fig. 12[9]. A few evidences (based on the time and the space dependence of the H_β structure) suggest that the observed lines are unlikely to result from H_2 molecule or from other impurities, particularly in the plasma region far from the anode surface. Assuming the H_β structure is due to plasma satellites, the satellite wavelength shifts from the line centre give a frequency equal to within 20% to the plasma frequency obtained from n_e that is inferred from the FWHM of the main H_β line. The satellite intensity yields an amplitude of $\simeq 30$ kV/cm for the turbulent fields, which is significantly larger than the estimated thermal wave amplitude in the plasma ($\simeq 2$ kV/cm).

XIV. POSSIBLE NON-MAXWELLIAN ELECTRON ENERGY DISTRIBUTION

As discussed in Sec.IX level population ratios for CIII triplet levels, MgII, and AlIII, atomic systems in which the levels lie within $\simeq 10$ eV, indicate an electron temperature of 5-8 eV. However, level population ratios for other atomic systems, observed for the same experimental parameters, suggest that the electron energy distribution is not a pure Maxwellian[10]. The ratios for CIII singlet levels (2p^2, 2p3p, 2s5d, and 2s5f obtained from the intensity of the 2297-Å, 4326-Å, 4122-Å, and 4056-Å lines, respectively) and BIII (2p, 4s, 5d, and 5f obtained from the 2066-Å, 2234-Å, 4243-Å, and 4487-Å lines, respectively), systems in which the levels lie within $\simeq 25$ eV, indicate a temperature of 20-50 eV. This can be explained by assuming a higher-energy component for the 5-8 eV electrons. For the data analysis, cross-sections for electron-excitations are integrated over the electron energy distribution and used in the time-dependent collisional-radiative calculations to yield the level population densities. Various electron energy distributions are examined. A possible explanation for this phenomenon is electron acceleration by electric fields in the plasma[18]. For this explanation we use an energy distribution function $f = f_M exp[\frac{E}{E_D}(\frac{mv^2}{2T_e})^2]$, where f_M is a Maxwellian energy distribution, E is the electric field in the plasma, v is the electron velocity, and $E_D = v_{th} m \nu / e$. Here, m, e, v_{th}, and ν are the electron mass, charge, thermal velocity, and collision frequency, respectively. It is found that using $E/E_D \simeq 0.07$, i.e., $E \simeq 15$ V/cm, provides a reasonable fit for the observed level population densities.

XV. PARTICLE FLUXES FROM THE ANODE SURFACE INTO THE PLASMA.

Detailed time-dependent calculations, based on the observed $n_e(t)$ and T_e, were used to analyze the line-intensity histories for various species and charge states. This analysis gave that neutral particles, singly charged ions, and doubly charged ions, produced at the immediate vicinity of the anode surface, are being continuously injected into the plasma[7]. Fluxes parabolic in time were used to fit the observed absolute line intensities. For this fit, level populations of each charge state were calculated using the direct flux of that charge state from the surface and the ionization in the plasma of the lower charge states. The satisfactory fits obtained for various elements and charge states gave, using the absolute system calibrations, the absolute time dependent fluxes for the various species as shown in Fig.13(a) for CI, CII, and CIII. The injection of neutral particles and singly charged ions of certain elements was found to continue also after the main pulse (for $t>100$ ns) as demonstrated in Fig.13(a) for CI. It is shown there that a parabolic CI flux, that drops to zero at the end of the pulse ($t \simeq 96$ ns), predicts a level population density after the pulse that is much lower than the measured one. On the other hand, a source that allows a continued injection for $t > 80$ ns (curve L(I) in Fig. 13(a)) fits the measurement reasonably well. This CI source is used in Figs. 13(b) and 13(c), together with CII and CIII parabolic sources, for the calculation of the CII and the CIII level populations. These are also found to be in good agreement with the data.

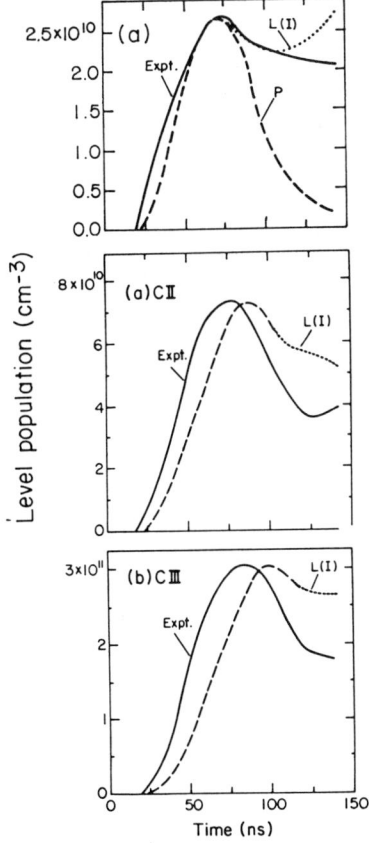

Fig. 13: Calculated level populations for CI, CII, and CIII using $T_e=7$ eV (dashed curves), together with the measured line intensities (solid curves).
(a) The CI 2p3s ^{1}P level population, calculated using the parabolic source A(CI,t)=A(CI) $((t-20)/19-(t-20)^2/38^2)$ atoms/cm^2 ns of CI at the ground state, is shown by the curve P. This source zeroes at $t=96$ ns. The population calculated for a source that is equal to S(CI,t) up to $t=80$ ns and that remains constant for $t >80$ ns, equal to S(CI, 80), is shown by the curve L(I). Here, A(CI)=7.1×10^{10} atoms/cm^2 ns.
(b) The CII 3p ^{2}P$_{3/2}$ level population calculated using a parabolic source for CII and the source for CI that corresponds to the curve L(I) given in (a). Here, A(CII)=8.3×10^{11} particles/cm ns.
(c) The CIII 2p^2 ^{1}D level population calculated using the CI L(I) source given in (a), the CII parabolic source given in (b), and a CIII parabolic source with A(CIII)=5.6×10^{11} particles/cm^2 ns.

The continuous particle injection into the plasma and the particle ionizations were found to be responsible for the observed rise in the plasma areal density[7].

Using the fluxes of the various charge states injected from the surface into the plasma and the velocity distributions of the various charge states observed in the plasma we deduced that ions injected into the plasma are accelerated by electric fields at the immediate vicinity (≤ 0.1 mm) of the anode surface[7]. This acceleration is believed to be responsible for the relatively high ion temperature seen[4] in the plasma (see further measurements described in Sec. XVII).

XVI. PARTICLE FLUXES FROM THE PLASMA INTO THE ACCELERATION GAP

Using the electron density (Sec. VIII), electron temperature (Sec. IX), and particle fluxes from the anode surface (Sec. XV), and the observed particle velocity distributions (Sec. VI), we calculated the ionization rates and the densities of ions and neutral particles as a function of time and position as they flow towards the outer (ion-emitting) region of the plasma[7]. For example, the equation for the CIII density is:

$$n(\text{CIII}, x, t) = \frac{2}{v(\text{CIII})\sqrt{\pi}} \int_0^t \frac{S(\text{CIII}, t')}{(t-t')} exp\left\{-\left[\frac{x/(t-t')}{v(\text{CIII})}\right]^2\right\} dt'$$

$$+ \frac{2}{v(\text{CII})\sqrt{\pi}} \int_0^t \frac{S(\text{CII}, t')}{(t-t')} exp\left\{-\left[\frac{x/(t-t')}{v(\text{CII})}\right]^2\right\}$$

$$\times \left[1 - exp\left\{-(t-t')/\tau(\text{CII})\right\}\right] dt'$$

$$+ \frac{2}{v(\text{CI})\sqrt{\pi}} \int_0^t \frac{S(\text{CI}, t')}{(t-t')} exp\left\{-\left[\frac{x/(t-t')}{v(\text{CI})}\right]^2\right\} \times \frac{1}{(\tau(\text{CII}) - \tau(\text{CI}))}$$

$$\times \left(\tau(\text{CI})\left[exp\left\{-(t-t')/\tau(\text{CI})\right\} - 1\right] - \tau(\text{CII})\right.$$

$$\left.\left[exp\left\{-(t-t')/\tau(\text{CII})\right\} - 1\right]\right) dt'$$

Here, S(CI), S(CII), and S(CIII) are the absolute fluxes for the different charge states, τ denotes the ionization time, and v denotes the average velocity in the x direction observed to be 1.1 cm/μs for CI and 1.8 cm/μs for CII and CIII[4]. The first term in this equation gives the contribution of the CIII continuous source at the surface. The other terms give the density of the CIII ions produced by the ionization of CII. CIII ionization is neglected due to its slow rate and the CIII ions that originate at CI in the plasma are assumed to retain the CI velocity.

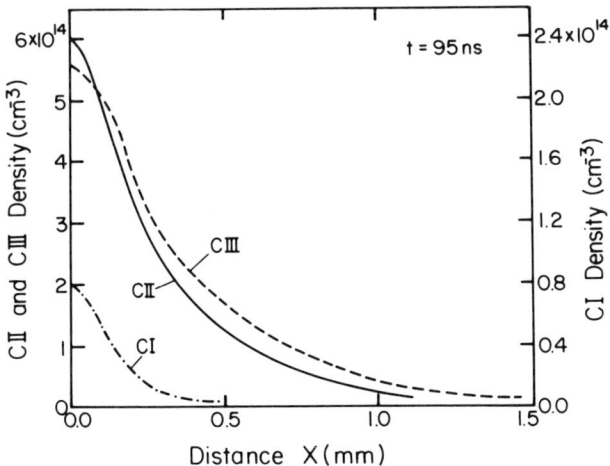

Fig. 14: Calculated axial density distributions of CI, CII, and CIII for $t=95$ ns obtained using the determined continuous particle fluxes from the anode surface into the plasma, the observed electron density, electron temperature, and particle velocity distributions.

These calculations predict that doubly charged ions and protons dominate the outer plasma region as shown in Fig.14 for the carbon particles. This is consistent with charge-collector measurements on the extracted ion beam.

Due to its relatively large concentration and high velocity a significant fraction of the neutral hydrogen reaches the plasma outer region. A quantitative analysis gives that the ionization of hydrogen in the plasma front contributes a proton flux that is similar to the density of the proton current extracted into the diode acceleration gap. This proton flux due to hydrogen ionization may be important if the proton flow in the plasma penetrated by the magnetic field[5] is limited.

Using equations similar to the above the fluxes of the various species at the outer the plasma region are obtained. The ratio between the CIII and proton fluxes at this region is predicted to rise from a low value to about unity towards the end of the pulse as shown in Fig.15. This rise is found to agree with the observed time dependent CIII to proton ratio in the extracted ion beam (see Fig. 15). This demonstrates that time-dependent measurements and analysis applied for plasma sources can be used to predict the composition of the extracted ion beams. The beam composition depends on the plasma atomic properties, on the plasma thickness, and on the particle velocities (see the equation above), which explains the diversity in the ion beam composition observed in various experiments using similar diode configurations.

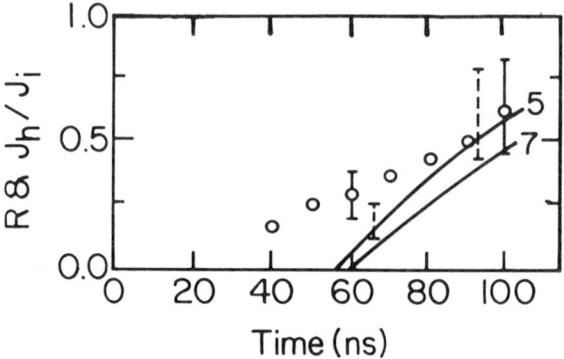

Fig. 15: The ratio R of the CIII charge flux to the total charge flux in the plasma for the distance $x_0=1.5$ mm from the anode surface calculated using $T_e=5$ and 7 eV as indicated (curves). Also given is the the measured ratio J_h/J_i of the non-protonic-ion current density to the total current density obtained from charge-collector measurements outside the diode (circles). Here, most of the non-protonic ions are CIII[7], and the uncertainties for the calculations and measurements are ±30% as shown.

XVII. DETERMINATION OF LOW-LYING LEVEL DENSITIES AND PARTICLE VELOCITIES USING LASER LIGHT ABSORPTION AND INDUCED FLUORESCENCE.

Since in pulsed-power systems most of the particles are at the ground state, the particle densities and velocity distributions can be known reliably only if the ground or low-lying level densities are determined. Spontaneous line emission can give misleading results since the determination of the ground state density from the excited level populations suffers from the usual poor knowledge of n_e and T_e (especially very close to surfaces) and from the continuous plasma replenishment.

We determine the particle density distributions as a function of time using laser absorption and laser-induced-fluorescence[8]. These methods allow the densities of the ground state, or of states that radiate in the VUV region (such as the hydrogen $n=2$ level), to be determined directly. The velocity distributions are obtained from the line absorption or the fluorescence spectral profiles. The laser-absorption technique is particularly advantageous since the results are not dependent on the plasma optical thickness or on the system absolute calibration, which usually increase the measurement uncertainties. In these measurements we achieved a spatial resolution of 30 μm. Unlike spontaneous emission, this allowed the region close to the anode surface to be investigated.

Fig.16 shows an absorption spectrum of the MgII $3s_{1/2} \rightarrow 3p_{3/2}$ and a

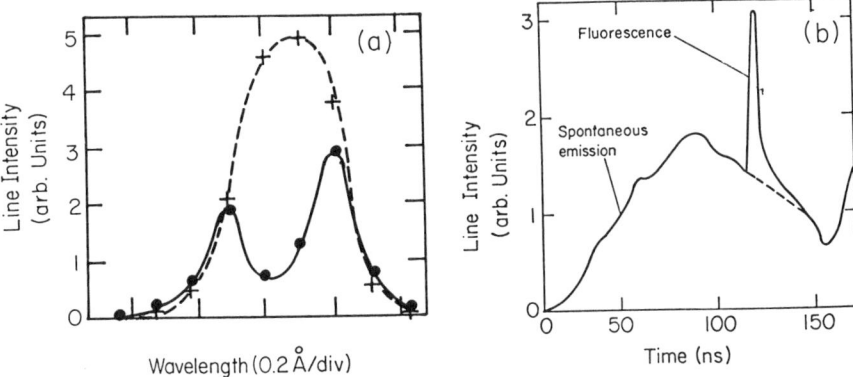

Fig. 16: (a) Absorption of the laser light by the MgII $3s_{1/2} \to 3p_{3/2}$ transition (2795.5Å) at $x \simeq 0$ and $t=75$ ns. The laser pulse is 6 ns long. The spectral profile of the laser light without plasma and after propagation through the plasma is shown by the dashed and solid curves, respectively.
(b) Spontaneous emission of the MgII $3p_{3/2} \to 3s_{1/2}$, together with the fluorescence induced by saturated laser absorption by the same transition.

laser-induced-fluorescence signal for the same transition. Fig.17 shows the MgII ground state density distribution obtained from such measurements. The data reveal a large density close to the anode surface that could not be seen by our previous measurements of spontaneous emission. The MgII total density within 0.1 mm near the anode surface is at least 10 times larger than the average density seen in the plasma. The 0.1-mm-wide layer close to the surface probably serves as the source of MgII and MgIII ions injected throughout the pulse into the plasma[7]. The absolute fluxes determined for the various charge states[7] are used to estimate particle ionization rates at the immediate vicinity of the anode surface. The ionization rate close to the surface is also obtained from the time dependent ratio of the MgI and MgII ground state densities determined from the laser-absorption measurements. The resulting estimate of n_e and T_e close to the surface is compared to the estimate obtained from the ratio between the ground and the first-excited level densities (the latter is also determined using the laser spectroscopy[8]). This ratio can also give information on the material release from the surface into the adjacent dense layer.

Similar measurements were performed for the hydrogen $n=2$ level. Hydrogen atoms were found to be distributed much farther from the anode surface than MgII. This is shown to result from the higher hydrogen velocity[4,9] and the slower hydrogen ionization[7].

The MgII Doppler broadened absorption profiles gave an MgII temperature $\simeq 20$ eV, showing that the MgII velocities seen in the anode plasma[4] are

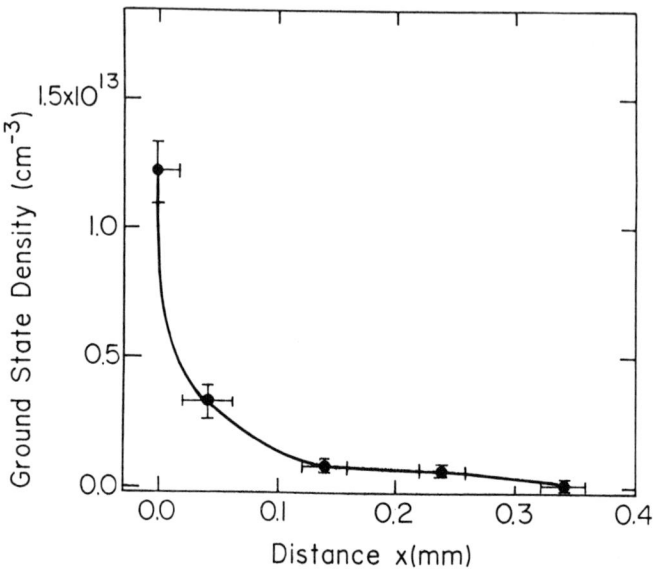

Fig. 17: The MgII ground state density (divided by the degeneracy) obtained from the laser absorption measurements for $t=60$ ns. The spatial resolution near the anode surface is $\simeq 30\mu$m. The line indicates the trend.

acquired by the ions within 0.1 mm from the anode surface. This is consistent with the findings[4,7] that suggest ion acceleration near the surface by relatively strong electric fields there (see Sec. XV). The significance of the results for the understanding of particle injection from a wall into a plasma is being studied.

XVIII. SUMMARY AND CONCLUSIONS

Diagnostic methods based on spontaneous emission and laser spectroscopy, aided by time-dependent collisional-radiative calculations, were developed to measure physical quantities inside intense ion diodes. We determined the electric field distribution in the diode acceleration gap, the ion transverse velocity distributions in the gap, the velocity distributions of ions and neutral particles in the anode plasma, the average ion flow velocity in the plasma, the magnetic field penetration into the plasma, the magnetic field induced on the anode side by the electrons drifting in the diode acceleration gap, the electron temperature and temperature-gradient in the plasma, the electron cooling rate due to inelastic collisions, turbulent electric fields and electron energy distribution in the plasma, the particle velocity and density distributions near the anode surface, the time-dependent fluxes of various species of different charge states from the anode surface into the plasma, the time-dependent plasma composition, and the particle fluxes and densities in the outer (ion-emitting) plasma region.

From these observations we concluded that electrons in the diode acceleration gap spread towards the anode beyond the theoretical electron sheath region and thus enhance the ion current densities and the magnetic field induced on the anode side. The ion transverse velocities in the diode gap are smaller than those outside the diode. Electron heating in the anode plasma is mainly due to the pressure-driven currents in the plasma and the electron elastic collisions with the hotter ions ($T_e \simeq 7$ eV vs. $T_i \simeq 20$-80 eV). A significant fraction of the electrons has a relatively high energy, 20-50 eV. The magnetic field penetration into the plasma suggests an anomalous plasma conductivity $\simeq 10\times$ lower than the classical one. Together with the observed plasma pressure gradient this explains the fast plasma expansion against the magnetic field. This conductivity also explains the observed uniformity of the electron temperature in the plasma, and the balance between the electron heating and cooling rates (the electron cooling is dominated by inelastic collisions). The amplitude of the turbulent fields is inferred to be significantly higher than that of the thermal fluctuations. Ionization processes within <0.1 mm from the anode surface determine the relative fluxes of various charge states injected into the plasma. The particles injected acquire their injection velocities by electric-field-acceleration within this distance from the the surface. The absolute fluxes of various species determined as a function of time and position in the plasma, together with the observed velocity distributions, yielded predictions on the composition of the generated ion beam, which were supported by charge-collector measurements outside the diode.

Spectroscopic observations have been recently started in Sandia National Laboratories to investigate the high-power PBFA-II ion diode the main application of which is to drive Inertial Confinement Fusion targets[19]. Spectroscopic systems were built to allow measurements in the visible and U.V. regions to be made in the hostile environment of this $\simeq 10$ MV, $\gtrsim 10$ TW diode. The electric fields studied ($\simeq 8$ MV/cm) are the highest ever spectroscopically determined, $\simeq 8\times$ stronger than in the previous measurements[2]. In these investigation, shifts, broadenings, and intensities of lines of various species are being used to study the plasma formation and behaviour and the properties of the ion beam inside the diode.

The studies described here can be used for the investigation of pulsed-power devices (high power diodes, microwave devices, plasma pinches, plasma switches, and transmission lines), for the design of plasma sources, and for studying plasma-wall interactions. The use of laser-spectroscopy with the geometry previously suggested[1] can be used to obtain better accuracy and to examine 2-D effects in the electric field distribution and in the plasma behavior.

REFERENCES

1. Y. Maron and C. Litwin, J. Appl. Phys. (1983).
2. Y. Maron, M.D. Coleman, D.A. Hammer, and H.S. Peng, Phys. Rev. Lett. 57, 699 (1986) and Phys. Rev. A36, 2818 (1987).
3. Y. Maron, M. Coleman, D.A. Hammer, and H.S. Peng, J. Appl. Phys. 61, 4781 (1987).
4. Y. Maron, E. Sarid, O. Zahavi, L. Perelmutter, and M. Sarfaty, Phys. Rev. **A39**,5842 (1989).
5. Y. Maron, E. Sarid, E. Nahshoni, and O. Zahavi, Phys. Rev. **A39**, 5856 (1989).
6. Y. Maron, M. Sarfaty, L. Perelmutter, O. Zahavi, M.E. Foord, and E. Sarid, Phys. Rev. **A40**, 3240 (1989).
7. Y. Maron, L. Perelmutter, E. Sarid, M.E. Foord, and M. Sarfaty, Phys. Rev. A. Jan. (1990).
8. L. Perelmutter, Ph.D.Thesis; L. Perelmutter, G. Davara, E. Nahshoni, A. Fruchtman, and Y. Maron, Investigation of a high-power-diode plasma using laser-induced-fluorescence, Bull. Am. Phys. Soc. **34**, 2063 (1989) to be published.
9. E. Sarid, Ph.D. Thesis; E. Sarid and Y. Maron, Measurements of anisotropic turbulent electric fields in a high power diode, Bull. Am. Phys. Soc. **34**, 2143 (1989) to be published.
10. M. Sarfaty, M.E. Foord, Y. Maron, and A. Fruchtman, Observation of non-Maxwellian electron energy distribution in a high-power-diode plasma, Bull. Am. Phys. Soc. **34**, 2063 (1989) to be published.
11. H. Griem, Plasma Spectroscopy (McGraw-Hill, New York, 1964).
12. C.W. Mendel and J.P. Quintenz, Comments on Plasma Phys. Controlled Fusion 8, 43 (1983).
13. T.M. Antonsen and E. Ott, Phys. Fluids 19, 52 (1976).
14. M.P. Desjarlais, Phys. Rev. Lett. **59**, 2295 (1987); Phys. Fluids **B1**, 1709 (1989).
15. C. Litwin and Y. Maron, Phys. Fluids, **B1**, 670 (1989).
16. H. Ehrich and D.E. Kelleher, Phys.Rev. **A21**, 319 (1980); D.H. Oza, R.L. Greene, and D.E. Kelleher, Phys. Rev. **A37**, 531 (1988).
17. E.K. Zavoiskii, Yu. G. Kalinin, V.A. Skoryupin, V.V. Shapkin, and G.V. Sholin, ZhETF Pis. Red. **13**, 19 (1971) [JETP Lett. **13**, 12 (1971)].
18. H. Dreicer, Phys. Rev. **115**, 23 (1959); A.V. Gurevich, Zh. Eksp. Teor. Fiz. **39**, 1296 (1960) [Sov. Phys. JETP **12**, 904 (1961)].
19. J. Bailey, A.L.Carlson, M.P. Desjarlais, D.J. Johnson, T.R. Lockner, R.L. Morrison, and Y. Maron, Time resolved visible spectroscopy of the PBFA-II Ion Diode. Bull. Am. Phys. Soc. **34**, 2108 (1989).

XUV Lasers

A QUEST FOR MORE EFFICIENT X-RAY LASERS: ATOMIC PARAMETERS INVOLVED AND RECENT EXPERIMENTS

R.C. Elton, E.A. McLean, J.A. Stamper, C.K. Manka and B.H. Ripin
U.S. Naval Research Laboratory
Washington, DC 20375-5000

ABSTRACT

The output energy of present x-ray lasers is $\lesssim 10^{-6}$ of that of the driver laser, for laser-produced plasma media. A significant improvement is essential for more compact, useful, and less costly devices. Some plasma/atomic processes that may bear on this goal and that deserve attention are described. Some experiments already indicate promise for improved efficiency. One recent one from the RIKEN laboratory (Japan) is described, along with initial attempts at the Naval Research Laboratory (NRL) to duplicate the results.

INTRODUCTION

In the last 5 years there have been significant advances in producing meaningful x-ray amplification of spontaneous emission (ASE) in elongated laser-produced plasmas. Most results have come from very large laser drivers, which are both cumbersome and expensive to operate. It is very important to improve the overall driver-to-output efficiency (presently $\lesssim 10^{-6}$). Associated with the relatively high gain coefficients measured (~ 5 cm^{-1}) are a number of questions that may depend on an improved understanding of atomic processes present in the lasing plasma, leading to improved overall efficiency. Some of these are listed in the next section according to the pumping process. Following this is highlighted one particular recent experiment in which similar gain coefficients are measured in a plasma created with >100-times less energy, i.e., with the potential for greatly increased overall efficiency at long lengths. A progress report on preliminary experiments underway at NRL to duplicate these results is described in some detail.

PLASMA/ATOMIC PROCESSES

Some current and vital questions of a plasma/atomic nature bearing on x-ray gain in plasmas are listed next, for a particular pumping mechanism. No attempt has been made to present a review with references to all sources. For such an in depth coverage, the reader less familiar with the various on-going efforts is referred to Chapters 3 and 4 of Ref. 1 and to Ref. 2.

Electron-collisional excitation pumping[3,4].

o Why are only Ne-like (n=6) ions found to lase for 3p-3s transitions in a $1s^2 2s^2 2p^n$ (n=1-6) configuration? E.g., why not F-like (n=5)? Is there an alternate decay mode for electrons in the 3p upper state (e.g. combined with 2s-2p promotion) for n<6?

- Why do two 3p J=2 upper levels lead to dominant gain and why are the gains essentially the same for the two transitions, when both upper and lower levels differ as well as the transition probabilities for the lasing transitions?

- Why does the 3p J=0 upper level continue to give lower gain than predicted. Both calculations and experiments indicate that the level kinetics are essentially correct. Are there plasma effects limiting the gain on this transtion?

Electron recombination pumping

- What is the mechanism of enhanced recombination in a solenoidal magnetic field in the experiments at Princeton Plasma Physics Laboratory[5]. Can it be applied to smaller geometries and higher fields (perhaps self generated) for shorter wavelengths?

- What is the best configuration and forced-cooling method?

In general:

- Are there plasma effects that are limiting the gain coefficient in all experiments to ~ 5 cm^{-1}?

- Can the overall gain be increased to the point of saturation of the medium? Or instead will refraction spread the beam excessively?

- Can the overall efficiency be improved by extending the length until saturation is achieved?

FIGURE OF MERIT FOR X-RAY LASERS

Short of consistent x-ray output power measurements for efficiency estimates, a useful figure-of-merit is the gain product GL divided by the input pumping-laser power P, where G is the gain coefficient and L the length of the laser medium. This is plotted in Fig. 1 for available data as a function of wavelength, as adapted and extended from a graph prepared by M. Key of Rutherford Laboratory[6] (again the reader is referred to Ref. 6 as well as to Ref. 1 for a review of the experiments involved). Besides the two pumping modes mentioned above, some long wavelength data from Auger decay into the upper level following innershell photoionization is included.

A RECENT HIGH-EFFICIENCY EXPERIMENT AT RIKEN[7]

While most data fall close to a single λ^{-2} fitted line in Fig. 1, there are recombination experiments showing figures-of-merit as much as 3 orders of magnitude higher (the Auger experiment shown high is a traveling-wave configuration). It is extremely important to understand the detailed processes responsible for such enhanced efficiencies, particular where they exceed expectations of existing

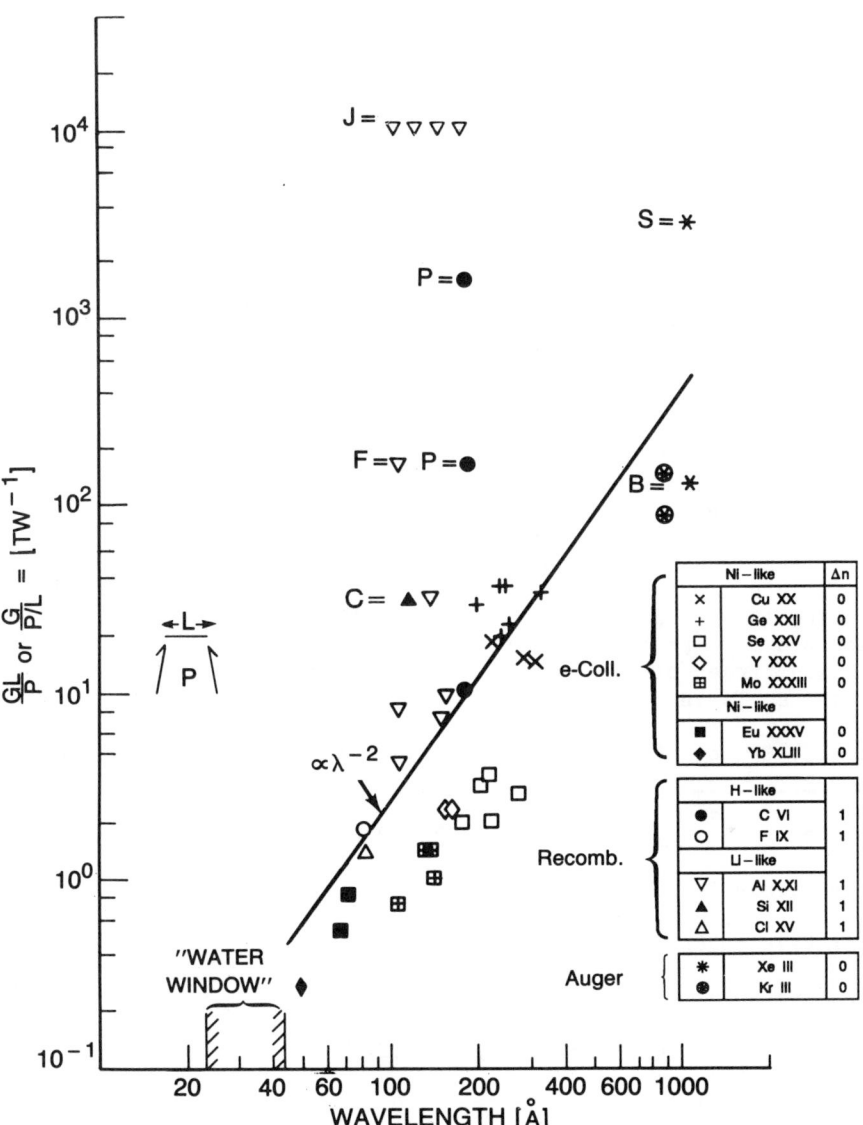

Fig. 1. Figures-of-merit vs. wavelengths. At high levels, J=Japanese (RIKEN) described herein, P=PPPL, F=France (Orsay), S=Stanford. (Adapted and expanded from Ref. 6).

models. The experiment showing by far the highest figure-of-merit designated in Fig. 1 by four ▽'s at GL/P ≳10⁴ was reported very recently[7]. The four laser wavelengths range from 105 to 177 Å and derive from n=4,5 to n=3 transitions in Be-like and Li-like aluminum (Al^{9+} and Al^{10+} ions, respectively) listed in Table I.

Table I. Reported[7] x-ray laser transitions.

Ion	Spectrum	Transition	Wavelength [Å]
Al^{9+}	Al X	4f-3d	177.8
Al^{9+}	Al X	5d-3p	123.49
Al^{10+}	Al XI	4f-3d	154.66
Al^{10+}	Al XI	5f-3d	105.7

Recombination-pumped lasing on the Li-like transitions listed in Table I has been reported elsewhere, as indicated in Fig. 1 and reviewed in Chapter 4 of Ref. 1. However, in the latest case, gain coefficients as large as 4.5 cm^{-1} at energies as low as 2 J (5 ns, 400 MW) are reported[8] and result in the large figures-of-merit shown. From basic ASE gain formulae[1], it can be shown that such a gain coefficient requires an upper laser state density of $\sim 4 \times 10^{15}$ cm^{-3}. This in turn implies an electron density of at least $N_e \sim 4 \times 10^{16}$ cm^{-3} (assuming 100% occupation in this upper state). It is more reasonable to expect an electron density of $N_e \sim 10^{18}$ cm^{-3} for an expanding target plasma at the specified distance of 800 μm from the target (see Table II).

Table II. Essential parameters in original[7,8] and NRL experiments.

Parameter	RIKEN	NRL
Driver Energy [J]	2 or 6	≥ 4 (9-16)
Driver Wavelength [μm]	1.05	1.05
Pulse Length [ns]	5	5
Line Focus Length on Target [mm]	3-12	3-12, 36
Focal Width [μm]	40	40
Irradiance [W/cm²]	$(0.8-2.5)10^{11}$	$\geq 2 \times 10^{11}$
Target	Al Slab	Al Slab
Expanding Plasma Viewed [μm]	200 @ 800	>600
Gain Measured [cm^{-1}]	3-4	0
Output Power [W]	??	~150

RESULTS OF PRELIMINARY ATTEMPTS
TO DUPLICATE THE RIKEN EXPERIMENT AT NRL

The RIKEN results are extremely interesting, and have led to several (so far unsuccessful) attempts[9] to verify and extend the results besides our own (NRL) to be described here. Our approach has been to duplicate as closely as possible[7,8] the original experiment, using the NRL Pharos III laser[4] as a driver (operated at low power). The original RIKEN and the NRL parameters are listed in Table II. A significant requirement according to the original authors is the tight ($\lesssim 40$ μm) line focus, which was achieved also at NRL.

In both experiments, the elongated plasma was created by focusing the driver beam with a combination of a concave cylindrical and a spherical lens. The focusing parameters are compared in Table III.

Table III. Spherical lens parameters at RIKEN[7,8] and NRL.

Parameter	RIKEN	NRL
Focal Length (F.L.) [cm]	40	120
Diameter D [cm]	6	20
f/#	6.7	6.0
Diffraction Limit λ/D [μrad]	17	5
Beam Divergence (Div.) [μrad]	120[a]	≤ 40[b]
Best Focus (Div.*F.L.) [μm]	48	≤ 50

[a]Private communication from T. Hara[8].
[b]Estimate of 8(λ/D) from S. Obenschain (NRL) for high power.

The NRL experiments were arranged similar to those at RIKEN and similar to a previous setup for gain experiments[4] in Cu and Ge. An aluminum slab was irradiated in a line focus, and soft x-ray spectra were obtained with a grazing incidence (photographic) spectrograph viewing along the axis of the line plasma. No focusing mirror was used before the slit of the spectrograph in the NRL experiments.

The NRL experiments were begun with a full 12-mm length plasma (maximum length used at RIKEN), and at a relatively high irradiance ($\sim 10^{12}$ W/cm^2) compared to that at RIKEN. Initially, the entire plasma (including the target region) was viewed. Not surprisingly, it was found that the spectrum was predominantly that of Li-like and Be-like aluminum. In particular, the four spectral lines reported by the RIKEN group to be lasing were observed to be strong. Also not surprisingly, the Al XI 4f-3d line at 154.66 Å (Al^{10+} ion) dominated. As a tracer, this line was observable to driver-laser energies as low as 5 J. This is plotted as the upper line (o—"TARGET VIEWED") in Fig. 2. This energy corresponds to an irradiance of 2x10^{11} W/cm^2, considered to be very low for such species. The corresponding

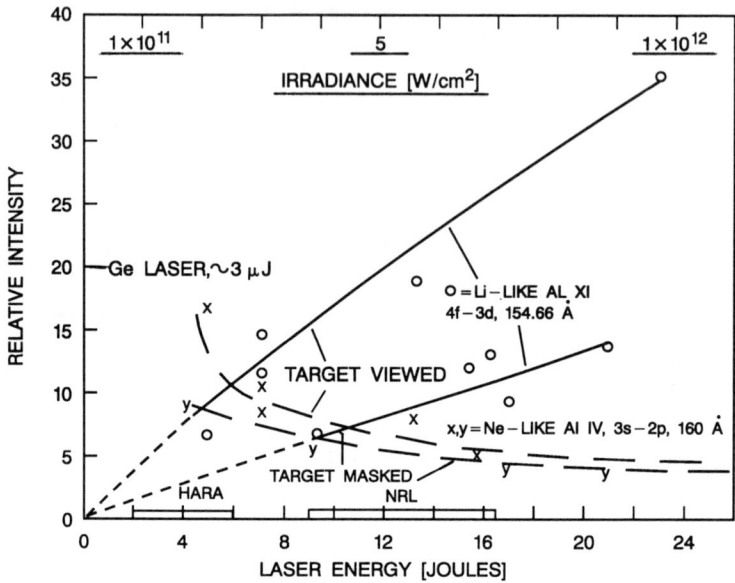

Fig. 2. X-ray intensity in two spectral lines versus driver energy. The upper solid (o) and dashed (x) plots for Al XI and IV, respectively, represent viewing of the entire plasma; the lower solid (o) and dashed (y) are when 500 μm near the target was masked from view.

relative intensity is 5 in Fig. 2 and relates to a photographic emulsion density of ~0.05 (note that photographic density has a linear dependence on intensity at this low level[10]). This suggests that there is an abundance of He-like ground-state ions formed at the target which remain "frozen in" that state and recombine to form Li-like (and then Be-like) ions in the cooler expansion region. The ionization time to the He-like state at the target is estimated to be <1 ns for an electron density and temperature of $N_e = 10^{21}$ cm^{-3} and $kT_e = 100$ eV, respectively. This is sufficiently shorter than the 5 ns driver pulse to render this model plausible.

Next a mask was added at the target to limit the field of view to distances >500 μm, in order to simulate the 200 m field of view at a distance of 800 μm for which lasing was reported[7,8]. Again the 154.66 Å line was observable to a relative intensity of 5 (photographic density of ~0.05) at an energy of 9 J and an irradiance of 4×10^{11} W/cm^2. This is also plotted (solid line, as o—"TARGET MASKED") in Fig. 2. It is interesting that there is only about a factor-of-2.5 decrease in intensity from the target region to the >500 μm region. This supports the "frozen-in" species model proposed above. An important result is that these data verify that indeed the Li-like lasing levels are significantly populated at energies at and below the 6 J originally reported[7].

Spectra representing lower ionization stages, namely Al IV, V and VI, were also observed. This was similar to the case at RIKEN, where the intensities were comparable to those for the gain lines at 12 mm plasma length. An Al IV, 3s-2p resonance line at 160 Å (Ne-like Al^{3+} ion) was found to increase rapidly at decreasing driver energies as indicated by "x" and "y" in Fig. 2 for the respective cases of total viewing and target masking (the two dashed lines). This indicates a rapid cooling at distances beyond 500 μm for energies lower than 8 J, as suggested by Hara[8].

From Fig. 2, it is apparent that the relative intensities in the 2-6 J region (for the lasing lines as well as Al IV) are ~5. Comparing the corresponding photographic density (~0.05) to that at which lasing was recorded for Ne-like Ge^{22+} in previous NRL experiments[4], a decrease of ~4-times was found (this germanium data point is included in Fig. 2). In the earlier experiments, an output energy of 3 μJ was quoted. Hence it is projected that the output power for the 154.66 Å aluminum line in the present case is ~150 W, based on a 5 ns pulse length. This is the origin of the value quoted in Table II above. In the germanium experiments, a driver energy of ~300 J was used, which is ~100-times larger than that in the RIKEN experiments. Hence, while the present efficiency might still lie in the 10^{-6} range, this regime of operation is potentially much more efficient providing that the intensity continues to exponentiate with increased plasma length.

As is now standard practice, the plasma length was varied in an effort to detect an exponentially varying output intensity associated with gain. Following the RIKEN example, the plasma length in this particular NRL experiment was varied by adjusting a mask on the focusing lens system. (This differs from the usual method which involves adjusting the target length without altering the driver beam, and has led to some concern.) At NRL, such a gain experiment was conducted with a driver energy in the nominally 12 J range (see the abscissa in Fig. 2), where it was felt that sufficient photographic exposure was available to detect an exponentially-decreasing spectral intensity at shorter lengths.

The first results obtained for the four lines reported[7] to be lasing by Hara, et al. are shown in Fig. 3. It is clear that no exponentiation of intensity with length is observed at our driver energy. Indeed, some decrease in intensity at extended lengths, particularly for the more intense lines, appears to be present. This may be a result of absorption, scattering or refraction of the beam.

Data for the Al XI resonance lines plotted in Fig. 4 show a strong saturation with length, which is associated with self absorption at high opacity. This effect was also apparent when we compared the Li-like spectrum for both a 12-mm line focus and a spot focus of equal area, as shown in Fig. 5. The overall ionic species present remained the same, as did the relative intensities for the 5f-3d and 5d-2p lines. However, the relative intensity of the 3d-2p resonance line decreased markedly for the point focus case, presumably because of such self absorption on the strong resonance line.

In contrast to the resonance line saturation with length in Fig. 4, the 5f-4d line (at 333 Å) intensity was found to increase linearly where it is was observable (as shown in Fig. 4 also). This shows a lack of self-absorption on such a transition between excited levels, to be expected. It might also be expected that this transition produce gain whenever the 5f-3d transition is in a lasing

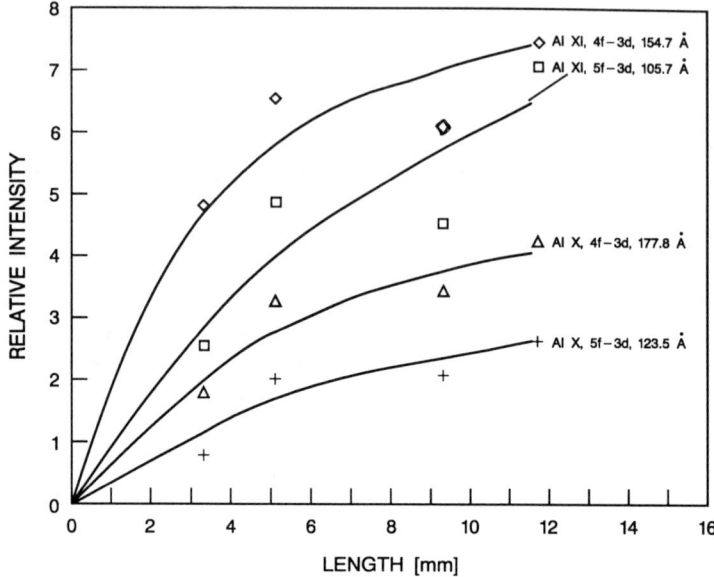

Fig. 3. NRL plasma length dependence of the four lasing lines reported[7]. The four symbols at 12 mm represent actual data points.

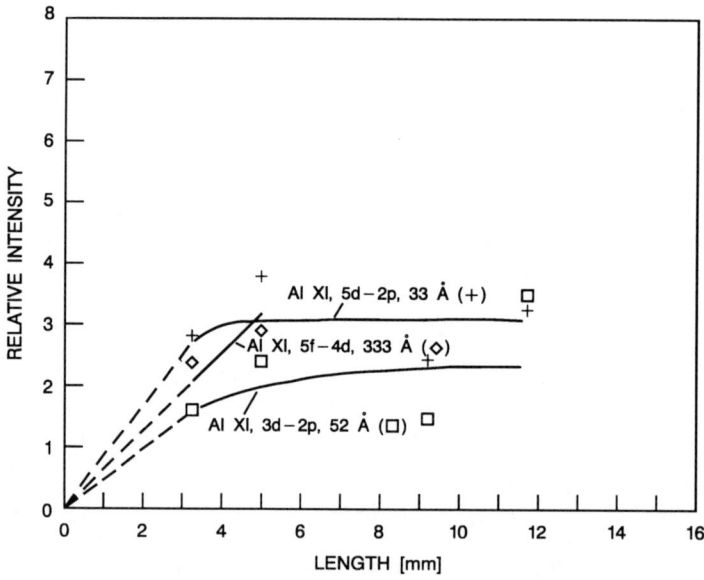

Fig. 4. NRL plasma length dependence of two resonance and one high-level transition.

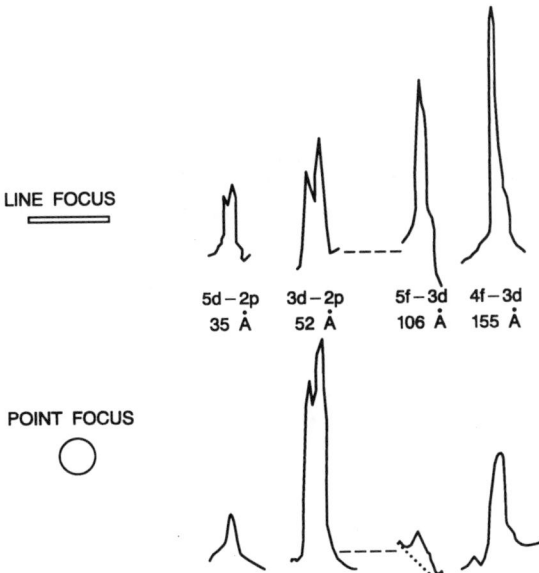

Fig. 5. Comparison of line and point focus spectral lines, showing dramatic self absorption in the first 3d-2p resonance line.

condition, because of the same upper level and a the longer wavelength (this assumes that the 4d level decays at a sufficiently rapid rate). While no gain is observed in the NRL experiments so far on this transition, it would be interesting to know whether gain is present in the RIKEN experiments.

The plasma length in the NRL experiments was increased to 36 mm, in hopes that the gain coefficient reported by the RIKEN group would lead to an exponential increase in output by a factor of ~10^4. Three attempts were made, at driver irradiances of $(3, 4\ \&\ 17)10^{11}$ W/cm² (still larger than the $(0.8-2.5)10^{11}$ W/cm² used in the RIKEN experiments--see Fig. 2). In general, the intensities for both Al XI and Al IV lines, as well as their ratios, remained similar to those in Fig. 2. For the shot at 4×10^{11} W/cm², which lies in the energy range of the length study shown in Fig. 3, the intensity of the Al XI 154.66 Å line was ~9 on the scale in that figure. This implies that the gradual increase of intensity with length shown in Fig. 3 continues to a length of 36 mm, rather than rising exponentially.

SUMMARY AND FUTURE PLANS

There are a number of anomalies that have arisen with the advent of successful x-ray gain experiments with high temperature plasmas that deserve attention. Generally the anomalies exist in differences between experiment and theory as well as in differences between experiments. Hence, it is important to have more experiments for comparisons and also to investigate the details of the atomic and plasma physics involved in the localized gain region. Some of this can be done on smaller scale experiments.

The initial experiments at NRL directed towards verifying the highly efficient recent experiments from the RIKEN laboratory have not as yet demonstrated measureable gain. By the same token, because of limited spectrograph sensitivity it has not been possible so far to perform them at the low driver energy (2 J) used at RIKEN, which is reportedly essential[8]. It is hoped that these experiments can be repeated at the prescribed conditions at NRL and elsewhere and that the unique atomic/plasma conditions leading to gain can be identified and understood. When this is accomplished, there may indeed be orders-of-magnitude increase in efficiency for developing a compact x-ray laser approaching the <50 Å region of interest.

ACKNOWLEDGEMENTS

One of us (RCE) is grateful for the details concerning the RIKEN experiments and the helpful advice freely given by Dr. T. Hara that laboratory. The technical assistances of J.L. Ford in performing the spectroscopic experiment and of H.R. Burris, Jr. and N.E. Nocerino in the configuration and operation of the laser are recalled with appreciation. Support for this project was provided by SDIO.

REFERENCES

1. R.C. Elton, "X-Ray Lasers" (Academic Press, NY, 1990).

2. R.C. Elton, Proceedings 1989 Lasers and Electro-Optics Society (LEOS) Conference, Orlando, FL, October 1989.

3. C.J. Keane, N.M. Ceglio, B.J. MacGowan, D.L. Matthews, D.G. Nilson, J.E. Trebes and D.A. Whelan, J. Phys. B $\underline{22}$, 3343 (1989).

4. T. Lee, E. McLean and R. Elton, Phys. Rev. Lett. $\underline{59}$, 1185 (1987).

5. C.H. Skinner, D. Dicicco, D. Kim, L. Meixler, C.H. Nam, W. Tighe and S Suckewer, IEEE Trans. Plasma Science $\underline{16}$, 512 (1988).

6. M.H. Key, J. de Physique, Colloque C1, $\underline{49}$, C1-135 (1988).

7. T. Hara, K. Ando, N. Kusakabe, H. Yashiro and Y. Aoyagi, Jap. J. Appl. Phys. $\underline{28}$, L1010 (1989); Proc. Japan Academy $\underline{65B}$, 60 (1989).

8. T. Hara, 9th International Conference on Vacuum-UV Radiation Physics, Honolulu, Hawaii, and private communication, July 1989.

9. P. Jaeglé (Orsay) and S. Suckewer (Princeton) in separate talks given at the Second Canadian Workshop on X-Ray Lasers, Ottawa, Canada, October 1989; also Lawrence Livermore National Laboratory and U. Rochester Laboratory for Laser Energetics (L. DaSilva and T. Boehly, respectively, private communications, November 1989).

10. B. Henke, S. Kwok, J. Uejio, H. Yamada and G. Young, J. Opt. Soc. Am. B $\underline{1}$, 818 (1984); and B. Henke, F. Fujiwara, M. Tester, C. Dittmore and M. Palmer, J. Opt. Soc. Am. B $\underline{1}$, 828 (1984).

IMPROVED PERFORMANCE OF RECOMBINATION BALMER-α LASER BY SHORT PULSE IRRADIATION

Y. Kato, H. Azuma, K. Yamakawa, M. Nishio, T. Tachi, H. Shiraga,
E. Miura, H. Takabe, K. Nishihara, S. Nakai
Institute of Laser Engineering, Osaka University, Suita, Osaka,
565 Japan

S. A. Ramsden[1], G. J. Pert[2], M. H. Key[3], S. J. Rose[3]
1 University of Hull, Hull, HU6 7RX, England
2 University of York, York, YO1 5DD, England
3 Rutherford Appleton Laboratory, Chilton OX11 OQX, England

C. P. J. Barty
Ginzton Laboratory, Stanford University, Stanford, Calif.,
94305, USA

ABSTRACT

Short pulse pumping is required to isoelectronically scale recombination Balmer-α laser to shorter wavelengths. By irradiating a planer NaF target with a 28 ps laser pulse, gain coefficient of 4 cm^{-1} has been attained on time- and space-integrated measurement. This gain is approximately a factor of 3 larger than the value we have obtained previously in target irradiation with a 130 ps laser pulse. Peak gain local in space and time of 18 cm^{-1} is suggested in computer simulation. A 3.5 ps, 8 TW pulse been generated by chirped pulse amplification which may be applicable to extention of the Balmer-α laser to shorter wavelengths.

INTRODUCTION

Current major objective of the x-ray laser research is to demonstrate saturated amplification at a short wavelength near to or within the water window of 43.8 Å - 23.3Å.[1] It will be desirable if the required pumping energy is modest so that the whole system can be installed within a room of a standard laboratory size. The recombination laser scheme[2,3] is attractive in this respect since it will be inherently more efficient than the collisional excitation scheme[4,5] comparing the ionization and excitation energies required to produce population inversion in the two schemes.[6]

Isoelectronic scaling of the recombination Balmer-α (Hα) laser in hydrogenic ions is obtained by considering similarity forms for the collisional radiative rate equations[7,8]:

laser wavelength $\quad \lambda \sim 6.55 \times 10^3 \, Z^{-2}$ (Å),
irradiation intensity $\quad I \sim 4 \times 10^{10} \, Z^4$ (W/cm^2),
characteristic time $\quad \tau \sim 5 \times 10^{-7} \, Z^{-4}$ (s).

Here the coefficient for the irradiation intensity was normalized to the condition for Na XI reported here. The characteristic time τ represents the plasma expansion time to the density at which gain occurs after the laser pulse is switched off, assuming laser pulse width is shorter than τ. This characterise time is 380 ps for Z=6 (C VI) and 34 ps for Z=11 (Na XI), respectively.

A gain coefficient of g=4 cm^{-1} was obtained in time-resolved measurement on C VI Hα at 182 Å by irradiating a fiber target with a 70 ps laser pulse.[3] This experiment was extended to observation of gain on F IX Hα at 80.9 Å.[9] Extension to shorter wavelengths was tested by using the glass laser facility at Osaka University by irradiating a stripe (and a foil) target with a 130 ps laser pulse (hereafter we call this "long pulse experiment").[10] This resulted in a time-integrated gain of 1.2 cm^{-1} on Na XI Hα at 54.2 Å and 1.9 cm^{-1} on Mg XII Hα at 45.5 Å. The time-resolved peak gain was 3-4 cm^{-1} on Na XI Hα. Since the pulse width in this experiment was long compared with $\tau \sim$34 ps for Na, we have recently implemented a revised experiment where a foil target was irradiated with a 28 ps pulse ("short pulse experiment"). The time-integrated gain of 4 cm^{-1} obtained in this experiment is approximately 3 times higher than the value obtained with 130 ps irradiation. This result is consistent with the concept that the target should be irradiated with a laser pulse shorter than the characteristic time τ in the "explosive mode" of recombination Hα laser.

PROCEDURES FOR GAIN DETERMINATION[11]

In both the long pulse and short pulse experiments, the gain coefficient g was determined by comparing the line intensities emitted along (Ia) and transverse (It) to the x-ray laser axis by using the relation Ia/It=[exp(gL)-1]/(gL) here L is the gain length.

The experimental arrangement is shown in Fig. 1. Two XUV spectrometers equipped with variable spacing gratings for flat-field imaging recorded axial and transverse spectra. A grazing incidence spherical mirror formed a 10:1 reduced astigmatic image of the source onto the slit which was wide opened to 500 µm. Since the source image was completely contained within the slit opening, a possibility for variations in light collection efficiency with different target lenghts and laser focusing conditions has been eliminated.

The relative sensitivity of the two instruments, which is required to determine the "true" Ia/It ratio from the experimentally obtained "apparent" value for Ia/It, was calibrated in each experiment by recording spectra from a narrow (1mm-wide) planer foil target whose target normal was oriented at an angle bisecting the two spectrometers. Since the two instruments viewed the small plasma from symmetric directions, the relative sensitivity is determined irrespective of the

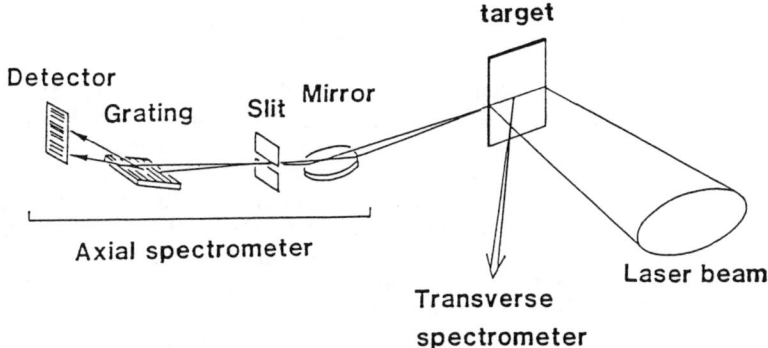

Fig. 1 Experimental arrangement for observing XUV spectra axial and transverse to the x-ray laser axis.

optical thickness of the calibration spectral lines. The characteristic response curve of the Kodak 101-07 film used for time-integrated recording of the spectra was calibrated for the batch of the film used in each experiment.

In these experiments, the gain length was limited to less than 1 cm in order to achieve high irradiation intensity required for producing fully stripped ions with the laser energies available at present. When the gain-length product is small, the present procedure for gain determination is especially suitable since the gain value is determined for each laser shot. In the case the gain is determined from the plot of the axial intensity vs the gain length, the gain coefficient is affected by shot-to-shot variations when the gL value is small.[12]

LONG PULSE IRRADIATION

In the first-phase experiment, a target was irradiated with a 351 nm, 130 ps laser pulse with the maximum energy of 152 J line-focused to a 7 mm length and 50 μm width. The maximum intensity was 3.3×10^{14} W/cm^2. The laser material of either NaF for Na XI Hα or MgF$_2$ for Mg XII Hα was coated over a 0.13 μm-thick CH foil in a stripe of 30 μm width and 1 μm thickness. Also a foil target was tested where NaF was coated to a 0.2 μm thickness over one surface of the foil.

In Fig. 2 we show the axial (a) and transverse (b) spectra of a NaF stripe target irradiated at 73 J. The Na XI Hα line at 54.2 Å shows stronger intensity in the axial spectrum compared with the transverse spectrum. The Ia/It values of various spectral lines determined from (a) and (b) and the relative spectral sensitivity of the two instruments are shown in (c). The Na XI Hα has Ia/It >1 showing it has gain. Some of the lines have Ia/It ~1 (optially transparent) such as Na X 4d-2p at

47.5 Å and 3d-2p at 66.3 Å and F IX 4d-2p at 60.0 Å. Other lines, especially resonance lines, are optically thick as expected.

The Na IX Hα is composed of a doublet (J=5/2-3/2 at 54.19 Å and J=3/2-1/2 at 54.05 Å) and is closely adjacent to Na IX 4p-2s at 53.86 Å and F IX 5d-2p at 53.5 Å. These lines were deconvolved and Ia/It for each line was determined. The result is shown in Fig. 3. The overall error in Ia/It arising from cross calibration, film response and line shape analyses was evaluated to be approximately ±30%. Table 1 shows the measured time-integrated gain coefficient of the Na XI Hα J=5/2-3/2 transition at different irradiation conditions. This shows that when the target is irradiated with ~70 J energy, the gain value of approximately 1 cm^{-1} is obtained. The target geometry does not strongly affect the gain. Increasing the laser energy may slightly reduce the gain. Interesting case is Na$_2$B$_4$O$_7$ which was intended to test dilution effect which will show higher gain if Ly-α trapping is the major factor for limiting gain. Our experimental data is opposite to this expectation, and consistent with the recent observations that simulation calculations are in better agreement with experiments when Ly-α trapping is neglected.[13]

Table 1 Measured time-integrated gain coefficients of the Na XI Hα J=5/2-3/2 transition at different irradiation conditions for target irradiation with a 130 ps pulse.

Target	Laser Energy [J]	Ia/It	g [cm^{-1}]
NaF Stripe	73	1.33±0.40	$0.9^{+0.8}_{-1.2}$
NaF Stripe	153	1.12±0.34	$0.4^{+0.8}_{-1.3}$
NaF Foil	63	1.47±0.44	$1.2^{+0.8}_{-1.1}$
Na$_2$B$_4$O$_7$ Stripe	69	0.74±0.22	$-1.1^{+0.9}_{-1.4}$

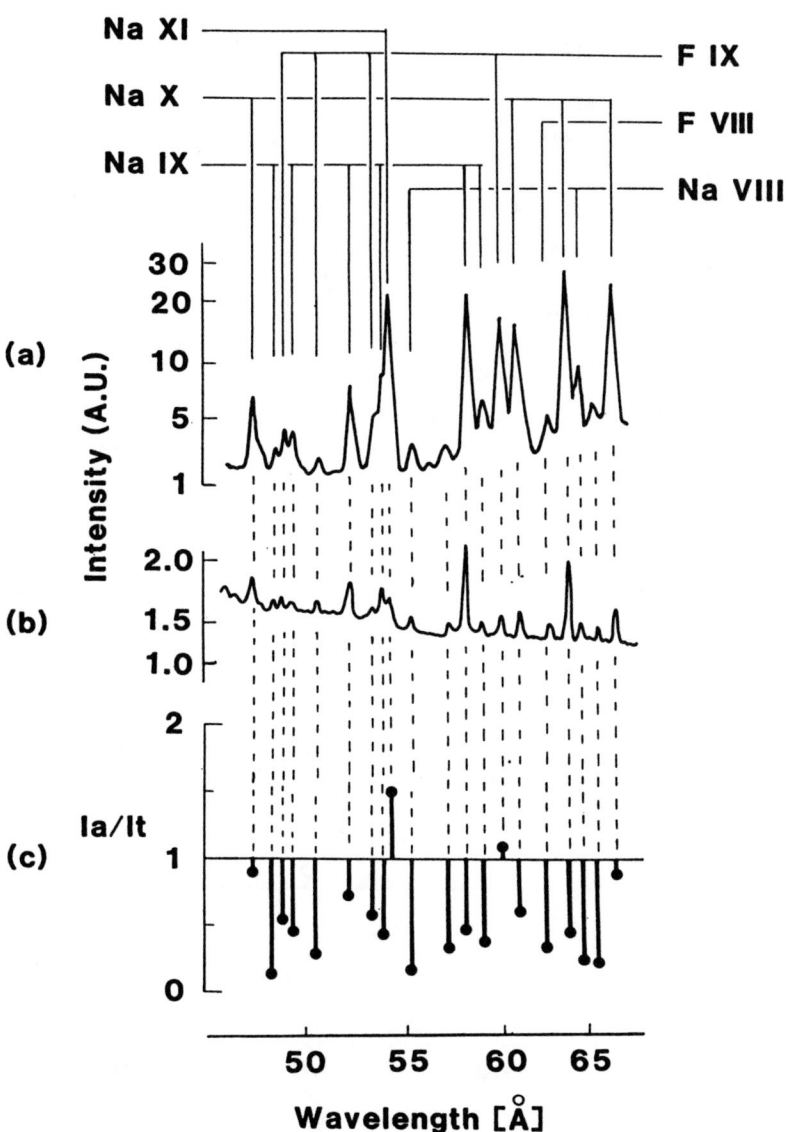

Fig. 2 Axial (a) and transverse (b) spectra of NaF. The axial-to-transverse intensity ratios for the spectral lines are shown in (c). The laser pulse duration was 130ps.

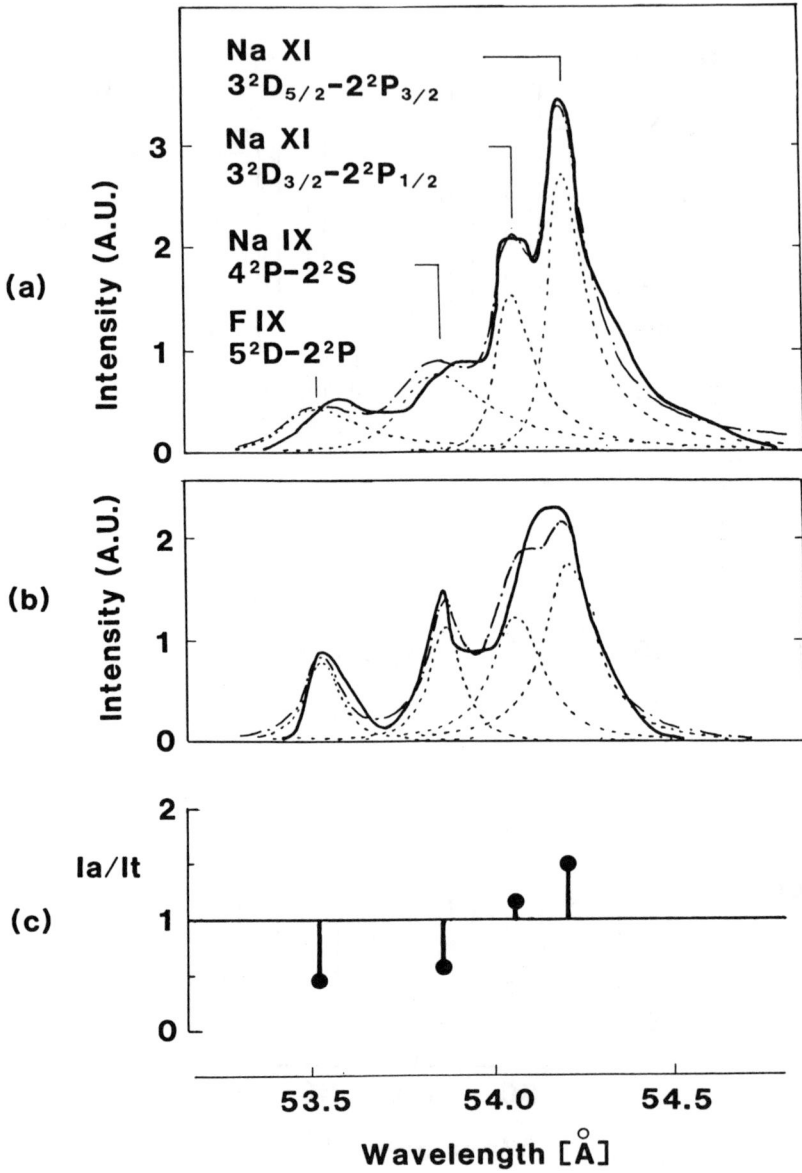

Fig. 3 Axial (a) and transverse (b) spectra of NaF near Na XI Hα line. Thick solid curves are the experimental data. Thin dotted curves are deconvolved lines for each component with the convolved spectra shown by line-dot curves. The Ia/It ratio for each component is given in (c). The laser pulse duration was 130 ps.

Determination of time-resolved gain was also made on Na XI Hα. Time-resolved axial and transverse spectra were taken with x-ray streak cameras for the NaF targets and the cross calibration targets. The time dependence of $I_a(t)/I_t(t)$ was used to determine $g(t)$. The gain sets on at ~100 ps after the laser peak and becomes maximum at 150-200 ps taking a value of $g=3.2\pm1.0$ cm^{-1}. This result is regarded to be consistent with the time-integrated gain value considering the gain reduction in time averaging.

Time-integrated measurements on Mg XII Hα at 45.5 Å were made on MgF$_2$ stripe targets using the same procedure as for Na XI Hα described above. The axial-to-transverse intensity ratio of $I_a/I_t=1.9\pm0.6$ corresponding to the gain of $g=1.9\pm1.2$ cm^{-1} has been obtained.

SHORT PULSE IRRADIATION

In this experiment, a 28 ps laser pulse was generated from a 100 ps pulse by using a fiber-grating pulse compressor. It was amplified to the beam diameter of 18 cm and frequency up-converted to 526 nm yielding a maximum energy of 25 J on target. The laser beam was focused to a line of 4.5 mm length and 30 µm width. The target was a foil target with NaF coating of 0.45 µm ~ 0.12 µm thickness. Gain determination was made in time-integrated measurements with the same procedure as in the long pulse experiment.

The axial and transverse spectra and the I_a/I_t values for the spectral lines near Na XI Hα are shown in Fig. 4. This data was obtained with NaF of 0.45 µm thickness with 23 J irradiation energy corresponding to the intensity of 6.0×10^{14} W/cm^2. Comparing this with Fig. 3, it is apparent that the Na XI Hα doublet have significantly higher axial-to-transverse intensity ratios in this short pulse irradiation experiemnt. This ratio corresponds to the time-integrated gain of $g=4.0\pm1.0$ cm^{-1}. This result was reproducibly obtained for the irradiation energies of 22-25 J and did not depend strongly on the coating thickness over 0.45-0.12 µm. The gain becomes strongly negative for the irradiation energy of 10 J. The intensity ratios for other spectral lines were similar to the one shown in Fig. 2.

Computer simulation of this experiment was made using a code which employs a simplified two dimentional hybrid model to represent the hydrodynamics. Figure 5 shows the spatial distribution of gain of Na XI Hα at different times after laser irradiation. The Ly-α trapping was not included in the calculation shown in Fig. 5. The gain sets on at 40 ps and becomes maximum at 60 ps taking a value of 18 cm^{-1}. The maximum gain region is 15 µm away from the initial target surface. Spatial averaging reduces the peak gain to 11 cm^{-1} and further time averaging reduces the gain to 5 cm^{-1}, which should be compared with the experimental value of 4 ± 1 cm^{-1}. This comparison indicates that a large amplification factor may be

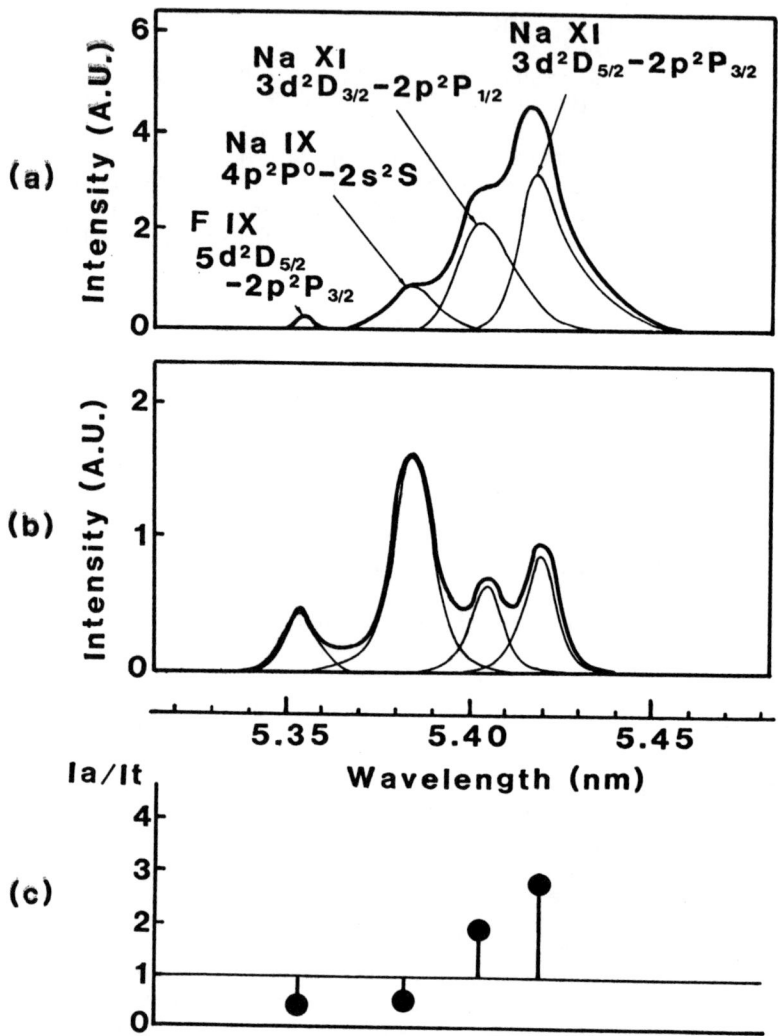

Fig. 4 The same as Fig. 3, but the laser pulse duration was 28 ps.

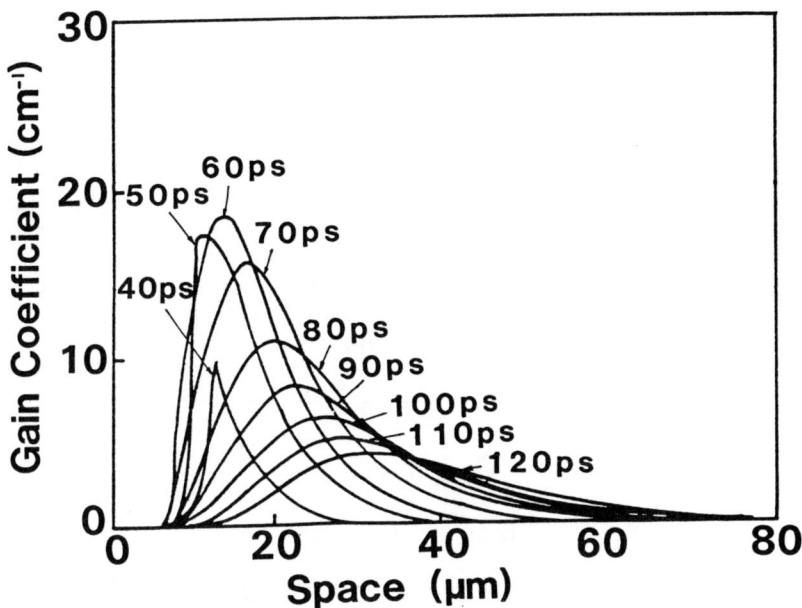

Fig. 5 The space and time resolved gain coefficient of Na XI Hα line calculated for target irradiation with a 28 ps pulse.

realized when a longer gain length is irradiated under the optimized irradiation condition.

This experiment has shown that the gain on Na XI Hα is strongly increased when a target is irradiated with a short pulse.

CHIRPED PULSE AMPLIFICATION

A laser pulse of higher energy and shorter duration is required in extending the recombination Hα laser to higher amplification factors and shorter wavelengths. Chirped pulse amplification (CPA) was proposed to produce such a pulse using solid state lasers.[14] We have tested the CPA technique in our large aperture glass laser system. Using a pair of large aperture Au-coated reflection gratings, a compressed pulse of 28 J energy and 3.5 ps duration with a beam diameter of 13.5 cm has beam generated. The peak power of 8 TW in this experiment is almost an order of magnitude higher than the value attained where a short pulse is amplified in this system with this beam diameter. A novel method has been introduced in this experiment to reduce the pedestal which contributes to formation of a low density plasma prior to irradiation by the main short pulse.

SUMMARY

Experimental procedures for determination of gain of Na XI Hα transition at 54.2 Å have been described. The gain was measured in two sets of experiments where a target was irradiated either by a 130 ps, ~70 J laser pulse, or by a 28 ps, 23 J laser pulse. The time-integrated gain of 4±1 cm^{-1} obtained in the short pulse experiment is approximately 3 times higher than the value obtained in the long pulse experiment. Computer simulation shows that the peak gain local in space and time reaches to 18 cm^{-1}. This is a very encouraging result for attaining strong amplification at short wavelengths by irradiation with modest laser energies. Future advances in laser development such as employment of the CPA technique may provide a laser system of a standard laboratory size meeting the practical requirements.

REFERENCES

1. R. A. London, M. D. Rosen and J. E. Trebes, Appl. Opt. 28, 3397 (1989).
2. S. Suckewer, et al., Phys. Rev. Lett. 55, 1753 (1985).
3. C. Chenais-Popovics, et al., Phys. Rev. Lett. 59, 2161 (1987).
4. D. L. Matthews, et al., Phys. Rev. Lett. 54, 110 (1985).
5. B. J. MacGowan, et al., Phys. Rev. Lett. 59, 2157 (1987).
6. M. H. Key, J. Mod. Opt. 35, 575 (1988).
7. R. W. McWhirter and A. G. Hearn, Proc. Phys. Soc. 82, 641 (1963).
8. G. J. Pert, J. Phys. B9, 3301 (1976) and B12, 2067 (1979).
9. M. Grande, et al., Opt. Commun., to be published.
10. Y, Kato, et al., OSA Proc. Short Wavelength Coherent Radiation: Generation and Applications, Vol.2 (Opt. Soc. Am., Washignton, D. C., 1988) pp.47-51; H. Nishimura et al., ibid. pp.137-140.
11. Y. Kato, et al., Appl. Phys. B, to be published.
12. A. Carillon, et al., Appl. Phys. B, to be published.
13. G. J. Pert and S. J. Rose, Appl. Phys. B, to be published.
14. D. Strickland and G. Mourou, Opt. Commun. 56, 219 (1985).

HARMONIC GENERATION IN RARE GASES : SINGLE ATOM RESPONSE AND PROPAGATION EFFECTS

L.-A. Lompré, A. L'Huillier, M. Ferray and G. Mainfray

Service de Physique des Atomes et des Surfaces,
C.E.N. Saclay, 91191 Gif-sur-Yvette, cédex, FRANCE.

ABSTRACT

The purpose of the present paper is to discuss the main experimental and theoretical results on VUV light emission in a gaseous medium exposed to a strong 1064 nm laser field. Our experimental results obtained in rare gases will be reviewed. They show for example up to the 33rd harmonic in Ar at 3×10^{13} Wcm^{-2}. New experimental results in Xe will help in understanding the influence of multiphoton ionization. The effect of the focalization on the conversion efficiency will also be discussed.

INTRODUCTION

With the exception of two experiments using a 1064 nm[1] and a 1315 nm[2] laser radiation, most of the previous experiments on harmonic generation in a gaseous medium use short incident wavelengths in order to obtain the shortest generated wavelengths[3-6]. A number of experiments have been performed by using the fourth harmonic (266 nm)[3,4] of the Nd-YAG laser, or the fundamental of an excimer, 308 nm (XeCl)[5] or 248 nm (KrF)[6]. McPherson et al have recently reported up to the 17th harmonic in Ne using a 248 nm excimer laser[6] at an intensity of 10^{16} Wcm^{-2}. The conversion seems to be strongly dependent on the medium.

The purpose of the present paper is to discuss experimental measurements of VUV light emission (355 - 30 nm) and in particular harmonic generation in a rare gas medium irradiated by an intense 1064 nm laser field in the 10^{13} Wcm^{-2} intensity range. Recent experiments[6-8] have showed the production of high-order harmonics in a rare gas medium, with photon energies much above the ionization energy. The most salient features of these results[7,8] are the following: the harmonic intensity distribution first decreases rather steeply from the third to the fifth or seventh harmonic, remains almost flat from the seventh to a high-order harmonic, which depends on the atom and on the laser intensity, and finally decreases again.

These results give rise to two essential questions : (i) What is the influence of multiphoton ionization on the conversion efficiency ? Does harmonic generation occur in a neutral medium or in an ionized medium ?. (ii) What is the role of the propagation effects ? The conversion efficiency depends both on the single atom response and on the many-atom response. Both questions are of the greatest interest in the perspective of the development of very high intense lasers, which would enable the study of harmonic generation in intense laser fields.

EXPERIMENTAL ARRANGEMENT

The experimental set-up has been described in details elsewhere and only a brief description will be given here[7,8]. The laser used is a 30 ps mode-locked Nd-YAG laser operated at 10 Hz. It is focused into the vacuum chamber by a lens. The focal length of this lens can be varied from 75 to 300 mm. The gaseous target is provided by a pulsed gas jet. At 0.5 mm from the 1 mm diameter nozzle, the gas density in the jet has been measured to be 15 Torr, for a backing pressure of 150 Torr. The low ratio between the backing pressure and the pressure measured at the output of the gas jet is very important in order to minimize the contribution of dimers. The complete description of the gas jet and the method to measure the pressure in the gas jet has been described elsewhere[9]. The use of a gas jet limits the reabsorption of the VUV light and enables the detection of very short wavelengths. The VUV radiation (from 350 to 10 nm) emitted along the laser axis is analyzed by a monochromator. It is composed of a toroidal grating which forms on the output slit an aberration free image of the interaction region for the spectral range of the grating. The advantage of this device is that the collection solid angle is imposed by the grating size and not by an entrance slit. The VUV radiation is detected by a photomultiplier or a windowless electron multiplier depending on the spectral range investigated.

EXPERIMENTAL RESULTS

Figure 1 shows typical experimental photon spectra obtained between 30 and 160 nm in Xe and Ar at a pressure of 15 Torr and using an electron multiplier as photon detector. The laser intensity which is the same for the two scans, is equal to 4.5×10^{13} Wcm^{-2}. Several major points come out at first sight : (1) only odd harmonics are observed. That is due to the fact that the harmonics

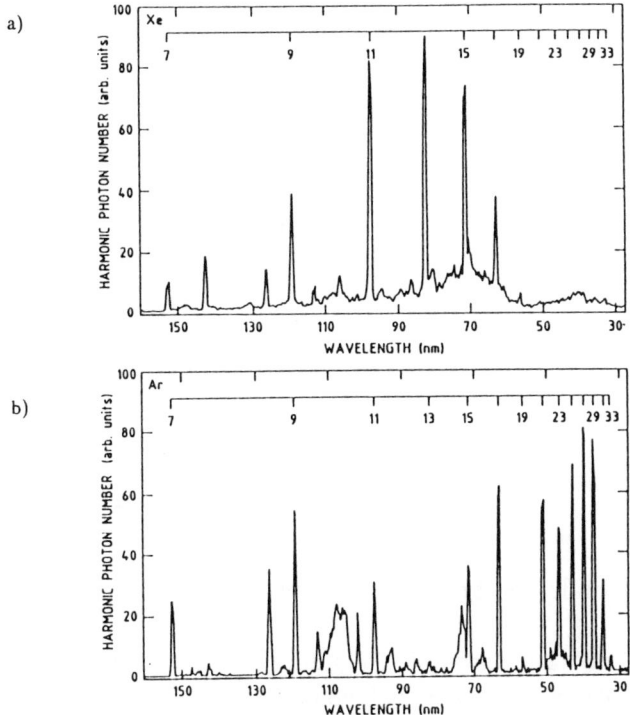

Fig.1 Photon spectra obtained at 1064 nm, 15 Torr, in a) Xe and b) Ar

are created in an isotropic gaseous medium with inversion symmetry which excludes even harmonics. (2) the maximum number of harmonics created at the maximum laser intensity is strongly dependent on the rare gas investigated; up to the 21st harmonic in Xe (Fig.1a) and up to the 33rd harmonic in Ar (Fig.1b). (3) a continuous background appears in Xe, which is much reduced in Ar. It must be pointed out that there is no convincing explanation about the origin of this background. (4) Other lines than harmonics seen in the first and second order of the grating can be observed : they correspond to transitions 6s → 5p (147 nm) and 6s' → 5p in neutral Xe (Fig.1a), or 4s → 3p in neutral Ar, and 4d → 3p in ionic Ar (Fig.1b)[8]. Over the whole spectrum extent (10-355 nm), a small number of lines are observed, corresponding to transitions from excited states to the fundamental of the neutral atom and also of the singly charged ion for our maximum laser intensity.

In order to understand the influence of ionization on harmonic generation, the intensity dependences of all harmonics have been studied, in particular in the vicinity of the saturation intensity. Figure 2 presents the number of photons

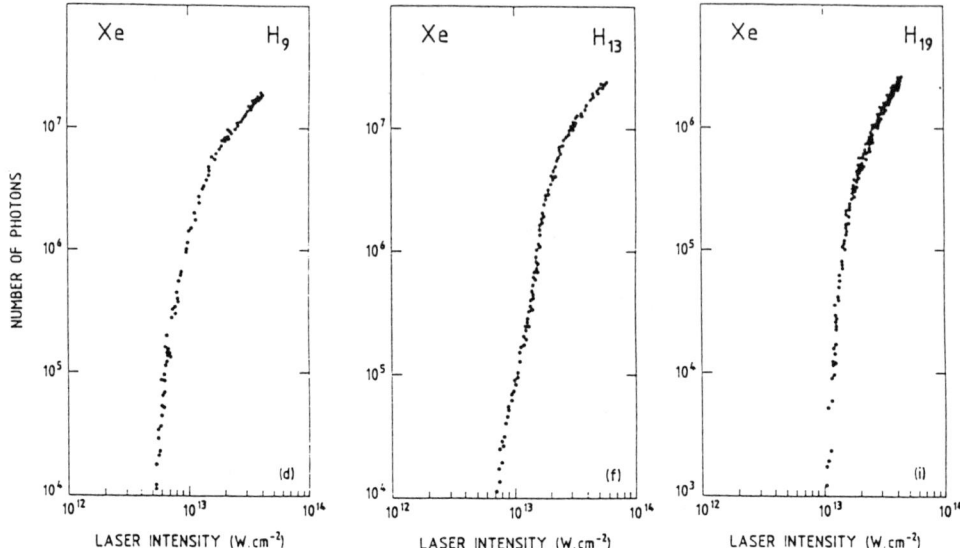

Fig.2 Intensity dependence of the 9th, 13th and 19th harmonics in Xe

at the 9th, 13th and 19th harmonic frequency in Xe as a function of the laser intensity in a double logarithmic plot. The behaviour of the harmonic power as a function of the laser intensity is very similar to the ion behaviour[10]. A saturation of the number of ions produced is observed at the same laser intensity (I_S). This observation is very important because it shows that the harmonic production is strongly influenced by the presence of ions in the interaction volume. This observation has two possible explanations, either the harmonic production efficiency is much smaller in an ionic medium than in an atomic vapor, or the presence of free electrons in the interaction volume seriously affects phase mismatching conditions, introducing an additional (positive) phase mismatch[11]. Unfortunately the experiment does not allow to conclude which effect is the major one.

Furthermore, these intensity dependences show another interesting feature. The harmonics produced below the ionization limit (3, 5, 7, 9) closely follow a I^q power law for laser intensity lower than I_S. The 11th and 13th harmonics have a much more complicated behavior, which could have several explanations : from resonant processes, non-perturbative effects, phase matching modifications, to absorption. Finally, all the higher order harmonics (15-21) vary approximately as I^{12} for laser intensities below I_S. These results have an important consequence for comparison with theoretical calculations in atoms. They show that data must

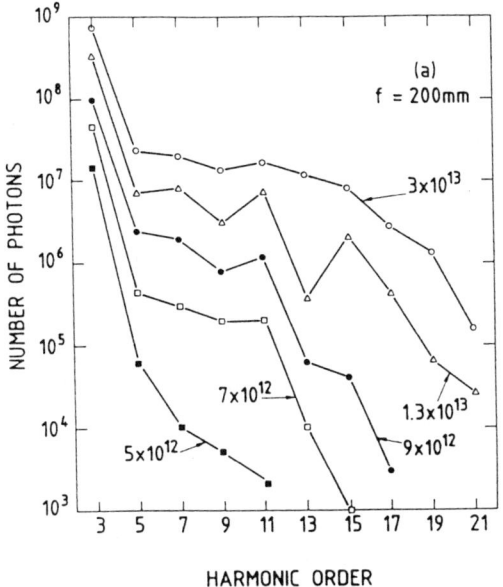

Fig.3 Number of photons at each harmonic at different laser intensity, in Xe

be taken for laser intensities below or close to the saturation intensity, but in any case not for laser intensities larger than I_S for which a complicated situation has to be taken into account, including ionization of the medium.

From these intensity dependences, it is possible to plot the number of photons produced at each harmonic at different laser intensities. Figure 3 shows this plot in xenon at several laser intensities from 5×10^{12} Wcm^{-2} to 3×10^{13} Wcm^2. The vertical scale has been calibrated in absolute value by estimating the collection efficiency, in particular by taking into account the spectral responses of the grating and the photon detectors for the different harmonics. The experimental uncertainty is estimated to a factor of 5. Most of the modulations in Fig.3 are therefore not significant. As the intensity increases, a plateau appears (at 7×10^{12} Wcm^{-2}) and extends up to the saturation intensity, estimated to 1.3×10^{13} Wcm^2. Beyond this intensity, the distribution becomes smoother, but the maximum order that can be detected remains constant. Most of the high order harmonics are created in an intensity such that the medium is not or weakly ionized. This ruled out plasma effects as responsible for the production of high order harmonics[12]. This maximum number of harmonics, 21 in Xe, 29 in Kr, 33 in Ar is directly related to the saturation intensity. Let us point out that, in the case of Ne and He which have higher saturation intensity about 2×10^{14} and 3×10^{14} Wcm^{-2} respectively, we may expect the creation of the 59th and the 65th harmonics of the incident laser radiation at 1064 nm. This would correspond to generated radiations at 18.3 and 16.8 nm. But these radiations may be more difficult to observe because of the poor efficiency, compared to the heavier rare gases.

DISCUSSION

These experimental results[6-8] have stimulated a number of theoretical works[13-18] either within the framework of perturbation theory (in H)[13-16] or going beyong it, using Floquet method (also in H)[17] or performing time-dependent calculations[18-19]. It must be pointed out that all calculations are based on a single atom response. Figure 4a shows for example the result obtained by Potvliege and Shakeshaft[17]. The square of the Fourier transform of the dipole moment is plotted in solid line as a function of the harmonic order. The dashed line corresponds to the results obtained in the framework of lowest-order perturbation theory : important non-perturbative effects appear beyond 10^{13} Wcm^{-2}. In Fig.4b the result obtained in Xe by Kulander and Shore is shown[19]. They numerically solve the time-dependent frozen-core Hartree-Fock equations. Besides the harmonics, there is a pronounced resonance and a continuous background.

If we assume that the number of photons that could be detected in an experiment is directly proportional to the rate shown in fig.4, then both (non-perturbative) theoretical results show a quite similar behavior to the experimental results : a rapid decrease followed by a plateau and another decrease. A direct comparison between experimental and theoretical resuslts can be made in Xe : both results behave identically. But is it meaningful to compare in this way experimental and theoretical results, or is this apparent agreement accidental ? The theoretical results include neither the ionization of the gaseous medium which occurs in the experiment, nor propagation effects which could significantly affect the overall harmonic distribution. Generally speaking, the coherent radiation field is created from the laser induced polarization of the medium. This is a collective process due to the response of the whole medium to the incident field.

In order to extract from the experimental results some informations on the microscopic response of the atomic medium, it is necessary to briefly recall the main equations which govern nonlinear optical phenomena[11,20]. An electromagnetic field propagating in a dielectric medium characterized by an electronic polarization must satisfy the Maxwell propagation equation. If the incident laser field is assumed to be monochromatic, and considering only the

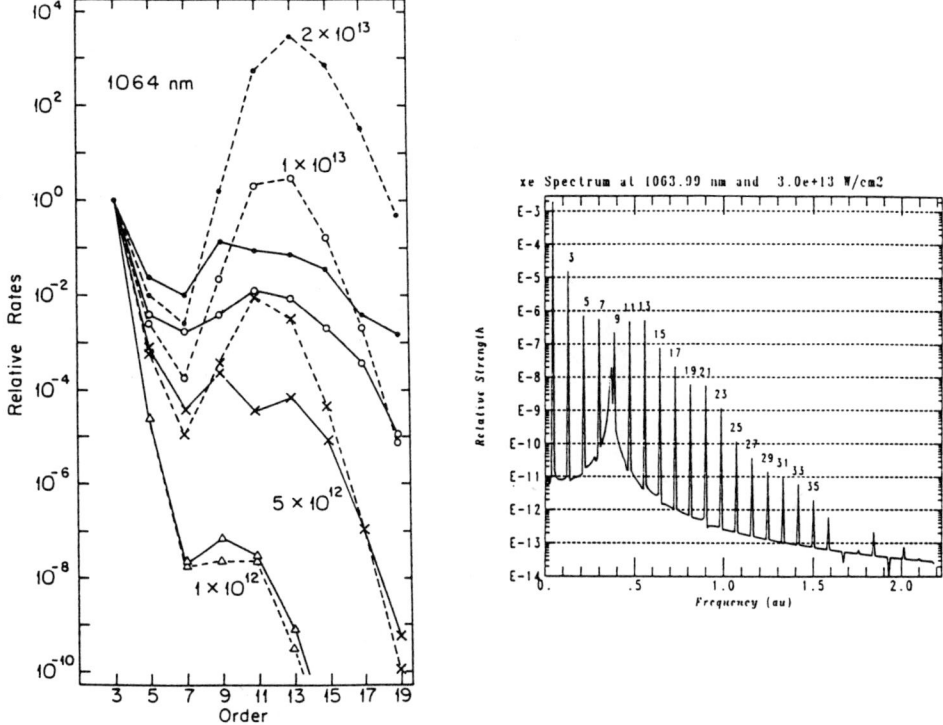

Fig.4 Theoretical calculations a) left, from ref.17, b) right, from ref.18

lowest-order in the interaction with the radiation field, then, for a gaussian beam focused into a nonlinear medium of finite length, the intensity of the qth harmonic field is[11]:

$$I_q(0,0,0) = (2^{q+1} b^2 q^2 \pi^{q+3} / n_1 n_q \lambda^2 c^{q-1}) \; I^q \; |F_q|^2 \; |N_a \chi^{(q)}|^2$$

where λ is the incident wavelength, c the speed of the light, n_1 and n_q are the refractive indices of the nonlinear medium at ω and $q\omega$ respectively and N_a the atomic density; b is the laser confocal parameter. The qth order susceptibilty $\chi^{(q)}$ characterizes the response of an individual atom to the radiation field. This expression involves two important terms, the square of the atomic nonlinear polarization at frequency $q\omega$ written here in a perturbative picture, which characterizes the response of each individual atom to the laser radiation field, and the phase matching factor F_q which defines the response of the whole medium and

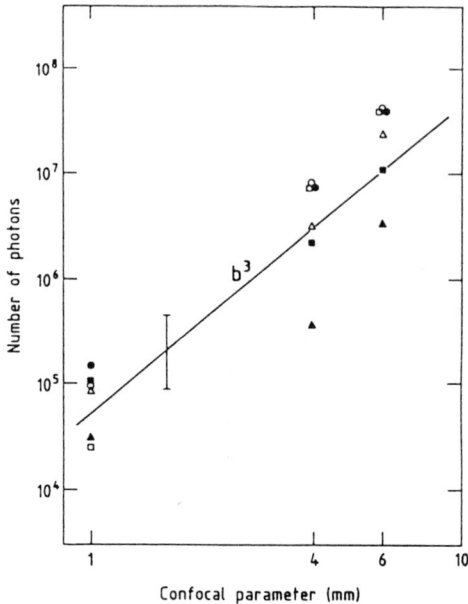

Fig.5 Number of photons versus confocal parameter b for different harmonics.

depends on macroscopic quantities such as the confocal parameter b or the atomic density N_a. Separating single-atom and many-atom contributions, the number of photons N_q detected in an experiment may be written as :

$$N_q = K [(2\pi I/c)^q |\chi^{(q)}|^2] * [b^3 N_a^2 |F_q|^2/q^{1/2}]$$

where K is a constant. This expression shows the main parameters which govern harmonic generation : the microscopic one, $\chi^{(q)}$ as well as the macroscopic ones, b and N_a. The quantity represented in Fig.4 is the non-perturbative generalization of the first bracket :

$$|d(q\omega)|^2 = (2\pi I/c)^q |\chi^{(q)}|^2$$

where $d(q\omega)$ is the Fourier transform of the dipole moment. As already mentioned, this calculation can be applied only for laser intensity below the saturation intensity I_s in order to eliminate any ionization effects. But is it possible to estimate the contribution of phase matching in our experimental conditions to the harmonic intensity distribution ? We have tried experimentally

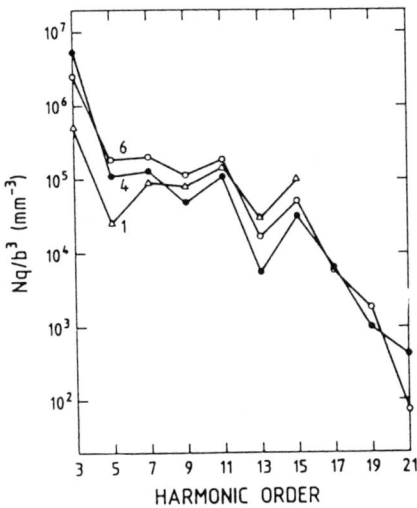

Fig.6 Number of photons divided by b^3 as a function of the harmonic order.

to study the influence of propagation effects by varying the confocal parameter, using different lenses to focus the laser beam. Some preliminary results are shown in Fig.5 [21]. The number of harmonics photons detected at the saturation intensity in xenon (1.3x10^{13} Wcm^{-2}) is plotted as a function of the confocal parameter, for three different values of b, 1, 4 and 6. The points for harmonics follow a b^3 power law.

Figure 6 shows the number of harmonic photons detected at I_s divided by b^3. Harmonic distributions behave very similarly within our experimental error bars. Clearly, apart from the b^3 power law, these results do not show an important influence of phase matching effects. Similar results are obtained, whatever the focalization of the incident laser beam is, as if the phase matching factor $|F_q|^2$ did not depend on b.

Assuming a I^q power dependence of the qth harmonic, $|F_q|^2$ has been calculated in Xe, neglecting ionization of the medium, using a 1 mm width lorentzian distribution of the atomic density of the gas jet, correctly reproducing the gas jet geometry[8,22]. Figure 7 shows the calculated results as a function of q, for a non dispersive medium, using Δk values obtained from a calculation of the dynamic polarizability of Xe[23]. For q > 11, the phase matching factor $|F_q|^2$ does not decrease any more, because, above the threshold, the real part of Δk becomes negative, and thus favors high order harmonic generation. We can observe on this figure a significant variation of the phase matching factor as a

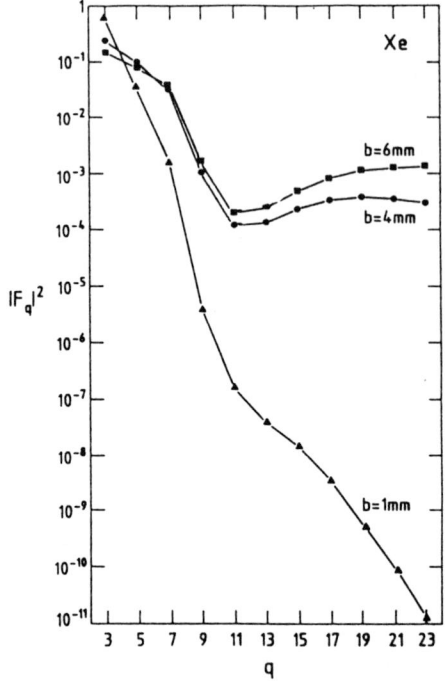

Fig.7 Calculated phase matching factor as a function of q for three values of b.

function of b. The huge difference between b=1 mm and b=4 or 6 mm, is not observed in the experiment. This calculation lies upon several approximations : absence of ionization of the medium and the weak field limit, which may not be valid in the present problem .

The interpretation of the experimental results and in particular the understanding of propagation of high-order harmonics in a nonlinear medium exposed to an intense laser field remains an open problem. Let us come back to the question raised at the beginning of the present paper : does the experimental harmonics distribution reflect the single-atom response to the nonlinear interaction ? The experimental results do not allow to give a definite answer to this question. The medium exposed to an intense infrared radiation field appears to behave as if the laser were a plane wave or as if the effective interaction length were extremely small, much smaller than the 1 mm of the interaction region, with a $|F_q|^2$ independent of b.

The interpretation of these results are very promising for the near future. With the development of new high power lasers working in the terawatt regime[24] laser intensities of the order of 10^{18} Wcm^{-2} can be reached and a new field of investigation is opened. At such a high laser intensity, a great number of electrons can be removed of the atom, leading to the production of very highly charged ions. We may expect that similar processes may occur in an ionized gas, leading to the production of higher order harmonics, and shorter wavelengths. The authors would like to thank Dr. C. Manus for many stimulating discussions, X.F. Li and A. Sanchez for their help in the experiments.

REFERENCES

1. M.G. Groseva, D.I. Metchkov, V.M. Mitev, L.I. Pavlov and K.V. Stamenov, Opt. Commun. 23, 77 (1977).
2. J. Wildenauer J. Appl. Phys. 62, 41 (1988).
3. J. Bokor, P.H. Bucksbaum and R.R. Freeman, Opt. Lett. 8, 217 (1983).
4. J. Reintjes, C.Y. She and R. Reckardt, IEEE J. Quatum Electron. QE-14, 581 (1978).
5. J. Reintjes, L.L. Tankersley and R. Christensen, Opt. Commun. 39, 334 (1981).
6. A. McPherson, G. Gibson, H. Jara, U. Johann, T.S. Luk, I. McIntyre, K. Boyer and C.K. Rhodes, J. Opt. Soc. Am. B 4, 595 (1987).
7. M. Ferray, A. L'Huillier, X.F. Li, L.A. Lompré, G. Mainfray and C. Manus, J. Phys. B 21, L31 (1988).
8. X.F. Li, A. L'Huillier, M. Ferray, L.A. Lompré and G. Mainfray, Phys. Rev. A 4, 5751 (1989).
9. L.A. Lompré, M. Ferray, A. L'Huillier, X.F. Li and G. Mainfray, J. Appl. Phys. 63, 1791 (1988).
10. A. L'Huillier, L.A. Lompé, G. Mainfray and C. Manus, J. Phys. B 16, 1363 (1983).
11. J.F. Reintjes, in "Nonlinear Optical Parameter Processes in Liquids and Gases", Academic Press, Inc. (1984).
12. R.L. Carman, C.K. Rhodes and R.F. Benjamin, Phys.Rev. A 24, 2649 (1981).
13. Y. Gontier and M. Trahin, IEEE, J. Quantum Electron. QE-18 ,1137 (1982).
14. B. Gao and A.F. Starace, Phys. Rev. A 39, 4550 (1989).
15. L. Pan, K.T. Taylor and C.W. Clark, Phys. Rev. A 39,4894 (1989).
16. J.H. Eberly, Q. Su and J. Javanainen, Phys. Rev. Lett 62, 881 (1989).
17. R.M. Potvliege and R. Shakeshaft, Phys. Rev. A 40, xx (1989).
18. K. Kulander and B.W. Shore, Phys.Rev. Lett. 62, 524 (1989).
19. K. Kulander and B.W. Shore, J.O.S.A. B in press (April 1990).
20. N. Bloembergen, in "Nonlinear Optics", Benjamin, New-York (1965).

21. L.A. Lompré, A. L'Huillier, M. Ferray, P. Monot and G. Mainfray, to be published.
22. A. Lago, G. Hilber and R. Wallestein, Phys. Rev. A **36**, 3827 (1987).
23. A. L'Huillier, X.F. Li and L.A. Lompré, J.O.S.A. B in press (April 1990).
24. M. Ferray, L.A. Lompré, O. Gobert, A. Gomez, A. L'Huillier, G. Mainfray and C. Manus, Opt. Comm. in press (January 1990).

Author Index

A
Azuma, H., 267

B
Barty, C. P. J., 267
Blaha, M., 177
Bollinger, J. J., 152

C
Casperson, D. E., 217
Cobble, J. A., 217
Comly, J. C., 217

D
Davis, J., 177
de Heer, F. J., 40
Den Hartog, E. A., 107

E
Elton, R. C., 257

F
Ferray, M., 277
Finkenthal, Michael, 95
Fonck, R. J., 122
Fujimoto, Takashi, 116

G
Gallagher, T. F., 48
Gilbert, S. L., 152
Griffin, D. C., 3

H
Heinzen, D. J., 152
Hoekstra, R., 40
Hooper, Jr., C. F., 204
Hulse, Russell A., 63

I
Iglesias, C. A., 82
Itano, W. M., 152

J
Jones, L. A., 217
Jones, R. R., 48

K
Kato, Y., 267
Key, M. H., 267
Kilcrease, D. P., 204
Kim, Yong-Ki, 19
Kyrala, G. A., 217

L
LaGattuta, K. J., 217
Lawler, J. E., 107
Lee, P. H. Y., 217
L'Huillier, A., 277
Lompré, L.-A., 277

M
Mainfray, G., 277
Mancini, R. C., 204
Manka, C. K., 257
Maron, Y., 232
McLean, E. A., 257
Miura, E., 267
Morgenstern, R., 40

N
Nakai, S., 267
Nishihara, K., 267
Nishio, M., 267

O
Olson, G. L., 217

P
Pert, G. J., 267
Pindzola, M. S., 3

R

Ramsden, S. A., 267
Ripin, B. H., 257
Rogers, F. J., 82
Rose, S. J., 267

S

Schappert, G. T., 217
Shemansky, D. E., 163
Shiraga, H., 267
Stamper, J. A., 257
Stratton, B. C., 135

T

Tachi, T., 267
Takabe, H., 267
Taylor, A. J., 217

W

Wilson, B. G., 82
Wineland, D. J., 152
Woltz, L. A., 204

Y

Yamakawa, K., 267
Younger, Stephen M., 193
Yu, Yan, 73

AIP Conference Proceedings

		L.C. Number	ISBN
No. 139	High-Current, High-Brightness, and High-Duty Factor Ion Injectors (La Jolla Institute, 1985)	86-70245	0-88318-338-2
No. 140	Boron-Rich Solids (Albuquerque, NM, 1985)	86-70246	0-88318-339-0
No. 141	Gamma-Ray Bursts (Stanford, CA, 1984)	86-70761	0-88318-340-4
No. 142	Nuclear Structure at High Spin, Excitation, and Momentum Transfer (Indiana University, 1985)	86-70837	0-88318-341-2
No. 143	Mexican School of Particles and Fields (Oaxtepec, México, 1984)	86-81187	0-88318-342-0
No. 144	Magnetospheric Phenomena in Astrophysics (Los Alamos, 1984)	86-71149	0-88318-343-9
No. 145	Polarized Beams at SSC & Polarized Antiprotons (Ann Arbor, MI & Bodega Bay, CA, 1985)	86-71343	0-88318-344-7
No. 146	Advances in Laser Science–I (Dallas, TX, 1985)	86-71536	0-88318-345-5
No. 147	Short Wavelength Coherent Radiation: Generation and Applications (Monterey, CA, 1986)	86-71674	0-88318-346-3
No. 148	Space Colonization: Technology and The Liberal Arts (Geneva, NY, 1985)	86-71675	0-88318-347-1
No. 149	Physics and Chemistry of Protective Coatings (Universal City, CA, 1985)	86-72019	0-88318-348-X
No. 150	Intersections Between Particle and Nuclear Physics (Lake Louise, Canada, 1986)	86-72018	0-88318-349-8
No. 151	Neural Networks for Computing (Snowbird, UT, 1986)	86-72481	0-88318-351-X
No. 152	Heavy Ion Inertial Fusion (Washington, DC, 1986)	86-73185	0-88318-352-8
No. 153	Physics of Particle Accelerators (SLAC Summer School, 1985) (Fermilab Summer School, 1984)	87-70103	0-88318-353-6
No. 154	Physics and Chemistry of Porous Media—II (Ridge Field, CT, 1986)	83-73640	0-88318-354-4
No. 155	The Galactic Center: Proceedings of the Symposium Honoring C. H. Townes (Berkeley, CA, 1986)	86-73186	0-88318-355-2

No. 156	Advanced Accelerator Concepts (Madison, WI, 1986)	87-70635	0-88318-358-0
No. 157	Stability of Amorphous Silicon Alloy Materials and Devices (Palo Alto, CA, 1987)	87-70990	0-88318-359-9
No. 158	Production and Neutralization of Negative Ions and Beams (Brookhaven, NY, 1986)	87-71695	0-88318-358-7
No. 159	Applications of Radio-Frequency Power to Plasma: Seventh Topical Conference (Kissimmee, FL, 1987)	87-71812	0-88318-359-5
No. 160	Advances in Laser Science–II (Seattle, WA, 1986)	87-71962	0-88318-360-9
No. 161	Electron Scattering in Nuclear and Particle Science: In Commemoration of the 35th Anniversary of the Lyman-Hanson-Scott Experiment (Urbana, IL, 1986)	87-72403	0-88318-361-7
No. 162	Few-Body Systems and Multiparticle Dynamics (Crystal City, VA, 1987)	87-72594	0-88318-362-5
No. 163	Pion–Nucleus Physics: Future Directions and New Facilities at LAMPF (Los Alamos, NM, 1987)	87-72961	0-88318-363-3
No. 164	Nuclei Far from Stability: Fifth International Conference (Rosseau Lake, ON, 1987)	87-73214	0-88318-364-1
No. 165	Thin Film Processing and Characterization of High-Temperature Superconductors	87-73420	0-88318-365-X
No. 166	Photovoltaic Safety (Denver, CO, 1988)	88-42854	0-88318-366-8
No. 167	Deposition and Growth: Limits for Microelectronics (Anaheim, CA, 1987)	88-71432	0-88318-367-6
No. 168	Atomic Processes in Plasmas (Santa Fe, NM, 1987)	88-71273	0-88318-368-4
No. 169	Modern Physics in America: A Michelson-Morley Centennial Symposium (Cleveland, OH, 1987)	88-71348	0-88318-369-2
No. 170	Nuclear Spectroscopy of Astrophysical Sources (Washington, D.C., 1987)	88-71625	0-88318-370-6
No. 171	Vacuum Design of Advanced and Compact Synchrotron Light Sources (Upton, NY, 1988)	88-71824	0-88318-371-4

No. 172	Advances in Laser Science–III: Proceedings of the International Laser Science Conference (Atlantic City, NJ, 1987)	88-71879	0-88318-372-2
No. 173	Cooperative Networks in Physics Education (Oaxtepec, Mexico 1987)	88-72091	0-88318-373-0
No. 174	Radio Wave Scattering in the Interstellar Medium (San Diego, CA 1988)	88-72092	0-88318-374-9
No. 175	Non-neutral Plasma Physics (Washington, DC 1988)	88-72275	0-88318-375-7
No. 176	Intersections Between Particle and Nuclear Physics (Third International Conference) (Rockport, ME 1988)	88-62535	0-88318-376-5
No. 177	Linear Accelerator and Beam Optics Codes (La Jolla, CA 1988)	88-46074	0-88318-377-3
No. 178	Nuclear Arms Technologies in the 1990s (Washington, DC 1988)	88-83262	0-88318-378-1
No. 179	The Michelson Era in American Science: 1870–1930 (Cleveland, OH 1987)	88-83369	0-88318-379-X
No. 180	Frontiers in Science: International Symposium (Urbana, IL 1987)	88-83526	0-88318-380-3
No. 181	Muon-Catalyzed Fusion (Sanibel Island, FL 1988)	88-83636	0-88318-381-1
No. 182	High T_c Superconducting Thin Films, Devices, and Application (Atlanta, GA 1988)	88-03947	0-88318-382-X
No. 183	Cosmic Abundances of Matter (Minneapolis, MN 1988)	89-80147	0-88318-383-8
No. 184	Physics of Particle Accelerators (Ithaca, NY 1988)	87-07208	0-88318-384-6
No. 185	Glueballs, Hybrids, and Exotic Hadrons (Upton, NY 1988)	89-83513	0-88318-385-4
No. 186	High-Energy Radiation Background in Space (Sanibel Island, FL 1987)	89-083833	0-88318-386-2
No. 187	High-Energy Spin Physics (Minneapolis, MN 1988)	89-083948	0-88318-387-0
No. 188	International Symposium on Electron Beam Ion Sources and their Applications (Upton, NY 1988)	89-084343	0-88318-388-9
No. 189	Relativistic, Quantum Electrodynamic, and Weak Interaction Effects in Atoms (Santa Barbara, CA 1988)	89-084431	0-88318-389-7

No. 190	Radio-frequency Power in Plasmas (Irvine, CA 1989)	89-045805	0-88318-397-8
No. 191	Advances in Laser Science–IV (Atlanta, GA 1988)	89-085595	0-88318-391-9
No. 192	Vacuum Mechatronics (First International Workshop) (Santa Barbara, CA 1989)	89-045905	0-88318-394-3
No. 193	Advanced Accelerator Concepts (Lake Arrowhead, CA 1989)	89-045914	0-88318-393-5
No. 194	Quantum Fluids and Solids—1989 (Gainesville, FL, 1989)	89-81079	0-88318-395-1
No. 195	Dense Z-Pinches (Laguna Beach, CA, 1989)	89-46212	0-88318-396-X
No. 196	Heavy Quark Physics (Ithaca, NY, 1989)	89-81583	0-88318-644-6
No. 197	Drops and Bubbles (Monterey, CA, 1988)	89-46360	0-88318-392-7
No. 198	Astrophysics in Antarctica (Newark, DE, 1989)	89-46421	0-88318-398-6
No. 199	Surface Conditioning of Vacuum Systems (Los Angeles, CA, 1989)	89-82542	0-88318-756-6
No. 200	High T_c Superconducting Thin Films: Processing, Characterization, and Applications (Boston, MA, 1989)	90-80006	0-88318-759-0
No. 201	QED Stucture Functions (Ann Arbor, MI, 1989)	90-80229	0-88318-671-3
No. 202	NASA Workshop on Physics From a Lunar Base (Stanford, CA, 1989)	90-55073	0-88318-646-2
No. 203	Particle Astrophysics: The NASA Cosmic Ray Program for the 1990s and Beyond (Greenbelt, MD, 1989)	90-55077	0-88318-763-9
No. 204	Aspects of Electron-Molecule Scattering and Photoionization (New Haven, CT, 1989)	90-55175	0-88318-764-7
No. 205	The Physics of Electronic and Atomic Collisions (XVI International Conference) (New York, NY, 1989)	90-53183	0-88318-390-0

APR 16 1991